THE
MATH WORKSHOP
ALGEBRA

THE MATH WORKSHOP ALGEBRA

Deborah Hughes-Hallett
HARVARD UNIVERSITY

W • W • NORTON & COMPANY • INC • New York • London

Library of Congress Cataloging in Publication Data

Hughes-Hallett, Deborah.
The math workshop : algebra.

Includes index.
1. Algebra.   I. Title.
QA154.2.H83   1980   512.9   79–29665

W. W. Norton & Company, Inc., 500 Fifth Avenue, New York, N.Y. 10110
W. W. Norton & Company Ltd., 37 Great Russell Street, London WC1B 3NU
1 2 3 4 5 6 7 8 9 0

ISBN 0-393-09030-2

To everyone who saw the sun rise over Route 2:
Barbara, Bruce, Ken, Rob, Ken

CONTENTS

x   *Contents*

PREFACE

Origin

The Math Workshop has grown out of eight years' experience in teaching algebra and precalculus at Harvard. Here, as at a great many colleges and universities, there are significant numbers of freshmen who need to learn algebra before going on in mathematics, as well as students who are stricken with "math anxiety" when required to undertake anything more than basic arithmetic. *The Math Workshop* evolved in response to the needs of such students. The book has been used very successfully in the classroom for several years, including use by teaching fellows who are initially inexperienced and therefore rely very heavily on it.

Purpose

The goal of this book is to show students that they *can* understand math—that whether they enjoy it or not, they can learn it. I write for ordinary people, not for mathematicians, and I attempt to talk to my readers, rather than lecture them. I hope that by the end the student will feel that math is an ordinary human activity—some parts are easy, some parts are not, but all of it is possible, and perhaps even enjoyable.

Audience

The Math Workshop: Algebra is for students who have not had algebra or who need a complete review. It was designed for a one-semester or two-quarter course for students expecting to go on into precalculus. (The companion text, *The Math Workshop: Elementary Functions*, completes the students' preparation for calculus). With certain topics omitted, *The Math Workshop: Algebra* can also be used for a terminal algebra course.

Approach

Many students enrolling in introductory college math courses approach the subject with anxiety and little confidence. I am convinced that the best way to restore such students' confidence is to start them on material they feel comfortable with and to explain new ideas in terms of ones they know well.

Therefore, the book assumes a command only of the addition, subtraction, multiplication, and division of positive whole numbers, and leads off with short chapters that review arithmetic. Algebra is then shown to be a generalization of arithmetic.

I am also convinced that in order to hold the students' attention (or suspend their disbelief), a book must not only explain what is correct but also dispel misconceptions and explain why what is correct has to be that way. I have tried to anticipate potential questions and discuss common mistakes, particularly in the "Things You Can and Can't Do" sections.

Exercises

The problems have been carefully designed and tested. They move gradually from routine ones designed to build the student's confidence to those that are in less standard form. Working through a reasonable selection of the problems in this book should build a thorough understanding of the subject.

The first eight chapters are short and consist mostly of review material, so the exercises are grouped at the end of each chapter. Thereafter, each chapter has problem sets at the ends of sections as well as a review at the end.

Treatment of Specific Topics

Arithmetic Review. This serves both to build confidence and to ensure that students from different backgrounds all start algebra with the same understanding of arithmetic. Numerical operations (for example, adding fractions) are laid out here in exactly the same way as the corresponding algebraic operations in later chapters, emphasizing that algebra is merely "arithmetic in disguise."

Factoring. This is taught by the systematic method—taking the guesswork out of math is extremely important in teaching students who are not comfortable with the subject.

Expressions versus Equations. It cannot be emphasized too often that expressions and equations are *different,* and should be treated differently. A less than clear understanding of the difference between them is probably responsible for more algebra mistakes than anything else; therefore, a short chapter on this distinction is included.

Literal Equations. The power of the viewpoint that "algebra is just arithmetic in disguise" comes to the fore again here. Students traditionally have a good deal of trouble with literal equations, even if they are perfectly at home with ordinary equations. In appropriate situations the text shifts to a two-column format in which each literal equation is shown solved in parallel with a similar equation with numerical coefficients. This approach has been very effective.

Word Problems. Students always complain that word problems are hard because there is no "method"; that is, they can't rely on memorization. *The Math Workshop* helps to structure and direct students' thinking, but without doing it for them. A set of general steps to be followed is outlined, and if a student follows these steps, his or her ability to do word problems improves dramatically.

Absolute Value. I have found that the geometric method for solving absolute value problems given in *The Math Workshop* is superior to the algebraic method in all but the most complicated problems. In some examples both methods are given side by side, for the benefit of those who have seen the algebraic method before, but for those who are having difficulty the geometric method can greatly simplify things.

Functions. The concept of a function is not treated explicitly in this text. I have found that students understand functions much better after considerable exposure to graphing techniques. A brief introduction to the abstract notion of a function is not likely to be retained, nor is the concept necessary when one is learning algebra. *The Math Workshop: Elementary Functions*, the companion text to this one, begins with graphing and then treats functions comprehensively.

Acknowledgments

To thank properly everyone who helped with *The Math Workshop* would take another book. But for their particular contributions of ideas, problems, coffee, and cheerful good humor in the middle of the night, I would like to thank Barbara Peskin, Ken Manning, Ken Argentieri, Rob Olian, Bruce Molay, Reed Eichner, Adele Peskin, Charlie Klippel, Brian Leverich, Steve Ballmer, Mike Graceffo, John Maggio, Mark Robbins, Caren Jahre, Paul Segel, Lynn Smolik, Jim Rhodes, Karen Fifer, Ellen Gravellese, Ann Ginsberg, Rich Nelson, Dave Moskowitz, Jill Einstein, John Dubaz, Steve Krow-Lucal, Grace Young, Dan Freed, Bruce Patton, John Engelman, Dan Feldman, Nat Kuhn, Kathy Rosenthal, and Barbara Alpert.

The publisher would like to thank the following reviewers for their helpful comments and criticism: Louise Raphael, Clark College, Atlanta; Douglas Burke, Malcolm X College; Andrew J. Berner, Alleheny College; A. W. Goodman, Univ. of South Florida; Stanley M. Lukawecki, Clemson Univ.; Deborah T. Haimo, Univ. of Missouri, St. Louis; Douglas Brown, Nassau Community College; Maurice Monahan, South Dakota State Univ.; Ignacio Bello, Hillsborough Community College; Calvin Lathan, Monroe Community College; Cleon R. Yohe, Washington Univ.; Ward Bouwsma, Southern Illinois Univ.; Robert Donaghey, Baruch College, CUNY; David Cohen, UCLA; Daniel Marks, Stanford Univ.; Jerry Karl, Golden West College.

THE
MATH WORKSHOP
ALGEBRA

1 NUMBER SYSTEMS REVISITED

1.1 ARITHMETIC IS THE FOUNDATION FOR ALGEBRA

Algebra is really arithmetic in disguise. In arithmetic you learn how to operate with different kinds of numbers—how to add fractions or multiply decimals, for example. In algebra you generalize what you learned in arithmetic: you do *exactly* the same operations, but you do them with letters instead of with specific numbers. Consequently, if you have all the arithmetical operations straight, algebra won't be hard, because it involves exactly these same operations, only in another context.

The first six chapters are a review of what you need to know about arithmetic. They are relatively short, and the exercises are grouped at the end of each chapter. Remember: in order to be able to use these operations in a different situation, you not only need to know them, but to know them *well*.

1.2 NUMBER SYSTEMS

The first numbers people started using were the "counting numbers"—1, 2, 3, 4, and so on—which in math language are called the *natural numbers*. It took some time for zero to be invented (the Romans didn't have one, although the Maya, who were in many ways less advanced, did, and used a drawing of a shell to symbolize it). The natural numbers together with zero are called the *whole numbers*.

The *negative numbers* did not come into use in Europe until the Renaissance, and even then people regarded them with the greatest suspicion—after all, it's easy enough to see what is meant by 3 trees or 5 trees, but −5 trees? The name "negative" comes from the Latin, *negare*, "to deny," because people denied that such numbers could exist. But, as often happens with a new mathematical idea, negative numbers got used anyway, even though people weren't sure they existed, because they were useful. So, if you find the behavior of negative numbers peculiar, don't be discouraged; a great many people have felt that before you, and history shows that working with them will eventually make them seem more reasonable. The negative numbers together with the positive numbers and zero are called the *integers* (symbolized by **Z**).

Long before negative numbers came into use, people had discovered a need for *fractions* to express a part of a number. One fifth, $\frac{1}{5}$, for example, means the number you get by dividing one into five pieces and taking one of them. Four thirds, $\frac{4}{3}$, therefore, means the result of dividing one into three pieces and taking four such pieces. Clearly this will involve cutting up two ones each into three pieces. When you put four of these pieces together, you will have more than one of the original ones, which is why $\frac{4}{3}$ is greater than 1. See Figure 1.1

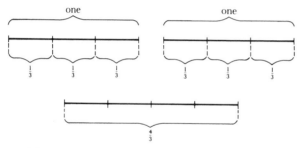

FIG. 1.1

The word "fraction" comes from the Latin word meaning "to break"; the word "fragment" comes from the same root. All fractions can be written as one number divided by another (called the *quotient* of two numbers). Thus, we can look at the natural numbers as fractions because they can be written as something divided by 1. For example, $2 = \frac{2}{1}$, and $5 = \frac{5}{1}$. There are also negative fractions, for example, $-\frac{1}{2}$ and $-\frac{5}{3}$, and the negative integers can also be written as fractions, because $-3 = \frac{-3}{1}$ and $-10 = \frac{-10}{1}$. All the fractions together (including the integers) are called *rational* numbers, symbolized by **Q**.

Two particular sorts of fractions deserve special mention: ones of the form

$$\frac{0}{\text{something}} \quad \text{or} \quad \frac{\text{something}}{0}$$

First, anything of the form $\dfrac{0}{\text{something}}$ is 0. For example, look at $\dfrac{0}{5}$. You can think of this as the result of taking the one, dividing it into five pieces and taking none of them. Alternatively, think of it as the number of times that 5 goes into 0. Either way, you have 0. On the other hand, anything of the form $\frac{\text{something}}{0}$ is *undefined*, which means that it has no meaning. For example, look at $\frac{3}{0}$. You can think of this as the result of dividing one into zero pieces, and taking three of them. But however you divide up one, you can't have zero pieces, and even if you could manage to divide it into zero pieces, how could you take three of them? Alternatively, think of it as the number of times the 0 goes into 3—again an impossible question. Either way, $\frac{3}{0}$ makes no sense, and therefore is said to be undefined. Finally, $\frac{0}{0}$ can have different meanings in different contexts and so is said to be *indeterminate*. In particular, $\frac{0}{0}$ is not necessarily 1

It turns out—though it is by no means obvious—that there are also numbers that cannot be written as fractions. These numbers are called *irrational*. This seems to me a rather rude name for a number: it gives the impression that they might behave rather badly, but actually they obey just the same rules as the rest of the numbers. In practice one actually meets very few irrational numbers. The number π, which comes up in the formulas for the area and circumference of a circle, is irrational (though I won't prove it), and so is e, a number that arises in calculus and comes into just about every population growth problem. Slightly closer to home, $\sqrt{2}$ (the square root of 2, meaning the number that when multiplied by itself gives 2) is irrational, and so is $\sqrt{3}$. The number $\sqrt{4}$ is not, since $\sqrt{4} = 2$, but $\sqrt{5}$ is, and so is the square root of any number that doesn't come out exactly. The Greeks first proved that $\sqrt{2}$ was irrational and hence that there are numbers that are not fractions. The rational numbers together with the irrational numbers are called the *real numbers*, symbolized by **R**.

It is interesting to look at the breakdown of the reals into the rationals and irrationals in a different way. Any real number can be written as a decimal—if you allow decimals that go on forever, that is. For example, $\frac{1}{3} = 0.333...$, where the 3's go on forever. This is usually written $\frac{1}{3} = 0.\overline{3}$, where the bar over the 3 means that it repeats forever (or, is "recurring"). Similarly, $\frac{1}{11} = 0.090909... = 0.\overline{09}$ (the bar means that the 09 keeps repeating) and $\frac{2}{9} = 0.22... = 0.\overline{2}$.

We also know that $1 = 1.0$ and $2\frac{1}{2} = 2.5$, so all the decimals which fractions give rise to either terminate or repeat. Look at $\frac{1}{7}$: it turns out that

$$\frac{1}{7} = 0.142857142857142857... = 0.\overline{142857}$$

So, although it takes six figures to do it, $\frac{1}{7}$ is still a "repeating decimal." On the other hand, if we were to work out π or e or $\sqrt{2}$ as a decimal we would find that they *never* repeated, and that there was no pattern whatsoever. You might therefore guess—and you'd be right—that the rational numbers are those that can be expressed by repeating or terminating decimals, and the irrational numbers are those whose decimals never repeat nor terminate.

The name "real number" might lead you to guess that somewhere there are some "unreal" numbers. In the nineteenth century it was realized that it is often useful to use a number called i, the square root of -1, or $\sqrt{-1}$. But since there is no number (that we know of) whose square is -1 (all real numbers have positive squares: $2^2 = 4$, $(-2)^2 = 4$, $(\frac{1}{3})^2 = \frac{1}{9}$, $(-\frac{1}{3})^2 = \frac{1}{9}$, etc.), i was said to be an *imaginary number*. The name shows that the nineteenth-century mathematicians felt about i much the same as earlier people felt about negative numbers—that such numbers did not really exist. They kept on using i because it was useful (it turns out to be particularly helpful in the study of electric circuits and of wave motion), and gradually people came to believe that imaginary numbers exist as much as any other kind do. Hence their old name, "imaginary numbers," has mainly been dropped in favor of *complex numbers*—again not a very polite name, since they are no more complex than any other kind of number, but it's one step better than 'imaginary'. The complex numbers, symbolized by **C**, consist of the real numbers together with the numbers that you get by adding i and all its multiples.

CHAPTER 1 EXERCISES

State whether each of the following numbers is real, complex, whole, natural, rational, integral, or irrational. Be sure to list all of the groups to which the number belongs.

1. 7
2. $-\dfrac{1}{2}$
3. 0
4. $\sqrt{2}$
5. -1
6. $\dfrac{-4}{2}$
7. $\dfrac{3}{8}$
8. 3
9. -6
10. $\sqrt{-3}$
11. $\dfrac{2}{5}$
12. 0.5
13. $0.\overline{132} = 0.132132132\ldots$
14. $\dfrac{12}{3}$
15. $\sqrt{9}$
16. $-\sqrt{16}$
17. $1.4282828\ldots$
18. 1.428
19. $\dfrac{20}{7}$
20. $\dfrac{\sqrt{25}}{-2}$

List all of the following:

21. The natural numbers less than 8.

22. The whole numbers greater than 2 and less than 11.

23. The natural numbers less than 31 that are whole-number multiples of 6.

24. The negative integers greater than -6 but less than the smallest natural number.

25. The irrational numbers that can be expressed by repeating decimals.

Are the following collections of numbers finite or infinite?

26. The natural numbers greater than 2.

27. The whole numbers less than 100.

28. The positive integers less than 11.

29. The negative integers less than -11.

30. The rational numbers between 1 and 2.

An integer is *even* if it is twice some other integer, and *odd* if it is one more than an even integer.

31. If n represents a member of the integers, write expressions in terms of n for (i) an even integer, (ii) an odd integer.

32. Show that the sum of any two even integers is an even integer.

33. Show that the sum of any two odd integers is an even integer.

34. Show that the product of any two even integers is an even integer.

35. Is the product of two odd integers even or odd? Can you show this?

36. Using pennies, nickels, and quarters, how can you make exactly 28 cents in change?

37. Using 4-cent, 6-cent, and 8-cent stamps, how can you make exactly 43 cents postage?

38. If the sum of four integers is odd, how many of the four integers are odd?

39. If the sum of four integers is even, how many of the four integers are odd?

In Problems 40–50, fill in "All," "Some," or "No".

40. _____ integers are whole numbers.

41. _____ natural numbers are rational.

42. _____ irrational numbers are real.

43. _____ whole numbers are irrational.

44. _____ rational numbers are whole numbers.

45. _____ whole numbers are natural numbers.

46. _____ natural numbers are whole numbers.

47. _____ real numbers are rational.

48. _____ real numbers are complex.

49. _____ complex numbers are irrational.

50. _____ complex numbers are real.

2 THE INTEGERS REVISITED

2.1   THE NUMBER LINE

Having described what kinds of numbers there are, we will now discuss what we can do with them. It turns out to be very useful for doing things with numbers to be able to draw a picture of them, so that you can "see" what you're doing as well as calculate. We will draw a picture only of the real numbers, although one can be drawn for the complex numbers also. The picture of the real numbers is called a *number line* (see Figure 2.1). It is constructed by taking a point, called the *origin*, to represent 0, drawing a horizontal line through it, and marking off another point at a small distance to the right of 0 to represent 1. The same distance is marked off to the right of 1 to give 2, and so on; the negative numbers are marked off to the left of 0. The arrows at either end of the line indicate that it goes on forever in either direction.

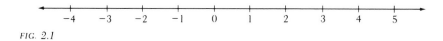

FIG. 2.1

2.2   ADDING INTEGERS

Adding two positive numbers can be done easily without the number line; for example, $3 + 2 = 5$. However, it will be much easier to generalize to adding negative numbers if we work it out on the number line as well.

How do we interpret $3 + 2 = 5$ on the number line? Think of 3 as the number represented by going three steps to the right from 0 on the number line; as in Figure 2.2. The number 2 is represented by going two steps to the right from 0 on the number line; as in Figure 2.3. To add the 2 to the 3, take the two steps representing the 2 from where you left off with the three steps representing the 3 (see Figure 2.4). You will end up five steps to the right of the origin, at the point representing 5. Hence we see that on the line, 2 added to 3 gives 5. Now I know that you can figure out $3 + 2$ without the line, but it is necessary to understand how $3 + 2$ works on the line in order to figure out $3 + (-2)$, $(-3) + 2$, and $(-3) + (-2)$.

FIG. 2.2

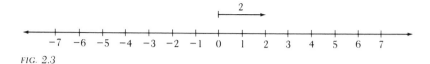

FIG. 2.3

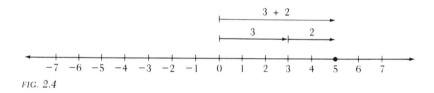

FIG. 2.4

What about $3 + (-2)$? First, we have to see that -2 means two steps to the left from 0 on the number line; as in Figure 2.5. Now, to add 3 and 2 we did the two steps representing 2 from where we left off with the three steps representing 3. We'll do the same in this case: we take the three steps to the right from 0 representing the 3; this leaves us at the point 3. Then we take the 2 steps to the left representing the -2; this leaves us at 1 (see Figure 2.6). Hence we shall say that $3 + (-2) = 1$.

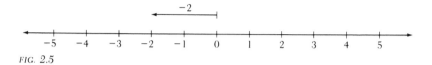

FIG. 2.5

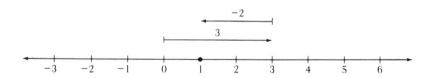

FIG. 2.6

By the same argument, $(-3) + 2$ is three steps to the left (representing -3), and then two to the right (representing 2), leaving us at -1 (see Figure 2.7). So $(-3) + 2 = -1$.

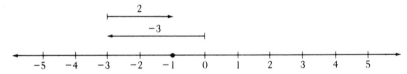

FIG. 2.7

What about $(-3) + (-2)$? Here we have to take three steps to the left (for the -3), followed by two more steps to the left (for the -2), and we end up at -5 (see Figure 2.8). So $(-3) + (-2) = -5$.

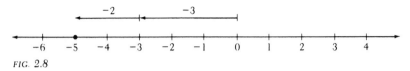

FIG. 2.8

There are a great many other ways of adding integers, and most books end up with some rule about it. However, the rule is itself usually so complicated, and so unintuitive, that it is often hard to remember it correctly, never mind apply it correctly. As a result, it is very much safer to work out each problem by the number-line method, which is at least straightforward, rather than to try and cut corners by learning a complicated "rule."

2.3   SUBTRACTING INTEGERS

As with addition, we will first do a problem with an answer that we know so that we can see what is happening on the number line.

How should we interpret $4 - 1 = 3$ *on the number line?* Since $4 - 1$ means go up to 4 and then down by 1, we should first take four steps to the right (representing the 4). Then, instead of taking one step to the right (representing 1) as in addition, we must take one step to the left, i.e. one step *in the opposite direction.* We end up at 3 (see Figure 2.9).

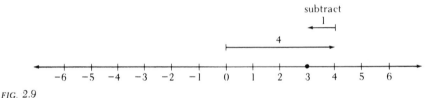

FIG. 2.9

The subtraction *10 − 7 can be done the same way.* Take ten steps to the right (for the 10), and then, since 7 is represented by seven steps to the right, take seven steps in the opposite direction, i.e. to the left; this will leave us at 3 (see Figure 2.10). So $10 - 7 = 3$.

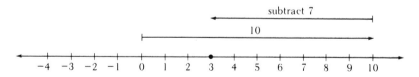

FIG. *2.10*

What about 4 − 7? That means take four steps to the right, and then seven to the left (as in the previous example), leaving us at −3 (see Figure 2.11). So $4 - 7 = -3$.

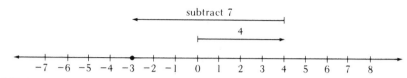

FIG. *2.11*

How about (−1) − 4? First, remember that −1 is represented by one step to the left, and 4 by four steps to the right. So we first take one step to the left, and then four steps to the left (to the left because we are subtracting 4). This leaves us at −5, as shown in Figure 2.12. So $(-1) - 4 = -5$.

FIG. *2.12*

The subtraction *(−4) − 3* works the same way; see Figure 2-13. Thus, $(-4) - 3 = -7$.

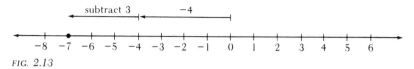

FIG. *2.13*

What about 4 − (−3)? This looks a bit more alarming than the previous cases because of the two − signs in a row, but it works by *exactly* the same method. First, we take four steps to the right, representing the 4. The −3 is represented by three steps to the left, but since we are subtracting we must go in the opposite direction, and take three steps to the right. This will set us down at 7, as shown in Figure 2.14. Thus, $4 - (-3) = 7$.

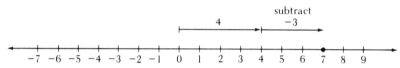

FIG. 2.14

You should notice that subtracting −3 comes to the same thing as adding 3. This happens because −3 is three steps to the left, and so when you go in the opposite direction you go three steps to the right, just as though you were adding 3. This will always happen when you are subtracting a negative number, and it gives rise to the saying that "two minuses make a plus."

For another example; look at $5 - (-2) = 5 + 2 = 7$. On the line, take five steps to the right, and then, since −2 is represented by two steps to the left, take two to the right; as in Figure 2.15.

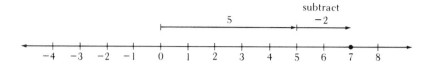

FIG. 2.15

Similarly, $10 - (-3) = 10 + 3 = 13$ and $0 - (-3) = 0 + 3 = 3$ (here we start by taking no steps and then take three steps to the right, leaving us at 3).

There's just one more example that will be useful to us: *what about* $(-2) - (-3)$? This means first go two to the left (for the −2) and then, since −3 is represented by 3 to the left, go three to the right (the opposite direction because we are subtracting); this leaves us at 1; as shown in Figure 2.16. So $(-2) - (-3) = 1$.

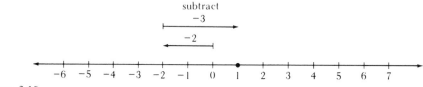

FIG. 2.16

Alternatively, $(-2) - (-3) = (-2) + 3$ (subtracting −3 is the same as adding 3), and −2 and 3 add to give 1. By the same reasoning, $(-4) - (-7) = (-4) + 7 = 3$ and $(-4) - (-1) = (-4) + 1 = -3$.

It is helpful to notice that an addition or subtraction problem with several signs between two numbers can be turned into a problem with only one sign between them. If you look back over the examples in this section you will see that they follow these rules:

$$
\begin{array}{ll}
+\,(+) = + & \qquad 5 + (+3) = 5 + 3 \\
+\,(-) = - & \qquad 5 + (-3) = 5 - 3 \\
-\,(+) = - & \qquad 5 - (+3) = 5 - 3 \\
-\,(-) = + & \qquad 5 - (-3) = 5 + 3
\end{array}
$$

2.4   PARENTHESES

The parentheses surrounding the -2 in $(-2) + 3$ are not really necessary: $-2 + 3$ means just the same thing; in either case you add -2 and 3. However, what does $5 - 1 + 2$ mean? Do you subtract 1 from 5 (giving 4) and then add 2 (giving 6)? Or do you add 1 and 2 (giving 3) and then subtract that from 5 (giving 2)? It looks as though either of these could perfectly well be right, and to eliminate confusion mathematicians had to decide which $5 - 1 + 2$ should mean. The decision was that the $-$ sign should apply only to the 1, not to the $1 + 2$, and that we should work from left to right. Hence there's no ambiguity about it: $5 - 1 + 2 = 4 + 2 = 6$.

If we want to denote 1 added to 2 and then subtracted from 5, we write $5 - (1 + 2)$. The parentheses mean that whatever is inside them should be done first. So $5 - 1 + 2 = 6$ could be written $(5 - 1) + 2 = 6$. So,

$$5 - (1 + 2) = 5 - 3 = 2$$
$$7 - (1 + 2) = 7 - 3 = 4$$

but

$$7 - 1 + 2 = 6 + 2 = 8$$

Again,

$$(5 + 3) - (1 + 2) = 8 - 3 = 5$$
$$6 - (3 - 1) = 6 - 2 = 4$$
$$7 - (2 - 4) = 7 - (-2) = 7 + 2 = 9$$

If you have several sets of parentheses, the convention is that you start with the innermost and work outwards. For example, $14 - [20 - (5 + 3)]$ means that you add the 5 and the 3 first, giving 8, and then subtract that from 20, giving 12, and then subtract that from 14, giving 2. Writing all that mathematically looks like this

$$14 - [20 - (5 + 3)] = 14 - [20 - 8] \qquad \text{(doing inner parentheses)}$$
$$= 14 - 12 \qquad \text{(doing outer parentheses)}$$
$$= 2$$

Just to see how necessary the parentheses are, try working out $14 - 20 - 5 + 3$:

$$14 - 20 - 5 + 3 = -6 - 5 + 3$$
$$= -11 + 3$$
$$= -8$$

So $14 - [20 - (5 + 3)]$ and $14 - 20 - 5 + 3$ are very different!

2.5   MULTIPLYING INTEGERS

The first thing we must decide is how to express multiplication. Arithmetic usually uses $\times$ and sometimes $\cdot$ Algebra, on the other hand, seldom uses $\times$ because it is too easily confused with the letter x. In preparation for algebra, therefore, we will represent all multiplication by $\cdot$

For positive integers, we know what happens when we multiply them. For example, $2 \cdot 4 = 8$, because if you have two 4's and add them, you get 8. For the same reason, $2 \cdot (-4) = -8$, because if you add two (-4)'s you get -8.

What about $(-2) \cdot 4$? One of the things we know is that $(-2) \cdot 4$ had better be the same as $4 \cdot (-2)$, because we don't want it to matter which order we multiply two numbers in—the result should always come out the same. Now $4 \cdot (-2) = -8$, because if you lay out four (-2)'s in a row and add them, you'll have -8. So $(-2) \cdot 4 = 4 \cdot (-2) = -8$. Since $2 \cdot (-4) = -8$ and $(-2) \cdot 4 = -8$, clearly the quickest way to multiply two numbers when one is negative is to just do the multiplication ignoring the signs, and then change the sign of the answer, making it negative. For example:

$$(-3) \cdot 2 = -6$$
$$3 \cdot (-7) = -21$$

How can we use this for $(-2) \cdot 4$? When you have just one $-$ sign you ignore that sign, do the multiplication, and change the sign of the answer. For two $-$ signs, then, you must ignore both signs, do the multiplication, and change the sign of the answer *twice*, which takes you back to 8. So $(-2) \cdot (-4) = 8$.

Similarly,

$$(-3) \cdot (-4) = 12$$
$$(-1) \cdot (-5) = 5$$

We therefore end up with the following scheme for multiplication:

+ times + = +
+ times − = −
− times + = −
− times − = +

$$2 \cdot \quad 4 = \quad 8$$
$$2 \cdot (-4) = -8$$
$$(-2)\cdot \quad 4 = -8$$
$$(-2)\cdot (-4) = \quad 8$$

2.6   DIVIDING INTEGERS

Division is closely connected to multiplication, and so it is reasonable to expect that we should be able to get the rules for division from the rules for multiplication. For example, $\frac{6}{3} = 2$ because 2 is the number that you must multiply 3 by to get 6. In other words,

$$6 = 3 \cdot \text{②} \quad \text{and so} \quad \frac{6}{3} = \text{②}$$

What is $\frac{-6}{3}$? It must be −2, because −2 is the number that you must multiply 3 by to get −6. In other words,

$$-6 = 3 \cdot \text{(−2)} \quad \text{and so} \quad \frac{-6}{3} = \text{(−2)}$$

What about $\frac{6}{-3}$? We know that

$$6 = (-3)\cdot -2 \quad \text{so} \quad \frac{6}{-3} = -2$$

The last case is $\frac{-6}{-3}$.

$$-6 = (-3) \cdot 2 \quad \text{so} \quad \frac{-6}{-3} = 2$$

More generally, we end up with the following scheme:

+ over + = +
− over + = −
+ over − = −
− over − = +

Note that the multiplication and division rules can be summarized together:

> Multiplying ⎱ two numbers of the *same*
> Dividing ⎰ sign gives a + answer.
>
> Multiplying ⎱ two numbers of *opposite*
> Dividing ⎰ sign gives a − answer.

2.7　POWERS OF INTEGERS

Just as $2 + 2 + 2 + 2$ can be written more compactly as $4 \cdot 2$, so $2 \cdot 2 \cdot 2 \cdot 2$ can be written more briefly as 2^4 where 2^4 or *two to the fourth power*, is defined to be *four 2's multiplied together*:

$$2^4 = 2 \cdot 2 \cdot 2 \cdot 2$$

Similarly, 2^3, or two to the third power, usually called two *cubed*, is three 2's multiplied together:

$$2^3 = 2 \cdot 2 \cdot 2$$

Two *squared*, the usual name for two to the second power, or 2^2, is two 2's multiplied together:

$$2^2 = 2 \cdot 2$$

2^1, therefore, should mean one 2 multiplied together, or just 2. So:

$$2^1 = 2$$

EXAMPLE:　2^5, or two to the fifth power, means five 2's multiplied together:

$$2^5 = 2 \cdot 2 \cdot 2 \cdot 2 \cdot 2 = 32$$

EXAMPLE:　$(-3)^2$, or −3 squared, means two (-3)'s multiplied together:

$$(-3)^2 = (-3) \cdot (-3) = 9$$

EXAMPLE:　$(-1)^3 = (-1) \cdot (-1) \cdot (-1) = -1.$

EXAMPLE:　Notice that 2^2 *and* $(-2)^2$ are the same, since

$$2^2 = 2 \cdot 2 = 4$$
$$(-2)^2 = (-2) \cdot (-2) = 4$$

EXAMPLE:　2^3 *and* $(-2)^3$, however, are not the same:

$$2^3 = 2 \cdot 2 \cdot 2 = 8$$
$$(-2)^3 = (-2) \cdot (-2) \cdot (-2) = -8$$

CHAPTER 2 EXERCISES

1. On a large sheet of paper, draw the number line. Plot the following numbers on the number line:

(a) 0   (b) 1   (c) 2   (d) 3   (e) -1 (f) -2

(g) -4 (h) $\dfrac{1}{2}$ (i) $\dfrac{3}{2}$ (j) -2.5 (k) -1.5 (l) -6.1

(m) $\sqrt{2}$ (n) π

2. Each of the points A, B, C, D, E, F, G in Figure 2.17 represents one of the following numbers.

$$\frac{-20}{9}, \; -0.5, \; 0.9, \; -0.9, \; \frac{-12}{10}, \; \frac{11}{10}, \; \sqrt{2}$$

Indicate which number is represented by each point.

FIG. 2.17

3. Each of the points H, I, J, K, L, M in Figure 2.18 represents one of the following numbers.

$$\frac{-27}{8}, \; 2.8, \; \frac{9}{10}, \; -\sqrt{2}, \; -\sqrt{8}, \; \pi$$

Indicate which number is represented by each point.

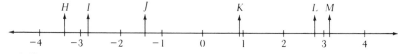

FIG. 2.18

Add the following on the number line. Draw a separate number line for each problem.

4. $1 + 4$ 15. $(-12) + (-2)$

5. $4 + 1$ 16. $4 + 3 + 1$

6. $1 + (-4)$ 17. $4 + 1 + 3$

7. $(-4) + 1$ 18. $3 + 2 + (-1)$

8. $4 + (-1)$ 19. $3 + (-2) + (-1)$

9. $1(-1) + 4$ 20. $(-2) + 4 + (-3)$

10. $(-1) + (-4)$ 21. $(-7) + 3 + 3$

11. $(-4) + (-1)$ 22. $(-2) + (-3) + (-1)$

12. $2 + 2$ 23. $1 + (-2) + 2$

13. $6 + 1$ 24. $3 + 4 + (-5) + (-1)$

14. $7 + (-3)$ 25. $(-2) + (-1) + (-3) + (-6)$

Subtract the following on the number line. Draw a separate number line for each problem.

26. $3 - 2$
27. $3 - 4$
28. $6 - 3$
29. $-3 - 2$
30. $-5 - 1$
31. $-2 - (-1)$
32. $8 - (-3)$
33. $2 - 9$
34. $2 - (-9)$
35. $-3 - (-4)$

36. $6 - (-2)$
37. $(-2) - 2 - 2$
38. $(-3) - (-1) - (-4)$
39. $6 - (-5) - 12$
40. $(-1) - 1 - (-1)$
41. $2 - (-3) - (-4)$
42. $(-2) - 3 - (-1)$
43. $3 - 8 - (-9)$
44. $3 - (-8) - 9$
45. $9 - 2 - (-3)$

Rewrite each of the following so that there is only one sign between two adjacent numbers, and then solve the problem on a number line.

46. $8 + (+4)$
47. $8 + (-4)$
48. $8 - (+4)$
49. $6 - (-3)$
50. $(-6) - (+3)$
51. $7 + (-2)$
52. $(-3) - (-11)$

53. $(-2) - (-6)$
54. $(-1) - 6 - (-7) - (+8)$
55. $(-1) - (-2) - (-3) - (+4)$
56. $0 - (-1) + (-2) + (-5) - (-7)$
57. $2 - (-3) + (-1)$
58. $(-2) - (+3) + (+2)$

Evaluate the following:

59. $(2 - 5) - (6 - 3)$
60. $(2 - 5) - (3 - 6)$
61. $2 - 5 - 6 - 3$
62. $2 - (5 - 6 - 3)$
63. $2 - (5 - 6) - 3$
64. $-4 - (3 + 2)$

65. $-4 - 3 + 2$
66. $-4 - (3 - 2)$
67. $1 - 4 - 2 - (-7)$
68. $-(1 - 4) + -(2 - 7)$
69. $(-1 - 3) - (2 - 6)$
70. $12 - [8 - (6 + 4)]$

71. $(12 - 8) - [6 + 4 - (2 + 1)]$
72. $1 - \{1 - [3 + (-1 - 2)]\}$
73. $[12 - (2 + 6)] - (3 - \{4 + [5 - (-1)]\})$
74. $-9 - (3 - 4 - \{2 + 3 - 4 + [6 - (-1)]\} + 6 - 2) + 7$

Evaluate the following:

75. $2 \cdot (-3)$
76. $(-2) \cdot 3$
77. $(-2) \cdot (-3)$
78. $3 \cdot (-2) \cdot (-4)$
79. $(-3) \cdot (-3) \cdot (-2)$
80. $(-3) \cdot 2 \cdot 6$
81. $(-1) \cdot 2 \cdot (-6) \cdot (-8)$
82. $(-3) \cdot 7 \cdot (-4) \cdot (-2) \cdot 2$

83. $\dfrac{3 \cdot 4}{(-2) \cdot (-3)}$

84. $\dfrac{(-3) \cdot (-4)}{(-2) \cdot 3}$

85. $\dfrac{3 \cdot 4 \cdot 8}{(-12) \cdot (-4)}$

86. $\dfrac{(-3) \cdot (-4) \cdot 8}{(-12) \cdot 4}$

87. $\dfrac{(-6) \cdot (-8) \cdot (-10)}{(-5) \cdot (-4)}$

Evaluate the following:

88. 1^3	95. 4^3	
89. 1^4	96. $(-3)^3$	102. $(-2)^6$
90. $(-1)^2$	97. $(-4)^2$	103. 2^6
91. $(-1)^4$	98. $(-4)^3$	104. $4 \cdot 4^2$
92. 3^2	99. 0^2	105. $2 \cdot 2^5$
93. 3^3	100. $(-2)^4$	106. $2^3 \cdot 2^3$
94. 4^2	101. $(-2)^5$	107. $2^2 \cdot 2^4$

108. Federal income tax laws permit a net loss resulting from the sale of stocks to be deducted from ordinary income (i.e., wages and salaries), subject to a $2000 ceiling on such deductions in any one year. Mrs. Dow Jones sells four stocks during the year with the following results:

Stock A	$7000 gain
Stock B	$8100 loss
Stock C	$7300 loss
Stock D	$6000 gain

(a) What is Jones's net loss?

(b) If her income before the stock loss deduction is $32,000, what will it be after the deduction?

(c) How much, if any, of Jones's loss is nondeductible because of the $2000 ceiling?

109. Gullett Razor Company is figuring out its overall profits for the year before deciding whether or not to build a new headquarters; if profits are less than $12 million, the company will make do with its old building. Gullett's various divisions report the following results for the year:

Razor division	$ 8 million profit
Hair dryer division	$10 million loss
Pen division	$ 6 million profit
Shaving cream division	$ 5 million profit
Razor blade division	Unreported

Will the company build a new headquarters on the basis of the profits reported by the first four divisions? How much profit does the razor blade division have to show before the company will go ahead with its building plans?

110. The Nailo Hardware Company made a profit of $125 on its opening day of business, lost $250 (because of a theft) on its second day, and made $175 profit on the third day.

(a) If the Nailo Company started with $50 cash on hand on opening day, how much did it have at the end of the third day?

(b) At the end of the fourth day of business the Nailo Company had $25 more than it had at the beginning of the first day. What was the outcome of the fourth day of business?

111. Ken made the following successive trips on Interstate 80, an east–west road: 150 miles to the east, 170 miles to the west, 20 miles to the west, 43 miles to the east.

(a) At the end of the fourth trip, is Ken east or west of his starting point?

(b) How far must Ken go to return to his starting point?

112. The Chicago Bears football team gained 4 yards on the first play, lost 5 yards on the second play, gained 7 yards on the third play, and gained 3 yards on the fourth play. If they started at the 50-yard line, where were they at the completion of the fourth play?

113. The boiling point of water is 100° Celsius (centigrade), that of alcohol is 78°C, and that of helium (a gas at normal temperatures) is −269°C.

(a) Through how many degrees must the temperature of a substance be raised if its temperature is to be increased from that of boiling alcohol to that of boiling water?

(b) If it is to be raised from that of boiling helium to that of boiling alcohol?

114. Star football player Joe led the conference in rushing last year, gaining 589 yards. Stan was not so fortunate; he lost yardage every time he got the ball and ended up losing 112 yards for the year. To make things even worse, Stan and Joe had made a friendly wager before the season, the loser agreeing to pay the winner $1 for every yard he was outgained. How much does Stan owe Joe?

115. At a recent Weight Watchers meeting, the members reported the following changes from the week before:

Member	Result
A	gained 2 pounds
B	lost 3 pounds
C	lost 2 pounds
D	lost 7 pounds
E	gained 6 pounds
F	gained 1 pounds
G	lost 4 pounds

What was the total gain or loss of the club?

116. Quabbin Reservoir, located in the western part of Massachusetts, supplies most of the water for the Boston area. The water from Quabbin is pumped into several holding reservoirs in the immediate vicinity of Boston, and Boston and its suburbs draw water from these holding reservoirs. At the end of each day, Quabbin refills the holding reservoirs to replace exactly the water drawn out during the day. Holding reservoirs A, B, and C start with

300 million gallons, 250 million gallons, and 150 million gallons at the beginning of each day. Assume that suburbs a_1 and a_2 each take 125 million gallons daily from A, suburbs b_1, b_2, and b_3 each take 25 million gallons daily from B, and Boston takes 137 million gallons daily from C. If Quabbin starts with 7000 million gallons on Monday morning and receives 400 million gallons of rain on Tuesday morning, how much water is left in Quabbin after it refills the holding reservoirs at the end of the day on Tuesday?

117. Ned the newspaper dealer opened his stand on Monday morning with 75 magazines on hand. After selling 38 magazines, the delivery truck dropped off 50 copies of the *Newsweek*. Ned sold 55 more magazines before the truck returned and dropped off 20 copies of the *Time*. If Ned always goes home once he has sold all but 10 magazines, how many more magazines must he sell before he can leave for the day?

118. The bottom of Death Valley is 289 feet below sea level. Telescope Peak (a mountain on the edge of the valley) is 10,811 feet high (measured from sea level).

(a) How far is it from the bottom of Death Valley to the top of Telescope Peak?

(b) If the temperature at the bottom of Death Valley is 60°C and drops by 1°C for every 300 feet as you climb up the mountain, what is the temperature at the top?

3 THE ORDER OF OPERATIONS AND THE DISTRIBUTIVE LAW

3.1 THE ORDER OF OPERATIONS AND PARENTHESES

Let us look at $2 \cdot 3 + 5$. This could mean multiply 2 and 3, giving 6, and add 5, giving 11. Alternatively, it could mean add 3 and 5, giving 8, and multiply by 2, giving 16. A decision is needed to settle which of these operations, the multiplication or the addition, should be done first.

Again, $3 \cdot 2^2$ could mean multiply 3 by 2, giving 6, and then square, giving 36. Alternatively, it could mean square 2, giving 4, and then multiply by 3, giving 12.

In order to decide such questions, mathematics resorts to a convention:

The order of operations

1. Parentheses, innermost first
2. Exponents (squaring, taking the cube, etc.)
3. Multiplication and division, working from left to right
4. Addition and subtraction, working from left to right

Hence, $2 \cdot 3 + 5 = (2 \cdot 3) + 5 = 6 + 5 = 11$, because multiplication comes before addition. Similarly, $3 \cdot 2^2 = 3 \cdot (2^2) = 3 \cdot 4 = 12$, because squaring comes before multiplication. If we have $5 - 2 \cdot (4 - 3)$, we must first do the parentheses:

$$5 - 2 \cdot (4 - 3) = 5 - 2 \cdot 1$$

then the multiplication:

$$= 5 - 2$$

and then the subtraction:

$$= 3$$

Working out the parentheses may itself involve using the order of operations within the parentheses.

EXAMPLE: $10 + (2 \cdot 3^2 - 8) \cdot 3 = 10 + (2 \cdot 9 - 8) \cdot 3$ (Inside the parentheses, exponents first, then

$$= 10 + (18 - 8) \cdot 3$$ multiplication, then subtraction)

$$= 10 + (10) \cdot 3$$ (Parentheses are now unnecessary. Finish

$$= 10 + 30$$ problem using order of operations again)

$$= 40$$

EXAMPLE: *Find* $2 \cdot [4 - (2 \cdot 3^2 - 17)]$.

Remember to work from the innermost parentheses outward:

$2 \cdot [4 - (2 \cdot 3^2 - 17)] = 2 \cdot [4 - (2 \cdot 9 - 17)]$ Inner parentheses: first exponents, then multiplication,

$$= 2 \cdot [4 - (18 - 17)]$$ then subtraction,

$$= 2 \cdot [4 - 1]$$ now outer parentheses.

$$= 2 \cdot 3$$

$$= 6$$

EXAMPLE: *Simplify* $\dfrac{2 \cdot 4 + 7}{3}$.

Here the fraction bar does the work of parentheses. $\dfrac{2 \cdot 4 + 7}{3}$

means that all of $2 \cdot 4 + 7$ must be divided by 3, so

$$\frac{2 \cdot 4 + 7}{3} = \frac{(2 \cdot 4 + 7)}{3}$$

$$= \frac{8 + 7}{3}$$

$$= \frac{15}{3}$$

$$= 5$$

EXAMPLE: *Simplify* $\dfrac{2 \cdot 3^2 - 8}{5} + 2.$

The 5 is to be divided into the whole of $2 \cdot 3^2 - 8$, so parentheses around this are understood:

$$\frac{2 \cdot 3^2 - 8}{5} + 2 \;=\; \frac{(2 \cdot 3^2 - 8)}{5} + 2$$

$$=\; \frac{(2 \cdot 9 - 8)}{5} + 2$$

$$=\; \frac{(18 - 8)}{5} + 2$$

$$=\; \frac{(10)}{5} + 2$$

$$=\; 2 + 2$$

$$=\; 4$$

Please remember: the order of operations is a convention—it is something that has been arbitrarily decided, not reasoned out. So there's no way to figure it out—you just have to learn it.

3.2   THE DISTRIBUTIVE LAW

The order of operations tells you that

$$2 \cdot (3 + 4) = 2 \cdot 7 = 14$$

But there's another way you could have reasoned this out: if you multiply 2 and 3, giving 6, and multiply 2 and 4, giving 8, and then add, you get 14. And so,

$$2 \cdot (3 + 4) = 14 = 2 \cdot 3 + 2 \cdot 4$$

The reason this works is that if you add 3 and 4, you have 7, which looks like this:

••• ••••

Multiplying the 7 by 2:

••• ••••
••• ••••

But this could be regrouped:

here is $2 \cdot 3$ ⟶ ⟨∴ ∴⟩   ⟨∴∴ ∴⟩ ⟵ here is $2 \cdot 4$

So $2 \cdot 7 = 2 \cdot 3 + 2 \cdot 4$

or

$$2 \cdot (3 + 4) = 2 \cdot 3 + 2 \cdot 4$$

In general, if a, b, and c are any numbers:

$$a \cdot (b + c) = a \cdot b + a \cdot c$$

This is called the *distributive law of multiplication over addition* because the a is distributed over the b and the c. It is being included in the arithmetic chapters so that you can see where it comes from and why it is true. However it is not particularly useful in arithemetic because you can work out $2 \cdot (3 + 4)$ just as easily as $2 \cdot 3 + 2 \cdot 4$, but it is of the greatest importance in algebra.

Multiplication distributes over subtraction in the same way. For example,

$$4 \cdot (5 - 3) = 4 \cdot 2 = 8 \quad \text{and} \quad 4 \cdot 5 - 4 \cdot 3 = 20 - 12 = 8$$

so

$$4 \cdot (5 - 3) = 4 \cdot 5 - 4 \cdot 3$$

Therefore, in general,

$$a \cdot (b - c) = a \cdot b - a \cdot c$$

The distributive law is often used to remove parentheses following a minus sign. For example,

$$-(3 + 8) \quad \text{means} \quad (-1) \cdot (3 + 8) = (-1) \cdot 3 + (-1) \cdot 8$$
$$= -3 + (-8)$$
$$= -3 - 8$$

In general,

$$-(b + c) = -b - c$$

This is often called "distributing minus signs."

You may remember being told that when you take away parentheses that have a minus sign in front, you must change the signs of *all* the terms inside the parentheses. That is simply another way of saying $-(b + c) = -b - c$.

CHAPTER 3 EXERCISES

Evaluate the following:

1. $-4 \cdot (3 - 8)$

2. $\dfrac{4 - (-5)}{3}$

3. $\dfrac{(-4 - 5)}{3}$

4. $\dfrac{1 - (-1 - 3)}{-4 - (9 - 18)}$

5. $-\left(\dfrac{-288 + (-3) \cdot (-100)}{1 - 2}\right)$

6. $(4 \cdot 2^2 - 1) \cdot 3 - 2$

7. $2 \cdot 3^2 - (2 \cdot 3)^2$

8. $\dfrac{-1 \cdot \{[-1 + (-1)^2]^2 - 1\}^2 + 2}{3^2 - 2^2 - 1^2}$

9. $4 + \dfrac{2 \cdot [(2^2 - 1)^2 - 6]}{7^2 - 6^2 + 4 \cdot (-3)} - 1$

10. $\dfrac{[2 \cdot (19 \cdot 2 - 3 \cdot 12)^2 + 5^2 - 4^2 - 3 \cdot 2^2]^3}{\{[(1^2 - 2)^2 + 3]^3 - 4^2 \cdot 3 - 2^2 \cdot 3\}^2}$

For each of problems 11–15 you may pick any integral number you wish to fill in the blanks, and then evaluate. (The answers will not depend on the number you pick!)

11. $\dfrac{2 \cdot (\underline{\quad} + 5) - 4}{2} + (-\underline{\quad})$

12. $\dfrac{2 \cdot (\underline{\quad} + 3) + 4}{2} + (-\underline{\quad})$

13. $\dfrac{[\underline{\quad} + (\underline{\quad} + 1) + 7]}{2} + (-\underline{\quad})$

14. $\dfrac{[(2 \cdot \underline{\quad}) + 9 + \underline{\quad}]}{3} + 4 + (-\underline{\quad})$

15. $\dfrac{[(3 \cdot \underline{\quad}) + (\underline{\quad} + 1) + 11]}{4} - (\underline{\quad})$

Evaluate the following, first using the distributive law and then without it.

16. $5 \cdot (3 - 2)$

17. $(2 - 3) \cdot 2$

18. $4 \cdot (3 - 5) - 2 \cdot (1 - 7)$

19. $5 \cdot [5 + (-4) - 1]$

20. $20 - \{2 \cdot [-(4 - 1)] - 3 \cdot (-4^2)\}$

21. Suppose you own two stocks, one of which pays you a dividend of 85¢ a share, the other $1.15 a share. If you own 35 shares of each stock, you will receive (0.85 · 35) dollars in dividends from the first stock and (1.15 · 35) dollars from the second. Use the distributive law to figure out your total dividends.

22. Use the distributive law to find the total area of two rectangular regions, one of which is 6 inches by 8 inches, and the other 12 inches by 8 inches.

Use the distributive law to solve the following problems:

23. Tickets for loge seats at Symphony Hall cost $10.25 apiece, while tickets for the mezzanine cost $9.75 apiece. What is the total cost of 77 seats in each section?

24. Mr. Footloose and his wife drove from Chicago to St. Louis at an average speed of 57 miles per hour for 4 hours. Then his wife drove back to Chicago at an average speed of 43 miles per hour. How far from Chicago was she after 4 hours?

25. My Datsun gets 33 miles per gallon. Last week, I used 4 gallons of gas on Tuesday, 3 on Wednesday, 1 on Thursday, 3 on Friday, and 9 on Saturday. How many miles did I drive?

4 FRACTIONS REVISITED

4.1 MULTIPLYING FRACTIONS

It may sound strange to talk about multiplication of fractions before addition, but multiplication is much easier than addition, so it makes sense to do it first.

The Word 'of' and Multiplication

If we have two *of* some number, it means the same as two *times* that number, and if we have four *of* some other number, then we have four *times* that number. By the same reasoning, half *of* some number should mean $\frac{1}{2}$ *times* that number. We will use this to work out $\frac{1}{2} \cdot \frac{1}{3}$.

$\frac{1}{2} \cdot \frac{1}{3}$ means half of a third. If we take the length from 0 to 1 on the number line and divide it into 3, then each length is $\frac{1}{3}$. See Figure 4.1. Let us now cut each third in half, as in Figure 4.2. This cuts the length from 0 to 1 into six pieces, and so each one must have length $\frac{1}{6}$. Hence $\frac{1}{2} \cdot \frac{1}{3} = \frac{1}{6}$.

FIG. 4.1

FIG. 4.2

Notice that you can get this answer by multiplying the tops and the bottoms of the original fractions separately:

$$\frac{1}{2} \cdot \frac{1}{3} = \frac{1 \cdot 1}{2 \cdot 3} = \frac{1}{6}$$

In general,

> To multiply fractions, multiply the tops and multiply the bottoms.

EXAMPLE: $\quad \dfrac{1}{2} \cdot \dfrac{3}{4} = \dfrac{1 \cdot 3}{2 \cdot 4} = \dfrac{3}{8}$

EXAMPLE: $\quad \dfrac{2}{5} \cdot \dfrac{5}{2} = \dfrac{2 \cdot 5}{5 \cdot 2} = \dfrac{10}{10} = 1 \qquad [10 \text{ into } 10 \text{ goes once, so } \dfrac{10}{10} = 1]$

Integers and fractions can be multiplied this way too.

EXAMPLE: $\quad 2 \cdot \dfrac{1}{3} = \dfrac{2}{1} \cdot \dfrac{1}{3} = \dfrac{2 \cdot 1}{1 \cdot 3} = \dfrac{2}{3}$

So can negative fractions.

EXAMPLE: $\quad \left(\dfrac{-2}{5}\right) \cdot \left(\dfrac{-1}{3}\right) = \dfrac{(-2) \cdot (-1)}{5 \cdot 3} = \dfrac{2}{15}$

4.2 CANCELLING

To get $\frac{3}{6}$ we cut the number line from 0 to 1 into 6 pieces and take three of them, (see Figure 4.3.) Clearly $\frac{3}{6}$ is half the length from 0 to 1, so $\frac{3}{6} = \frac{1}{2}$.

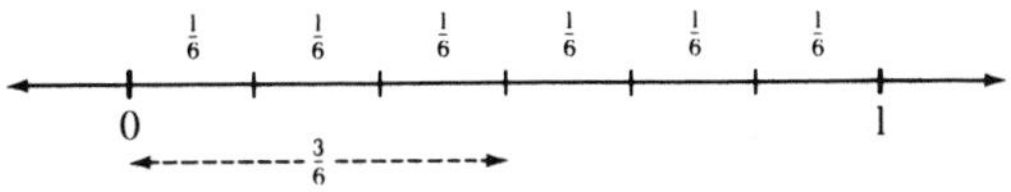

FIG. 4.3

Another way to do this is to *"cancel"* a 3 from the top and bottom of the fraction—meaning to divide 3 into the top and bottom of the fraction—leaving a 1 on top, and a 2 on the bottom. This is usually written:

$$\frac{\overset{1}{\cancel{3}}}{\underset{2}{\cancel{6}}} = \frac{1}{2}$$

But why can we do this (apart from the fact that it gives the right answer)? It is very important that you see why cancelling works. People's most frequent algebra mistakes are cancelling mistakes, which wouldn't happen if they knew why and how cancelling worked.

So here is the reason. Since $3 = 3 \cdot 1$ and $6 = 3 \cdot 2$, you can think of $\frac{3}{6}$ as the product of $\frac{3}{3}$ and $\frac{1}{2}$:

$$\frac{3}{6} = \frac{3 \cdot 1}{3 \cdot 2} = \frac{3}{3} \cdot \frac{1}{2} \qquad \text{(this is the rule for multiplying fractions written backwards)}$$

But $\frac{3}{3}$ is 1, so

$$\frac{3}{6} = 1 \cdot \frac{1}{2} = \frac{1}{2}$$

Now let's look at $\frac{120}{80}$. We can remove a 10 from both the top and bottom like this:

$$\frac{120}{80} = \frac{10 \cdot 12}{10 \cdot 8} = \frac{10}{10} \cdot \frac{12}{8} = 1 \cdot \frac{12}{8} = \frac{12}{8}$$

But we can go even further and take a 4 out of the top and bottom, too:

$$\frac{12}{8} = \frac{4 \cdot 3}{4 \cdot 2} = \frac{4}{4} \cdot \frac{3}{2} = 1 \cdot \frac{3}{2} = \frac{3}{2}$$

This whole process is usually written:

$$\frac{\overset{\overset{3}{\cancel{12}}}{\cancel{120}}}{\underset{\underset{2}{\cancel{8}}}{\cancel{180}}} = \frac{\overset{3}{\cancel{12}}}{\underset{2}{\cancel{8}}} = \frac{3}{2}$$

Cancelling can be continued as long as there are numbers that divide into both the top and the bottom of the fraction. Such numbers are called *common factors*. When there are no longer any common factors, the fraction cannot be *reduced* or simplified any more. It is then said to be in *lowest terms*.

Note on Negative Fractions

The fractions $-\frac{2}{3}, \frac{-2}{3}, \frac{2}{-3}$ *are all the same.*

Since $-$ over $+$ and $+$ over $-$ are both $-$ (from the division scheme) $\frac{-2}{3}$ and $\frac{2}{-3}$ are both $-$, and so both must be $-\frac{2}{3}$.

Therefore

$$\frac{-2}{3} = -\frac{2}{3} = \frac{2}{-3}$$

Also

$$\frac{-2}{-3} = \frac{2}{3}$$

because $-$ over $-$ is $+$ (from the division scheme), or because cancelling -1 top and bottom in $\frac{-2}{-3}$ gives $\frac{2}{3}$. Notice also that cancelling -1 top and bottom in $\frac{-2}{3}$ converts it to $\frac{2}{-3}$, thus: $\frac{-2}{3} = \frac{\cancel{-1} \cdot 2}{\cancel{-1} \cdot (-3)} = \frac{2}{-3}$.

4.3 ADDING FRACTIONS

How do we find out what $\frac{1}{2} + \frac{1}{3}$ *is?* The first thing you might think of is to add the tops and to add the bottoms, giving $\frac{2}{5}$. Unfortunately, $\frac{2}{5}$ is smaller than $\frac{1}{2}$ and so it couldn't be right, since if you add something (even something as small as $\frac{1}{3}$) to $\frac{1}{2}$, you must come out with something larger than $\frac{1}{2}$. So adding the tops and the bottoms is certainly wrong. If we try to add $\frac{3}{5} + \frac{1}{5}$, the problem is easy and the answer is $\frac{4}{5}$. To see this, cut the segment from 0 to 1 into five pieces, each of length $\frac{1}{5}$, and take three of them (representing $\frac{3}{5}$), and add another (representing $\frac{1}{5}$) onto the end; you will have four such pieces, or $\frac{4}{5}$ (see Figure 4.4.)

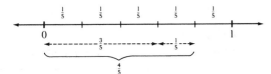

FIG. 4.4

The problem is easy here because the "pieces" are all the same size (namely $\frac{1}{5}$) and so adding them simply means adding the tops (since the top of the fraction tells you how many such pieces you have).

What about $\frac{1}{2} + \frac{1}{4}$? Here the pieces are not of the same length, but if you cut the $\frac{1}{2}$ in half, and make it into two pieces of length $\frac{1}{4}$, then you can add these two pieces to the other $\frac{1}{4}$ and get $\frac{3}{4}$, (see Figure 4.5.) This shows that $\frac{1}{2} + \frac{1}{4} = \frac{3}{4}$.

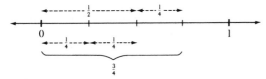

FIG. 4.5

Mathematically, cutting $\frac{1}{2}$ into two pieces, each of length $\frac{1}{4}$, means writing $\frac{1}{2}$ as $\frac{2}{4}$ (cancelling shows that $\frac{2}{4}$ and $\frac{1}{2}$ are exactly the same). So what we did was this:

$$\frac{1}{2}+\frac{1}{4}=\frac{2}{4}+\frac{1}{4}=\frac{2+1}{4}=\frac{3}{4}$$

In adding $\frac{1}{2}+\frac{1}{3}$ we have the same problem, that the "pieces" are not the same size. However, if we cut the $\frac{1}{2}$ into three pieces, each will be $\frac{1}{6}$; and if we cut the $\frac{1}{3}$ into two pieces, each will be $\frac{1}{6}$ (see Figure 4.6.) So altogether, the $\frac{1}{2}$ and the $\frac{1}{3}$ give us five pieces, each of length $\frac{1}{6}$. Thus we have shown that $\frac{1}{2}+\frac{1}{3}=\frac{5}{6}$.

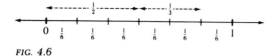

FIG. 4.6

In mathematical language, what we did was to replace $\frac{1}{2}$ by $\frac{3}{6}$, and $\frac{1}{3}$ by $\frac{2}{6}$ (cancelling shows that $\frac{3}{6}=\frac{1}{2}$, and $\frac{2}{6}=\frac{1}{3}$). So

$$\frac{1}{2}+\frac{1}{3}=\frac{3}{6}+\frac{2}{6}$$

$$=\frac{3+2}{6} \qquad \text{(since the ``pieces'' are the same size, we can just add the tops)}$$

$$=\frac{5}{6}$$

A method for adding fractions is now clear: first you have to convert the fractions into new fractions all with the same denominators by "uncancelling" (the opposite of cancelling), and then you can add the numerators. The main problem in this, obviously, is getting all the fractions over the same denominator (called a *common denominator*), and so the next section will be devoted to that alone.

4.4 FINDING A COMMON DENOMINATOR

One possible common denominator is the product of the original denominators. For example, for

$$\frac{1}{6}+\frac{3}{8}$$

$6 \cdot 8 = 48$ is a possible denominator. Rewriting $\frac{1}{6}$ with 48 in the denominator involves multiplying $\frac{1}{6}$ by $\frac{8}{8}$, which does not change its value since $\frac{8}{8} = 1$.

$$\frac{1}{6} = \frac{1}{6} \cdot 1 = \frac{1}{6} \cdot \frac{8}{8} = \frac{8}{48}$$

Rewriting $\frac{3}{8}$ with 48 in the denominator involves multiplying $\frac{3}{8}$ by $\frac{6}{6}$, which again does not change its value since $\frac{6}{6} = 1$.

$$\frac{3}{8} = \frac{3}{8} \cdot 1 = \frac{3}{8} \cdot \frac{6}{6} = \frac{18}{48}$$

So,

$$\left(\frac{1}{6}\right) + \left(\frac{3}{8}\right) = \left(\frac{8}{48}\right) + \left(\frac{18}{48}\right) = \frac{26}{48}$$

But

$$\frac{26}{48} = \frac{13}{24}$$

by cancelling a 2 from top and bottom, so

$$\frac{1}{6} + \frac{3}{8} = \frac{13}{24}$$

In general therefore:

> Fractions can be added by rewriting them all with a common denominator, and then adding the numerators.

4.5   FINDING AND USING A LEAST COMMON DENOMINATOR

In the example above, 48 is not the only possible common denominator. First notice that the way we convert a fraction into a new one with the same value is to multiply by something of the form

$$\frac{6}{6} \quad \text{or} \quad \frac{8}{8} \quad \text{or} \quad \frac{a}{a}$$

and hence any new denominator is a multiple of the old one. For example,

$\frac{1}{2} = \frac{3}{6}$, and 6 is a multiple of 2

$\frac{1}{6} = \frac{8}{48}$, and 48 is a multiple of 6.

Therefore, any possible common denominator is a multiple of each of the original denominators.

Since $24 = 4 \cdot 6 = 3 \cdot 8$, 24 is a multiple of both 6 and 8, and will do as a common denominator.

To convert $\frac{1}{6}$ into 24th's, we must multiply by $\frac{4}{4}$, because the factor 4 is what the 6 needs to be multiplied by to make it into 24.

$$\frac{1}{6} = \frac{1}{6} \cdot 1 = \frac{1}{6} \cdot \frac{4}{4} = \frac{4}{24}$$

To convert $\frac{3}{8}$ into 24th's, multiply top and bottom by 3, because that will make the denominator 24.

$$\frac{3}{8} = \frac{3}{8} \cdot 1 = \frac{3}{8} \cdot \frac{3}{3} = \frac{9}{24}$$

Then

$$\left(\frac{1}{6}\right) + \left(\frac{3}{8}\right) = \left(\frac{4}{24}\right) + \left(\frac{9}{24}\right) = \frac{13}{24}$$

So we can add the fractions from the preceding section using a denominator smaller than 48. In general, multiplying the denominators together will always give a denominator that will work, but not always the least one. In many cases it doesn't matter whether or not you have the *least* common denominator, but in some cases, unfortunately, it does matter. Hence this is a section on the *least common denominator*. But first we must define *factors* and *prime factors*.

If we write a number as a product of several integers, that number is said to be *factored*. For example, by writing

$$21 = 3 \cdot 7$$

we have factored 21; the 3 and the 7 are called *factors* of 21. But if we write

$$21 = 2 \cdot 10 + 1$$

then 21 has not been factored, because the 1 that has been added means that 21 has not been written simply as a product.

If a factor cannot be factored any further, then it is called a *prime factor*. For example, 3 is a prime factor of 24, but 8 is not because

$$8 = 2 \cdot 4 = 2 \cdot 2 \cdot 2 = 2^3$$

However, if we write

$$24 = 3 \cdot 2^3 = 3 \cdot 2 \cdot 2 \cdot 2$$

then each of the factors 3 and 2 is prime. (Note: writing $3 = 3 \cdot 1$ or $2 = 2 \cdot 1$ does not count as factoring.)

To factor a number into primes, write the number as a product of two integers and then try to factor those integers. Keep going until none of the factors will factor any further—then they must all be prime and you are done.

EXAMPLE: *Factor 36*

Write ⟋ prime

$36 = 2 \cdot 18 \longleftarrow$ not prime

$= 2 \cdot 2 \cdot 9 \longleftarrow$ not prime

$= 2 \cdot 2 \cdot 3 \cdot 3 \longleftarrow$ all factors are prime, so could stop here

$= 2^2 \cdot 3^2 \longleftarrow$ answer usually left in this form

$36 = 2^2 \cdot 3^2$ *is called the prime factorization of 36.*

Note that we could just as well have started with

$$36 = 4 \cdot 9$$

and this would have led to the same prime factorization:

$$36 = 2^2 \cdot 3^2$$

**To Add Fractions Using
a Least Common Denominator (L.C.D.)**
(with $\frac{1}{6} + \frac{3}{8}$ as an example):

1. Factor each denominator into prime factors.

    $6 = 2 \cdot 3 \qquad 8 = 2^3$

2. List all the different prime factors occurring in any of the denominators. For each of these factors, pick the highest power to which it occurs in any one denominator, and write the factor to that power. Multiply together the factors to these powers. This is the L.C.D.

    Factors occurring are 2 and 3 ($6 = 2 \cdot 3$, $8 = 2^3$)

    Highest power of 2 occurring is 3 (in $8 = 2^3$)
    Highest power of 3 occurring is 1 (in $6 = 2 \cdot 3^1$)
    So write down 2^3 and 3^1 and multiply these together to give
    L.C.D. $= 2^3 \cdot 3^1 = 24$

3. Convert each fraction into one with the L.C.D. for a denominator. For each fraction, pick out those factors in the L.C.D. but not in the denominator and multiply top and bottom by these factors.

Since $6 = 2 \cdot 3$ and   L.C.D. $= 2^3 \cdot 3$

the factors occurring in the L.C.D. but not in 6 are 2^2. So take $\frac{1}{6}$ and multiply top and bottom by 2^2 or 4:

$$\frac{1}{6} = \frac{1}{6} \cdot \frac{4}{4} = \frac{4}{24}$$

Now $8 = 2^3$, so 3 is the only thing in the L.C.D. but not in 8. Therefore we multiply top and bottom by 3:

$$\frac{3}{8} = \frac{3}{8} \cdot \frac{3}{3} = \frac{9}{24}$$

4. Replace old fractions by new ones with the same value, but over the L.C.D. Since the denominators are now all the same, we can add the numerators.

$$\frac{1}{6} + \frac{3}{8} = \frac{4}{24} + \frac{9}{24} = \frac{13}{24}$$

EXAMPLE: *Find $\dfrac{1}{2100} + \dfrac{7}{90}$.*

STEP 1   Factor into prime factors:

$$2100 = 2^2 \cdot 3 \cdot 5^2 \cdot 7$$

$$90 = 2 \cdot 3^2 \cdot 5$$

STEP 2   Different prime factors are 2, 3, 5, 7.
Highest power of 2 occurring is 2 (in $2100 = 2^2 \cdot 3 \cdot 5^2 \cdot 7$)
Highest power of 3 occurring is 2 (in $90 = 2 \cdot 3^2 \cdot 5$)
Highest power of 5 occurring is 2 (in $2100 = 2^2 \cdot 3 \cdot 5^2 \cdot 7$)
Highest power of 7 occurring is 1 (in $2100 = 2^2 \cdot 3 \cdot 5^2 \cdot 7$)
So write down 2^2, 3^2, 5^2, 7^1 and multiply them to give

L.C.D. $= 2^2 \cdot 3^2 \cdot 5^2 \cdot 7$ (= 6300 if you want to multiply it out, but that's not necessary)

STEP 3   Since $2100 = 2^2 \cdot 3 \cdot 5^2 \cdot 7$ and the L.C.D. $= 2^2 \cdot 3^2 \cdot 5^2 \cdot 7$, the only factor in the L.C.D. but not in 2100 is a 3. Therefore you must multiply $\frac{1}{2100}$ top and bottom by 3 so that the denominator becomes the L.C.D.

$$\frac{1}{2100} = \frac{1}{2100} \cdot \frac{3}{3} = \frac{3}{6300}$$

Since $90 = 2 \cdot 3^2 \cdot 5$ and the L.C.D. $= 2^2 \cdot 3^2 \cdot 5^2 \cdot 7$, the factors in the L.C.D. but not in 90 are 2, 5, and 7. Now, $2 \cdot 5 \cdot 7 = 70$, and therefore you must multiply $\frac{7}{90}$ top and bottom by 70 so that the denominator becomes the L.C.D.

$$\frac{7}{90} = \frac{7}{90} \cdot \frac{70}{70} = \frac{490}{6300}$$

STEP 4   Now add:

$$\frac{1}{2100} + \frac{7}{90} = \frac{3}{6300} + \frac{490}{6300} = \frac{493}{6300}$$

4.6   SUBTRACTING FRACTIONS

Subtracting fractions works exactly like adding fractions except that after all the fractions have been rewritten over a common denominator, the numerators are subtracted.

EXAMPLE: *Find* $\dfrac{1}{7} - \dfrac{1}{9}.$

Factoring denominators: 7 is prime, $9 = 3^2$, so the L.C.D. $= 7 \cdot 3^2 = 7 \cdot 9 = 63$.

$$\frac{1}{7} = \frac{1}{7} \cdot \frac{9}{9} = \frac{9}{63}$$

$$\frac{1}{9} = \frac{1}{9} \cdot \frac{7}{7} = \frac{7}{63}$$

So

$$\frac{1}{7} - \frac{1}{9} = \frac{9}{63} - \frac{7}{63} = \frac{9-7}{63} = \frac{2}{63}$$

In general:

Fractions can be subtracted by rewriting them a common denominator and subtracting the numerators.

4.7 DIVIDING FRACTIONS

By now you have probably noticed the "golden rule of fractions".

Golden Rule of Fractions

You may multiply a fraction by anything (except 0), *provided that you do it to both the top and the bottom.*

This rule holds because anything (except 0) divided by itself is 1, and multiplying by 1 doesn't change the value of the fraction. So far we have used this idea to put the fraction over a new denominator. For example,

$$\frac{2}{5} = \frac{2}{5} \cdot \frac{4}{4} = \frac{8}{20}$$

It is also perfectly possible to multiply top and bottom by a fraction. For example,

$$\frac{4}{12} = \frac{4}{12} \cdot \frac{\frac{1}{2}}{\frac{1}{2}} = \frac{4 \cdot \frac{1}{2}}{12 \cdot \frac{1}{2}} = \frac{2}{6}$$

or

$$\frac{4}{12} = \frac{4}{12} \cdot \frac{\frac{1}{4}}{\frac{1}{4}} = \frac{4 \cdot \frac{1}{4}}{12 \cdot \frac{1}{4}} = \frac{1}{3}$$ (this is another way of looking at cancelling)

In order to see how this applies to the division of fractions, there is one more thing that you must notice: If you multiply a fraction by the same thing turned upside down, you will always get 1. For example,

$$\frac{2}{3} \cdot \frac{3}{2} = \frac{2 \cdot 3}{3 \cdot 2} = \frac{6}{6} = 1$$

$$\frac{5}{111} \cdot \frac{111}{5} = \frac{555}{555} = 1$$

$\frac{3}{2}$ is called the *reciprocal* of $\frac{2}{3}$, and $\frac{111}{5}$ the reciprocal of $\frac{5}{111}$.

The *reciprocal of a fraction* $\frac{a}{b}$ *is the fraction turned upside down, i.e.* $\frac{b}{a}$.

From the examples above, you can see that the reciprocal of a fraction is the number which, when multiplied by that fraction, gives 1. This property is sometimes used as the definition of a reciprocal, and it is certainly what makes them useful here.

What about the reciprocal of 3? Thinking of 3 as $\frac{3}{1}$ shows that the reciprocal of 3 is $\frac{1}{3}$. Then, as we would expect, the product of 3 and its reciprocal is 1:

$$3 \cdot \frac{1}{3} = \frac{3}{3} = 1$$

Now for dividing fractions. Suppose we have $\frac{\frac{5}{8}}{\frac{2}{3}}$. This looks like a real mess until you realize that if you multiply top and bottom by $\frac{3}{2}$, the bottom will become 1:

$$\frac{\frac{5}{8}}{\frac{2}{3}} = \frac{\frac{5}{8}}{\frac{2}{3}} \cdot \frac{\frac{3}{2}}{\frac{3}{2}} = \frac{\frac{5}{8} \cdot \frac{3}{2}}{\frac{2}{3} \cdot \frac{3}{2}} = \frac{\frac{5}{8} \cdot \frac{3}{2}}{1} = \frac{5}{8} \cdot \frac{3}{2}$$

Now $\frac{5}{8} \cdot \frac{3}{2}$ is a perfectly reasonable multiplication problem, which can be done in the usual way. So,

$$\frac{\frac{5}{8}}{\frac{2}{3}} = \frac{5}{8} \cdot \frac{3}{2} = \frac{5 \cdot 3}{8 \cdot 2} = \frac{15}{16}$$

In general, multiplying top and bottom by the reciprocal of the bottom fraction makes the problem into a multiplication; it leaves you with the top fraction multiplied by the reciprocal of the bottom. In other words:

> To divide two fractions, flip the bottom one over and multiply the top fraction by it.

As with cancelling, please remember where this rule comes from; otherwise you can end up with some startling (and wrong) results!

EXAMPLE: *Simplify* $\dfrac{\frac{4}{3}}{\frac{2}{3}}$

Since the top of this fraction is twice the bottom, the answer should come out to 2. (When using a new method, it's always a good idea to do a problem you already know the answer to, just as a check on the method.)

$$\frac{\frac{4}{3}}{\frac{2}{3}} = \frac{\frac{4}{3}}{\frac{2}{3}} \cdot \frac{\frac{3}{2}}{\frac{3}{2}} = \frac{\frac{4}{3} \cdot \frac{3}{2}}{\frac{2}{3} \cdot \frac{3}{2}} = \frac{\frac{12}{6}}{\frac{6}{6}} = \frac{\frac{12}{6}}{1} = \frac{12}{6} = 2$$

EXAMPLE: *Simplify* $\dfrac{\frac{2}{45}}{3}$

This doesn't look like division by a fraction; however, if you think of 3 as $\frac{3}{1}$, then it can be done by the same method as before:

$$\frac{\frac{2}{45}}{3} = \frac{\frac{2}{45}}{\frac{3}{1}} = \frac{2}{45} \cdot \frac{1}{3} = \frac{2}{135}$$

4.8   FORMULAS FOR WORKING WITH FRACTIONS

Multiplication

Two fractions are multiplied by multiplying the tops and the bottoms separately. Therefore the general formula is

$$\boxed{\frac{a}{b} \cdot \frac{c}{d} = \frac{a \cdot c}{b \cdot d}}$$

Addition

To add

$$\frac{a}{b} + \frac{c}{d}$$

we must first put both fractions over a common denominator. Without knowing b and d we cannot write a formula for the L.C.D, but we can use the fact that $b \cdot d$ is a possible common denominator (if not the least). Then

$$\frac{a}{b} = \frac{a}{b} \cdot \frac{d}{d} = \frac{a \cdot d}{b \cdot d}$$

and

$$\frac{c}{d} = \frac{c}{d} \cdot \frac{b}{b} = \frac{b \cdot c}{b \cdot d}$$

so

$$\frac{a}{b} + \frac{c}{d} = \frac{a \cdot d}{b \cdot d} + \frac{b \cdot c}{b \cdot d} = \frac{a \cdot d + b \cdot c}{b \cdot d}$$

Therefore:

$$\boxed{\frac{a}{b} + \frac{c}{d} = \frac{a \cdot d + b \cdot c}{b \cdot d}}$$

Subtraction

The formula for subtraction is exactly like that for addition:

$$\frac{a}{b} - \frac{c}{d} = \frac{a \cdot d - b \cdot c}{b \cdot d}$$

Division

Two fractions are divided by multiplying top and bottom by the reciprocal of the denominator, which is equivalent to inverting the bottom and multiplying.

$$\frac{\dfrac{a}{b}}{\dfrac{c}{d}} = \frac{\dfrac{a}{b} \cdot \dfrac{d}{c}}{\dfrac{c}{d} \cdot \dfrac{d}{c}} = \frac{\dfrac{a}{b} \cdot \dfrac{d}{c}}{1} = \frac{a}{b} \cdot \frac{d}{c}$$

So:

$$\frac{\dfrac{a}{b}}{\dfrac{c}{d}} = \frac{a}{b} \cdot \frac{d}{c} = \frac{a \cdot d}{b \cdot c}$$

CHAPTER 4 EXERCISES

Find the prime factorization of the following numbers:

1. 990
2. 7860
3. 30^2
4. 396
5. 135
6. 273
7. 152
8. $24^2 \cdot 36$
9. 2652
10. 2160

Find the least common denominator of the following fractions:

11. $\frac{1}{8}$ and $\frac{1}{24}$
12. $\frac{1}{9}$ and $\frac{1}{21}$
13. $\frac{1}{18}$ and $\frac{1}{54}$
14. $-\frac{1}{8}, \frac{1}{20},$ and $-\frac{1}{36}$
15. $\frac{1}{12}, \frac{1}{18}, \frac{1}{28},$ and $\frac{1}{36}$
16. $\frac{1}{40}, \frac{1}{60}, \frac{1}{72},$ and $\frac{1}{120}$
17. $\frac{1}{77}, -\frac{1}{84},$ and $\frac{1}{24}$
18. $\frac{1}{36}, \frac{1}{42},$ and $\frac{1}{49}$
19. $\frac{1}{12}, \frac{1}{18},$ and $\frac{1}{112}$
20. $\frac{1}{26}, \frac{1}{29},$ and $\frac{1}{32}$

Find:

21. $\frac{1}{18} - \frac{1}{60}$
22. $\frac{1}{2} \cdot \frac{1}{4} - \frac{1}{4} \cdot \frac{1}{4} - \frac{1}{32}$
23. $\left(-\frac{1}{4}\right) \cdot \left(\frac{1}{3}\right) \cdot \left(-\frac{1}{2}\right)$
24. $\dfrac{\frac{5}{8}}{-\frac{15}{2}}$

25. $(\frac{1}{3}) \cdot 2 \cdot (-\frac{6}{7}) - (1 - 2)$

26. $2 \cdot [1 + \frac{1}{3} \cdot (-\frac{7}{2}) - 2 \cdot (\frac{1}{24} - \frac{1}{12})] - \frac{1}{3}$

27. $\dfrac{[(\frac{1}{7} - \frac{1}{10}) - \frac{1}{35}]}{-\frac{1}{5}}$

28. $\dfrac{4 - \frac{3}{4}}{2 + \frac{1}{2}}$

29. $\frac{2}{3} \cdot (\frac{3}{4} + \frac{3}{5})$

30. $\frac{3}{2} \cdot [1 - (\frac{1}{6} + \frac{1}{3}) + \frac{2}{3} \cdot (\frac{3}{5} - \frac{1}{10})]$

Evaluate:

31. $(1 - \frac{1}{4}) \cdot (1 - \frac{1}{5}) \cdot (1 - \frac{1}{6}) \cdots (1 - \frac{1}{20})$

32. $\cfrac{1}{1 + \cfrac{2}{1 + \cfrac{3}{1 + \frac{4}{5}}}}$

33. The Department of Transportation is ordering emergency phone boxes to go along the turnpike from Boston to Sturbridge. The phones are to be placed every 1056 feet on both sides of the road. If the distance between the cities is 48 miles, and if no phones are needed right at the end of the road, how many phones should be ordered? [1 mile = 5280 feet]

34. One-fourth of a half-gallon carton of ice cream has been eaten. The remainder is divided among five people. What fraction of a gallon does each person get?

35. Of the people entering high school, $\frac{2}{3}$ graduate. Of the people graduating high school, $\frac{5}{8}$ are admitted to college. Of the people admitted to college, $\frac{1}{2}$ graduate.

(a) What fraction of the people entering high school graduate from college?

(b) If 80 people graduate from high school, how many of these do *not* finish college?

36. If standard blueprints use a scale of 1 foot $= \frac{3}{4}$ inch, what is the size of the smallest sheet of blueprint necessary to depict a room that is 32 feet by 20 feet?

37. Three students bought two sausage pizzas. The anthropology major ate $\frac{2}{3}$ of one of them and the biology major ate $\frac{3}{8}$ of the other. How much pizza is left for the math major to eat?

38. The stock of Fifer Flute Company is offered for sale. Michael buys $\frac{1}{2}$ of the total number of shares; Rob takes $\frac{1}{3}$ of what is left after Michael's purchase; Karen buys $\frac{1}{4}$ of the remainder after Michael and Rob take their shares; and Janet buys the remainder. What fraction of the total shares does each person own?

39. A water purification plant works by passing the dirty water through three filters in turn. The first takes out all but $\frac{1}{4}$ of the dirt in the water, the second leaves $\frac{1}{5}$ of the dirt reaching it and the last takes out $\frac{1}{2}$ of what's left.

(a) What fraction of the dirt is left in the water after it has gone through the plant?

(b) What fraction of the dirt has the plant removed?

40. In a referendum, $\frac{2}{3}$ of a town's population voted to have the roads improved, and $\frac{7}{12}$ of the population voted to have a sewage system built.

(a) What is the smallest fraction of the population that could have voted for both?

(b) What is the largest fraction that could have voted for both?

41. Two-thirds of the people in Switzerland speak German, $\frac{7}{12}$ speak French, and $\frac{1}{4}$ speak Italian.

(a) What is the smallest fraction of the population that could speak all three languages?

(b) What is the largest fraction that could speak all three languages?

42. A pizza costs $2.75; three friends decide to split one. Tom takes a third, and then Dick takes a third of the remainder, and then Harry takes half of what's left.

(a) How should they split the total cost so that each pays an amount proportional to what he has eaten so far?

(b) Can they split the remaining pizza so that everyone ends up having eaten the same total amount? If so, how?

43. An earthquake leaves an isolated Red Cross post with an inadequate supply of gamma globulin to inject the people of the local village against hepatitis, which is common in the area. If the post originally had enough gamma globulin for two injections per person, and if the quake destroyed five-sixths of their supplies:

(a) What fraction of an injection (per person) have they left?

(b) If anything less than $\frac{5}{12}$ of a normal injection is useless, is it worthwhile for the Red Cross to inject the entire village?

44. What is the length of the smallest box that can be exactly filled with either 6-inch or 8-inch chocolate bars, placed end to end?

45. A recent survey of surgeons and family practitioners revealed that there are twice as many surgeons as family practitioners.

(a) What fraction of the doctors surveyed are family practitioners?
Suppose that $\frac{3}{4}$ of the doctors surveyed earn more than $100,000 per year, and that all surgeons earn more than $100,000 per year.

(b) What fraction of the doctors surveyed are the family practitioners who earn more than $100,000 per year?

(c) What fraction of the family practitioners earns more than $100,000 per year?

5   DECIMALS REVISITED

5.1   DEFINITION OF DECIMALS AND CONVERSION TO FRACTIONS

Fractions are essential in everyday life and in scientific work. However, they are a great nuisance to manipulate, because of the bother involved in changing the denominators every time you want to add them. What we really need, therefore, is some way of always using fractions with the same denominator. What is actually done is to use *decimals*, which are fractions whose denominators are always a power of ten (for any terminating decimal), and for which there is a convention so that one never has to bother writing the denominators at all.

In the number 213, the 2 gives the number of hundreds, the 1 the number of tens, and the 3 the number of units. In the number 0.213, the 2 gives the number of tenths, the 1 the number of hundredths, and the 3 the number of thousandths. Thus 0.213 means

$$\frac{2}{10} + \frac{1}{100} + \frac{3}{1000}$$

which is the same as

$$\frac{200}{1000} + \frac{10}{1000} + \frac{3}{1000} = \frac{213}{1000}$$

Any terminating decimal (one that doesn't go on forever) can be written as a fraction this way.

Conversely, a fraction like $\frac{525}{10,000}$ is written as 0.0525 as a decimal, because

$$\frac{525}{10,000} = \frac{500}{10,000} + \frac{20}{10,000} + \frac{5}{10,000}$$

$$= \frac{5}{100} + \frac{2}{1000} + \frac{5}{10,000}$$

$$= 5 \text{ hundredths, 2 thousandths, and 5 ten-thousandths}$$

$$= 0.0525$$

EXAMPLE: *Convert 12.75 into a fraction.*

$$12.75 = 12 + \frac{7}{10} + \frac{5}{100}$$

$$= \frac{1200}{100} + \frac{70}{100} + \frac{5}{100}$$

$$= \frac{1275}{100}$$

You can see that the middle stages of this conversion are really un-necessary. 12.75 can be thought of as 1275 of whatever the column furthest to the right indicates (in this case hundredths), and so

$$12.75 = \frac{1275}{100}$$

hundredths

EXAMPLE: *Convert $\frac{5138}{1000}$ into a decimal.*

$$\frac{5138}{1000} = \frac{5000}{1000} + \frac{100}{1000} + \frac{30}{1000} + \frac{8}{1000}$$

$$= 5 + \frac{1}{10} + \frac{3}{100} + \frac{8}{1000}$$

$$= 5.138$$

Again, the middle steps are unnecessary. $\frac{5138}{1000}$ means 5138 thousandths, which can be written as a decimal by putting the 8 three places to the right of the decimal point (because that's the thousandth's place). Hence,

$$\frac{5138}{1000} = 5.138$$

thousandths

EXAMPLE: $$0.0351 = \frac{351}{10,000}$$

EXAMPLE: $$\frac{12001}{1000} = 12.001$$

It is also possible to turn a fraction whose denominator is not a power of ten into a decimal. To do this, divide the denominator into the numerator, leaving the decimal point alone. (You may have to add zeros after the decimal point in the numerator to make this work.)

EXAMPLE: *Convert $\frac{2}{5}$ into a decimal.*

If we think of 2 as 2.0, the division will come out exactly:

$$
\begin{array}{r}
0.4 \\
5\overline{)2.0} \\
2.0 \\
\hline
0.0
\end{array}
$$

So $\frac{2}{5} = 0.4$.

EXAMPLE: *Convert $\frac{4}{3}$ into a decimal.*

Here you need to realize that 4 is the same as 4.0 or 4.00, or 4.00..., so

$$
\begin{array}{r}
1.33... \\
3\overline{)4.00...} \\
3.0 \\
\hline
1.0 \\
.9 \\
\hline
.10
\end{array}
$$

So $\frac{4}{3} = 1.33... = 1.\overline{3}$.

This fits in with what you may remember from Section 1.2, that fractions can all be written as terminating or repeating decimals.

5.2   ADDING AND SUBTRACTING DECIMALS

The wonderful thing about decimals (and in fact the reason they're around in such abundance) is that, provided you line up the decimal points, you can add and subtract them in columns, just as you do integers.   For example, $5.12 + 0.371$ becomes

$$\begin{array}{r} 5.12 \\ + \ 0.371 \\ \hline 5.491 \end{array}$$

so the answer is 5.491.   The reason for this is clear if you think what the numbers mean:

5.12   means 5 ones, 1 tenth, and 2 hundredths;

0.371 means 0 ones, 3 tenths, 7 hundredths, and 1 thousandth;

Altogether you have

5 ones, 4 tenths, 9 hundredths, and 1 thousandth

or

5.491

So, provided the decimal points are lined up, all the numbers in any given column represent numerators of fractions with the same denominator, and so they can be added.

EXAMPLE:
$$9.927 + 10.87 = 20.797 \text{ because } \begin{array}{r} 9.927 \\ + \ 10.87 \\ \hline 20.797 \end{array}$$

EXAMPLE:
$$0.532 - 0.411 = \ 0.121 \text{ because } \begin{array}{r} 0.532 \\ - \ 0.411 \\ \hline 0.121 \end{array}$$

5.3   MULTIPLYING DECIMALS

> To multiply two decimals, multiply the numbers ignoring the decimal points; then count the total numbers of places to the right of the decimal point in the original numbers—this will be the number of places after the decimal point in your answer.

EXAMPLE: *Find   1.2 · 0.2*

Multiply the numbers:

$12 \cdot 2 = 24$

The total number of places to the right of the decimal point is 2 (one place in 1.2 and one place in 0.2), so the answer is 0.24.

$$1.2 \cdot 0.2 = 0.24$$

The reason for this is best seen by turning everything back into fractions:

$$1.2 \cdot 0.2 = \frac{12}{10} \cdot \frac{2}{10} = \frac{24}{100} = 0.24$$

Each of the original numbers has one figure after the decimal point, and so has a 10 in the denominator when written as a fraction. The product therefore has $10 \cdot 10 = 100$ in the denominator, and so has two figures after the point when written as a decimal.

EXAMPLE: *Find* $(0.03)^2 = 0.0009.$

$$(0.03)^2 = (0.03) \cdot (0.03),$$

so multiplying the numbers:

$$3^2 = 9$$

The total number of places to the right of the point is 4, so

$$(0.03)^2 = 0.0009$$

The reason is again clear from looking at the fractions:

$$(0.03)^2 = \left(\frac{3}{100}\right)^2 = \frac{3}{100} \cdot \frac{3}{100} = \frac{9}{10000} = 0.0009$$

There is a particularly painless way of multiplying or dividing decimals by powers of ten, which deserves some space to itself.

To multiply by powers of 10, move the decimal point the number of places to the right equal to the power of 10.

EXAMPLE: $2.315 \cdot 100 = 2.315 \cdot 10^2 = 231.5$

because $2.315 \cdot 100 = \dfrac{2315}{1000} \cdot 100 = \dfrac{2315}{10} = 231.5$

EXAMPLE: $45.2 \cdot 1000 = 45.2 \cdot 10^3 = 45200$

> To divide by powers of 10, move the decimal point the number of places to the left equal to the power of 10.

EXAMPLE:

$$\frac{813.5}{100} = \frac{8\overset{\frown}{13.5}}{10^2} = 8.135$$

because $\dfrac{813.5}{100} = \dfrac{\frac{8135}{10}}{100} = \dfrac{8135}{10} \cdot \dfrac{1}{100} = \dfrac{8135}{1000} = 8.135$

EXAMPLE:

$$\frac{0.256}{1000} = \frac{0.256}{10^3} = 0.000256$$

5.4   DIVIDING DECIMALS

> To divide decimals when the number that you are dividing by (the divisor) is not a decimal, carry out the division as usual: do not disturb the decimal point of the number you are dividing into (the dividend).

EXAMPLE:

$$\frac{2.48}{2} = 1.24$$

EXAMPLE:

Find $\dfrac{0.016}{5}$.

As it stands, this doesn't divide exactly. But if you realize that 0.016 can equally well be thought of as 0.0160 (or 0.01600 or 0.016000), then

$$\frac{0.016}{5} = \frac{0.0160}{5} = 0.0032$$

EXAMPLE:

Find $\dfrac{276.9}{13}$.

This needs long division (at least if you don't know your 13-times table, which I don't).

$$\begin{array}{r} 21.3 \\ 13\overline{)276.9} \\ \underline{26} \\ 16. \\ \underline{13.} \\ 3.9 \\ 3.9 \end{array}$$

Note: keep everything lined up, with the decimal point in the answer right over the decimal point in the dividend and the calculation.

So

$$\frac{276.9}{13} = 21.3$$

To divide when the divisor is a decimal, think of the division problem as a fraction, and multiply top and bottom by a power of 10 that will exactly get rid of the decimal in the denominator. Then continue as above.

EXAMPLE: *Find* $\dfrac{2.56}{0.2}$

Multiply top and bottom by 10, because

$$0.2 \cdot 10 = 2$$

(multiplying by 10 moves the decimal point one space to the right)

$$\frac{2.56}{0.2} = \frac{2.56 \cdot 10}{0.2 \cdot 10} = \frac{25.6}{2}$$

(multiplying top and bottom by 10 moves the decimal point one space to the right in each)

$$= 12.8$$

EXAMPLE: *Find* $\dfrac{132}{0.011}$

Multiply top and bottom by 1000:

$$\frac{132}{0.011} = \frac{132,000}{11} = 12,000$$

EXAMPLE: *Find* $\dfrac{0.03171}{0.21}$

Multiply top and bottom by 100:

$$\frac{0.03171}{0.21} = \frac{3.171}{21}$$

Now we need long division again:

$$
\begin{array}{r}
0.151 \\
21\overline{)3.171} \\
2.1 \\
\hline
1.07 \\
1.05 \\
\hline
0.021 \\
0.021 \\
\hline
\end{array}
$$

So

$$\frac{0.03171}{0.21} = \frac{3.171}{21} = 0.151$$

5.5   ROUNDING OFF DECIMALS

Suppose you live out in the country several miles from the last traffic light in the nearest town. A friend is coming over and you want to tell her how far beyond the traffic light to go before looking for your house. A rather detailed calculation tells you that you live 2.563 miles beyond the traffic light. You realize, however, that no car odometer reads to more than a tenth of a mile, and so the question becomes: To the nearest tenth of a mile, what is your distance from that traffic light?

Now 2.563 is between 2.5 and 2.6 and is actually closer to 2.6. Therefore, to the nearest tenth, the distance is 2.6 miles. (See Figure 5.1)

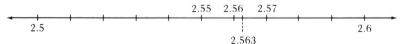

FIG. 5.1

The process of replacing 2.563 by 2.6 is called *rounding off to one decimal place*, or *correcting to one decimal place*.

To round off to n decimal places:

1. Throughout the calculation, carry at least one more decimal place than you want in the answer.

2. If the figure in the $(n + 1)$th place (the one after the last one you're interested in) is a 4 or less, drop the figures in the places that you don't need.

3. If the figure in the $(n + 1)$th place is a 5 or more, increase the figure in the nth place by 1 and drop the figures following it.

EXAMPLE: 2.563 rounded off to two decimal places is 2.56. (See Figure 5.1.)

EXAMPLE: 1.55 rounded off to one place is 1.6.

EXAMPLE: 5.103 rounded off to two places is 5.10. (You should leave in the 0 to show you didn't round off to one place, and that there actually is nothing in the second decimal place. This shows that the answer 5.10 is accurate to the nearest hundredth.)

EXAMPLE: 0.8895 rounded off to three places is 0.890. (Again, leave in the zero to show that the answer is accurate to the third decimal place.)

5.6   DEFINITION OF PERCENT

There is one kind of fraction that is used so frequently that it is given a special name. The fraction is the hundredth, and it is called a *percent* (from the Latin, *centum*, a hundred), and written %. So:

5%, or 5 percent, means 5|hundredths or $\dfrac{5}{100}$

0.2%, or 0.2 percent, means 0.2 hundredths or $\dfrac{0.2}{100}$

425%, or 425 percent, means 425 hundredths or $\dfrac{425}{100}$

Percents can also be written as decimals. For example,

$$59\% = \frac{59}{100} = 0.59$$

$$0.02\% = \frac{0.02}{100} = 0.0002$$

$$150\% = \frac{150}{100} = 1.50$$

The relationship between two numbers is often shown by expressing one as so many percent of the other. For example:

20 is 50% of 40

because 50% of 40 means 50% times 40 or $\dfrac{50}{100} \cdot 40$

which is $\dfrac{50 \cdot 40}{100}$ or 20

Similarly:

8 is 10% of 80

because 10% of 80 means $\dfrac{10}{100} \cdot 80$

and $\dfrac{10}{100} \cdot 80 = 8$

In general:

$$A \text{ is } p\% \text{ of } B \quad \text{means} \quad A = \frac{p}{100} \cdot B$$

p is the *percent*, so 20 is said to be 50 percent of 40. A is the *percentage* and B is the *base*, so 20 is a percentage of the base, 40.

EXAMPLE: *Find 5% of 250.*

5% of $250 = \dfrac{5}{100} \cdot 250$ (here $p = 5$, $B = 250$)

$\qquad\qquad = 12.5$ (so $A = 12.5$)

EXAMPLE: *Find $5\frac{1}{2}\%$ of 22.*

$5\frac{1}{2} = \dfrac{11}{2}$

so

$5\frac{1}{2}\%$ of $22 = \dfrac{\frac{11}{2}}{100} \cdot 22$ (here $p = \frac{11}{2}$, $B = 22$)

$\qquad\qquad = 1.21$ (so $A = 1.21$)

EXAMPLE: *What is 0.01% of 2?*

0.01% of $2 = \dfrac{0.01}{100} \cdot 2$ (here $p = 0.01$, $B = 2$)

$\qquad\qquad = 0.0002$ (so $A = 0.0002$)

EXAMPLE: *Find 110% of 90.*

$$110\% \text{ of } 90 = \frac{110}{100} \cdot 90 \qquad (\text{here } p = 110, B = 90)$$

$$= 99 \qquad (\text{so } A = 99)$$

EXAMPLE: *Find 100% of 83.*

$$100\% \text{ of } 83 = \frac{100}{100} \cdot 83 \qquad (\text{here } p = 100, B = 83)$$

$$= 83 \qquad (\text{so } A = 83)$$

If you look over these examples, you will see that:

If p is less than 100, then A is less than B.

If p is greater than 100, then A is greater than B.

For example, 12.5 is 5% of 250, and 99 is 110% of 90.

If p = 100, then A = B. Any number is 100% of itself.

For example, 83 is 100% of 83.

Percents come up all over the place in the "real world." In math books this means in word problems, so here are a couple.

EXAMPLE: *When Ford shut down its River Rouge plant in Detroit, it put ½% of Detroit's 1.2 million work force out of work. How many people did the plant employ?*

Number of people in plant $= \frac{1}{2}\%$ of 1.2 million

$$= \frac{\frac{1}{2}}{100} \cdot 1.2 \cdot 1,000,000$$

$$= 6000$$

EXAMPLE: *In singing "Oh Lord, won't you buy me a Mercedes Benz?", Janis Joplin was asking for a car with the unusual property that its value goes up as time goes on. If a model 280SL cost $14,000 when it was new, and sold for 160% of that price ten years later, what price did it fetch when sold, and how much profit did its owner make?*

We know:

$$\text{Selling price} = 160\% \text{ of cost}$$
$$= 160\% \text{ of } \$14,000$$
$$= \frac{160}{100} \cdot 14,000$$
$$= \$22,400$$

And:

$$\text{Profit} = \text{selling price} - \text{cost}$$
$$= \$22,400 - \$14,000$$
$$= \$8,400$$

Suppose now that instead of asking what a certain percent of a given number is (e.g., "What is 5% of 250?"), we want to know what percent one number is of another. For example, suppose we want to find what percent 45 is of 180. This means we want to look at 45 as a fraction of 180, i.e. $\frac{45}{180}$, and convert that fraction to one with 100 in the denominator. Now the easiest way to rewrite a fraction so as to have 100 in the denominator is first to convert the fraction to a decimal (because decimals automatically have powers of 10 in the denominator), and then to convert the decimal to hundredths. For this example;

first convert $\frac{45}{180}$ to a decimal:

$$\frac{45}{180} = 0.25$$

then convert to hundredths:

$$0.25 = 0.25 \cdot \frac{100}{100} = \frac{25}{100} = 25\%$$

So 45 is 25% of 180.

EXAMPLE: *What percent is 50 of 30?*

First look at the fraction:

$$\frac{50}{30} = \frac{5}{3}$$

Now convert to a decimal:

$$\frac{5}{3} = 1.666\ldots$$

Finally, convert the decimal to hundredths:

$$1.666\ldots = \frac{166.6\ldots}{100} = 166.6\ldots\% = 166.\bar{6}\%$$

So 50 is $166.\bar{6}\%$ of 30.

5.7   PERCENT INCREASE AND DECREASE

It is often useful to be able to compare the change in some quantity with the original value of that quantity. For example, a \$5 increase in a \$10 book (to \$15) is a lot more significant than a \$5 increase in a \$100 bicycle (to \$105). This comparison is usually made by expressing the change as a percentage of the original quantity. As in the two examples at the end of the last section, we can calculate that:

$$\$5 \text{ is } 50\% \text{ of } \$10$$

but

$$\$5 \text{ is } 5\% \text{ of } \$100$$

The book's price is said to have *increased by 50%*, whereas the bicycle's price has *increased by only 5%*. In general:

> *B is increased by r% if r% of B is added to B.*

If, however, the price of the book were to be lowered by \$5, then the price would be said to have *decreased by 50%*. Similarly, when the bicycle's price falls by \$5 to \$95, it is said to have *decreased by 5%*. So, in general:

> *B is decreased by r% if r% of B is subtracted from B.*

EXAMPLE: *What is 200 decreased by 10%?*

$$\text{Decrease} = 10\% \text{ of } 200 = \frac{10}{100} \cdot 200$$

$$= 20$$

So we must subtract 20 from 200, giving an answer of 180.

EXAMPLE: *What is 140 increased by 2%?*

Increase $= 2\%$ of $140 = \dfrac{2}{100} \cdot 140$

$= 2.8$

Therefore,

New value $= 140 + 2.8$

$= 142.8$

So 140 increased by 2% is 142.8.

Note 1: It is important to realize that the change is expressed as a percentage of the *original* quantity, not of the new value. For example, when 200 is decreased by 10%, the change is 20, and 20 is 10% of 200. But 20 is *not* 10% of 180, the final value. Similarly, when 140 is increased by 2%, the change is 2.8, which is 2% of 140, but not of 142.8 (the final value).

Note 2: Look again at the example above in which 200 is decreased by 10%. You will see that we could have arrived at the same answer by taking

$(100 - 10)\%$ of $200 = 90\%$ of 200

because

90% of $200 = \dfrac{90}{100} \cdot 200 = 180$

The reason for this is that decreasing 200 by 10% means subtracting 10% of 200 from 200. But 200 can be thought of as 100% of 200, and subtracting 10% of 200 leaves you with 90% of 200.

Similarly, 140 increased by 2% may also be found by taking 102% of 140, since

102% of $140 = \dfrac{102}{100} \cdot 140 = 142.8$

which is the same answer as we got before. This works because 140 can be thought of as 100% of 140, and adding 2% of 140 gives you 102% of 140.

EXAMPLE: *If inflation is running at 10% a year, what will a 40-cent package of margarine cost next year?*

Inflation at 10% a year means that prices increase by 10% a year. Next year the cost of margarine will therefore be 40 cents increased by 10%.

$$\text{New price} = (100 + 10)\% \text{ of } 40¢$$

$$= 110\% \cdot 40¢$$

$$= 44¢$$

Mark-up and Mark-down

When a store *marks an item up by 5%*, it increases the price of the item by 5%. During a sale, a sign saying *"Marked down by 20%"* means that the price has been decreased by 20%.

EXAMPLE: *After a good summer leading to a surplus in milk production, the Milk Commission decided on a 6% mark-down on milk prices. What is the new price (to the nearest cent) of a quart of milk that used to cost 60 cents?*

The price, 60 cents, is decreased by 6%.

$$\text{New price} = (100 - 6)\% \text{ of } 60¢$$

$$= 94\% \cdot 60¢$$

$$= 56.4¢$$

So, to the nearest cent, the new price = 56 cents.

Formulas for Percent Increase and Decrease

All the percent increase and decrease problems follow the same pattern, and so it is possible to find a formula that describes all of them.

Percent Increase   Suppose that after increasing B by $r\%$, we end up with A.

$$\text{Increase} = r\% \text{ of } B = \frac{r}{100} \cdot B$$

So we must add $\frac{r}{100} \cdot B$ to B to get A:

$$A = B + \frac{r}{100} \cdot B$$

Using the distributive law, you can check that this may be rewritten:

$$\boxed{A = \left(1 + \frac{r}{100}\right) \cdot B}$$

Percent Decrease   If A is obtained by decreasing B by $r\%$, then subtracting $r\%$ of B from B gives A.

$$\text{Decrease} = r\% \text{ of } B = \frac{r}{100} \cdot B$$

So

$$A = B - \frac{r}{100} \cdot B$$

This may also be rewritten:

$$A = \left(1 - \frac{r}{100}\right) \cdot B$$

CHAPTER 5 EXERCISES

Express each of the following as a decimal.

1. $\dfrac{23}{10}$

2. $\dfrac{4.5}{1000}$

3. $\dfrac{0.025}{10}$

4. $100\,(0.003)$

5. $\dfrac{1}{10 \cdot 10 \cdot 10}$

6. $0.02 + 1.23$

7. $1.003 + \dfrac{5}{100}$

8. $\dfrac{0.23}{10 \cdot 10} + 10(0.005)$

9. $(1.2) \cdot (0.02)$

10. $(0.04)\,(0.04)\,(2.1)$

11. $\dfrac{5.2}{2}$

12. $\dfrac{4.11}{0.03}$

13. $\dfrac{0.64}{3.2}$

Evaluate the following:

14. $\dfrac{0.01 \cdot (14 - 37)}{3.7 - 4.2}$

15. $\dfrac{1}{0.6}$

16. $(1.21) \cdot (1.21)$

17. $\dfrac{1}{1.2}$

Express each of the following as a decimal:

18. $\frac{1}{2}$ 21. $\frac{1}{5}$ 24. $\frac{1}{8}$ 27. $\frac{1}{25}$ 30. $\frac{3}{4}$

19. $\frac{1}{3}$ 22. $\frac{1}{6}$ 25. $\frac{1}{9}$ 28. $\frac{1}{50}$ 31. $\frac{4}{5}$

20. $\frac{1}{4}$ 23. $\frac{1}{7}$ 26. $\frac{1}{10}$ 29. $\frac{2}{3}$ 32. $\frac{7}{4}$

Evaluate the following:

33. $\dfrac{222}{7.4}$

34. $\dfrac{-236.73}{6.07}$

35. $\dfrac{2.9029}{1.001}$

36. $\dfrac{57.6}{0.24}$

37. $\dfrac{-40.74036}{800.4}$

Evaluate the following, rounding off the answer to three decimal places:

38. $(1.23) \cdot (4.56)$
39. $(3.01) \cdot (\frac{1}{9})$
40. $\frac{1}{15} \cdot \frac{1}{24}$
41. $\dfrac{1}{0.27} \cdot \dfrac{3}{6.9}$
42. $\dfrac{2 \cdot (0.5) + 0.8}{3 \cdot (0.5) - 0.6}$

43. $\dfrac{1.37 - 0.0742}{21.8 \cdot 3.17 + 4.06}$

44. $\dfrac{0.012 \cdot 18.34 - 0.03}{(1.14 - 0.026) \cdot 7.1}$

45. $3 \cdot \left[1 - 0.25 \left(\dfrac{7}{12} - \dfrac{1}{4} \right) + 0.6 \cdot 13 \right] - 2$

Express each of the following as a percent:

46. 3.12
47. 0.00065
48. 31.750
49. 0.00706
50. 2.23
51. 0.230
52. 0.060708
53. 12.0

Express each of the following as a decimal:

54. 11%
55. 0.003%
56. 1.7%
57. $\frac{3}{4}$ of 1%
58. 98.09%
59. 321%
60. 2% of 3%
61. 16%
62. 1.05%

Express each of the following as a percent:

63. $\frac{1}{2}$
64. $\frac{1}{3}$
65. $\frac{1}{4}$
66. $\frac{1}{5}$
67. $\frac{1}{6}$
68. $\frac{1}{7}$
69. $\frac{1}{8}$
70. $\frac{1}{9}$
71. $\frac{1}{10}$
72. $\frac{1}{25}$
73. $\frac{1}{50}$
74. $\frac{2}{3}$
75. $\frac{3}{4}$
76. $\frac{4}{5}$
77. $\frac{7}{4}$

78. What is 5% of 140?
79. What is 12% of 1.03?
80. What percent of 400 is 20?
81. 250% of 23 is what?
82. Find 0.64% of 15.
83. 300% of 25 is what?
84. What is 0.05% of 0.05?
85. What is 28 increased by 40%?
86. What is 38 decreased by 16%?
87. What is 17.4 increased by 8.1%?
88. What is 0.06 decreased by 212%?
89. What is 0.071 increased by 16%?

Evaluate the following (rounding to three decimal places):

90. $\frac{3}{8} \cdot (2.40\%)$

91. $\dfrac{3 \cdot 2.41 - 8\%}{\frac{3}{7}}$

92. $\dfrac{1.09 \cdot 4.17 - 220\%}{\frac{3}{8} - 12.5\% \cdot 7}$

93. 21% of 430% of 69

94. 13.7% of 12% of $\dfrac{7.17}{3.29}$

95. $\dfrac{6.08}{9\%}$ increased by 5%

96. 0.05% of 72, decreased by 11%

97. $\left(\dfrac{\frac{5}{8}}{62.5\%}\right)^2$

For each of the following, fill in "is more than," "is less than," or "is equal to."

98. 1 _____ 90.5%

99. $\frac{3}{8}$ _____ 40%

100. $\frac{1}{7}$ _____ 13%

101. 23% _____ $\frac{1}{4}$

102. $\frac{1}{6}$ _____ 17%

103. 15% _____ $\frac{1}{16}$

104. $\frac{1}{9}$ _____ 0.12

105. 0.34 _____ 32%

106. 0.04 _____ 0.9%

107. $\frac{1}{40}$ _____ 2.5%

108. $\frac{3}{13}$ _____ 23%

109. 0.16 increased by 2% _____ $\frac{1}{6}$

110. 20.9% _____ $\frac{1}{5}$

111. 0.076 _____ 11% of 0.69

112. A man who earns a million dollars a year wins another million in a lottery. If the income tax on $1 million is 40% and the tax on $2 million is 80%, is he any better off?

113. What percent of Detroit's 1.2 million labor force becomes unemployed if Ford lays off 3000 workers?

114. If a 25¢ candy bar goes up to 28¢, by what percent has its price increased?

115. (a) You make $110/week. Your pay is first cut by 10%; two weeks later you get a 10% cost-of-living raise. What are you making now?

(b) What percent cost-of-living increase would you have needed to get back to your $110/week salary?

116. At the moment, there are 14,000 students entering medical school each year. If the number goes up by 2%, how many will there be next year?

117. The Consumer Price Index (CPI) is set so that 100 represents the price level existing in 1967. Thus, if the same goods purchased in 1967 for $10.00 now cost $14.40, the CPI is 144.

(a) If prices increased by 7.8% in 1968, what was the CPI for that year?

(b) If the CPI last year was 143 and it went up 6 points (to 149) this year, what was the percent of the increase (i.e., the rate of inflation)?

(c) "Double-digit inflation" means that the CPI has gone up more than 10% from one year to the next. If the CPI is presently 149, what is the maximum level it can reach next year if we are to avoid double-digit inflation?

118.   At a certain school, 25% of humanities majors graduate with honors, 16% of social sciences majors graduate with honors, and 52% of math majors graduate with honors. Thirty percent of the students are humanities majors, 50% are social sciences majors and 20% are math majors. What percent of the overall student body graduates with honors?

119.   In baseball, slugging averages (commonly called "slugging percentages" even though they aren't expressed as percentages at all) are figured by dividing a hitter's total bases (1 base for a single, 2 bases for a double, 3 for a triple, 4 for a home run) by his number of official at-bats.

(a) What is the highest possible slugging average a player can have? (Express your answer in decimal form.)

(b) Does a player have a higher slugging average if he hits one triple and one single every eight times at bat, or if he hits one home run every five times at bat?

(c) In 480 at-bats, Louis V. Slugger hits 25 home runs, 22 doubles, 3 triples, and 111 singles. What is his slugging average?

120.   In Las Vegas, each slot machine has the following payoff schedule:
"Grand" payoffs occur 1% of the times the machine is played, and are worth $25.
"Very-grand" payoffs occur 0.5% of the times the machine is played, and are worth $50.
"Super-grand" payoffs occur 0.05% of the times the machine is played, and are worth $500.

(a) If U. N. Lucky plays the machine 6000 times at $1 a shot, and the machine pays off according to the above schedule, how much will he lose?

(b) How many payoffs of each type are there over the course of 100,000 plays of the machine?

(c) What percent of all payoffs are "super-grand" payoffs?

121.   One often hears the major oil companies say that they make a profit of only 1 or 2%, which sounds pretty bad; after all, you and I can get 5% interest on a savings account at a bank. However, the oil companies are quoting profits not as a percentage of investment, as most companies do, but rather as a percentage of sales. Thus, if an oil company puts up an original investment of

$10 million, which leads to sales of $40 million, and profits of $1 million, they say that their profit is

$$\frac{\$1 \text{ million}}{\$40 \text{ million}} = 2.5\%$$

while you and I might say that it's

$$\frac{\$1 \text{ million}}{\$10 \text{ million}} = 10\%$$

(a) If a company whose investment is 22% of sales has profits of 2% of sales, what is their profit expressed as a percentage of investment?

(b) What is the amount of profit for a company that has sales of $13.7 million and says that its profit is 1.8% (of sales)?

122.   In an attempt to refute recent claims that most cereals are only as nutritious as cardboard, the cereal companies surveyed a number of families. In the course of their research, they find that 50% of the families eat three boxes of cereal a week, 25% eat two boxes a week, 20% eat one box, and the rest eat none.

(a) What is the total number of boxes eaten per week by 1000 families?

(b) What is the average number of boxes eaten per week by those 1000 families?

(c) How many of those 1000 families eat any cereal at all?

(d) What is the average number of boxes eaten per week by the families that eat any? (The average may be defined as the total number of boxes eaten divided by the number of families. In general, the *average* of n numbers can be found by adding all the numbers together and then dividing that sum by n.)

6 ROOTS AND RADICALS

6.1 DEFINITIONS OF SQUARE AND CUBE ROOTS

The *square root* of a is defined to be a number that when squared gives a. For example, $2^2 = 4$ and $(-2)^2 = 4$, so both 2 and -2 are numbers that when squared give 4. Therefore 4 has two square roots, namely 2 and -2.

Convention says that the symbol $\sqrt{4}$ shall be reserved for the positive square root of 4, so $\sqrt{4}$ means 2. Therefore the two square roots of 4 are given by $\sqrt{4} = 2$ and $-\sqrt{4} = -2$.

In general, $\sqrt{a}$ is defined to be the positive square root of a, so

$$(\sqrt{a})^2 = a$$

Also

$$(-\sqrt{a})^2 = a \quad \text{since} \quad (-) \text{ times } (-) \text{ is } (+),$$
$$\text{and} \quad (-\sqrt{a})^2 = (-\sqrt{a}) \cdot (-\sqrt{a}).$$

Therefore $-\sqrt{a}$ is the negative square root of a.

EXAMPLE: $\sqrt{1} = 1$.

Now $1^2 = 1$ and $(-1)^2 = 1$, so

$$\sqrt{1} = 1 \quad \text{and} \quad -\sqrt{1} = -1$$

Therefore 1 and -1 are the square roots of 1.

EXAMPLE: $\sqrt{2}$ is an irrational number.

$\sqrt{2}$ can't be written as an integer, nor as a fraction, nor as a finite decimal. It can, however, be approximated by a decimal, and its value can be shown to be about 1.414. Therefore 2 has two square roots, one positive and one negative, which are approximately 1.414 and -1.414. Since the square root of 2 cannot be written down exactly either as a fraction or as a decimal, the symbol $\sqrt{2}$ will be used to mean the exact positive square root of 2. Since $\sqrt{2}$ is defined to be a number which when squared gives 2, we have

$$(\sqrt{2})^2 = 2$$

or, in other words,

$$\sqrt{2} \cdot \sqrt{2} = 2$$

EXAMPLE: $\sqrt{3}$ is an irrational number.

Like $\sqrt{2}$, the square root of 3 cannot be expressed exactly either as a decimal or as a fraction. The value of the positive square root is about 1.732, but $\sqrt{3}$ will be used to stand for the exact value; $-\sqrt{3}$ will stand for the negative square root, which is about -1.732.

Again, $\sqrt{3} \cdot \sqrt{3} = (\sqrt{3})^2 = 3$ because $\sqrt{3}$ was defined that way.

EXAMPLE: $\sqrt{0} = 0$.

Since $0^2 = 0$, $\sqrt{0}$ must be 0. This is the only example of a number with exactly one square root.

As you may have guessed from the above examples, the square roots of integers either come out exactly (as integers) or they don't come out neatly at all (and are irrational). The only integers that have rational square roots are those that are *perfect squares* (meaning those numbers that are the squares of another integer, for example, $1 = 1^2$, $4 = 2^2$, $9 = 3^2$, etc.).

The negative integers have no square roots, unless one allows complex numbers, which will be introduced in Section 14.5. Since there is no real number whose square is -1 or -4, $\sqrt{-1}$ and $\sqrt{-4}$ stand for complex numbers.

The *cube root* of a, $\sqrt[3]{a}$, is defined to be that number which when cubed gives a, so

$$\boxed{(\sqrt[3]{a})^3 = a}$$

EXAMPLE: $\sqrt[3]{8} = 2$

because $2^3 = 2 \cdot 2 \cdot 2 = 8$, so 2 cubed is 8.
8 has only one real cube root. -2 is not a cube root of 8 because

$$(-2)^3 = (-2) \cdot (-2) \cdot (-2) = -8.$$

So -2 is a cube root of -8 but not of 8; that is,

$$\sqrt[3]{-8} = -2$$

Thus, although they don't have real square roots, negative numbers do have real cube roots.

EXAMPLE: $\sqrt[3]{1} = 1$

because $1^3 = 1 \cdot 1 \cdot 1 = 1$, so 1 cubed is 1 and $\sqrt[3]{1} = 1$.

EXAMPLE: $\sqrt[3]{-1} = -1$

because $(-1)^3 = -1$

EXAMPLE: $\sqrt[3]{4}$ is an irrational number.

There is no integer that when cubed gives 4. You can see that 1 is too small because $1^3 = 1$, and 2 is too large because $2^3 = 8$. So the cube root of 4 must lie between 1 and 2, but it is irrational and so can't be written as a fraction. Since $\sqrt[3]{4}$ means the number that when cubed gives 4,

$$(\sqrt[3]{4})^3 = 4$$

Higher roots can be defined in the same way. The *fourth root* of a, $\sqrt[4]{a}$, means that (positive) number which when taken to the fourth power gives a. The *nth root* of a, $\sqrt[n]{a}$, means that number which when taken to the nth power gives a, so

$$\boxed{(\sqrt[n]{a})^n = a}$$

Convention has it that if there's a choice, $\sqrt[n]{a}$ is positive. Numbers such as $\sqrt{2}$, $\sqrt[3]{3}$, $\sqrt{\tfrac{1}{5}}$, $\sqrt[3]{4}$, are sometimes called radicals, from the Latin, *radix*, "a root." The $\sqrt{}$ sign is called the radical sign.

EXAMPLE: $\sqrt[4]{16} = 2$

because $(2)^4 = 2 \cdot 2 \cdot 2 \cdot 2 = 16$, so 2 to the fourth power is 16 and hence $\sqrt[4]{16} = 2$.

EXAMPLE: $\sqrt[4]{81} = 3$
because $(3)^4 = 3 \cdot 3 \cdot 3 \cdot 3 = 81$

EXAMPLE: $\sqrt[6]{64} = 2$
because $(2)^6 = 2 \cdot 2 \cdot 2 \cdot 2 \cdot 2 \cdot 2 = 64$

It is also perfectly possible to take the square, cube, or any other root of fractions, decimals, or other kinds of real numbers.

EXAMPLE: $\sqrt{\dfrac{4}{9}} = \dfrac{2}{3}$
because $(\frac{2}{3})^2 = \frac{2}{3} \cdot \frac{2}{3} = \frac{4}{9}$, so $\frac{2}{3}$ squared is $\frac{4}{9}$ and therefore $\sqrt{\frac{4}{9}} = \frac{2}{3}$.

EXAMPLE: $\sqrt{\dfrac{1}{4}} = \dfrac{1}{2}$
because $(\frac{1}{2})^2 = \frac{1}{2} \cdot \frac{1}{2} = \frac{1}{4}$, so $\frac{1}{2}$ squared is $\frac{1}{4}$ and therefore $\sqrt{\frac{1}{4}} = \frac{1}{2}$.

It is interesting to note that the squares of numbers larger than 1 are greater than the original number. For example,

$$2^2 = 4$$

$$3^2 = 9$$

$$4^2 = 16$$

Since $1^2 = 1$, the square of 1 is exactly equal to itself. However, the squares of anything between 0 and 1 are smaller than the original number. For example,

$$\left(\frac{1}{2}\right)^2 = \frac{1}{4}$$

$$\left(\frac{1}{3}\right)^2 = \frac{1}{9}$$

$$(0.1)^2 = 0.01$$

EXAMPLE: $\sqrt[3]{\dfrac{1}{27}} = \dfrac{1}{3}$
since $\left(\dfrac{1}{3}\right)^3 = \dfrac{1}{3} \cdot \dfrac{1}{3} \cdot \dfrac{1}{3} = \dfrac{1}{27}$

EXAMPLE: $\sqrt{\dfrac{1}{2}}$ is irrational, exactly as $\sqrt{2}$ is.

EXAMPLE: $\sqrt{\dfrac{-1}{4}}$ is not a real number, any more than $\sqrt{-1}$ and $\sqrt{-4}$ are; it can, however, be expressed as a complex number.

EXAMPLE: $\sqrt[5]{\dfrac{1}{32}} = \dfrac{1}{2}$

since $\left(\dfrac{1}{2}\right)^5 = \dfrac{1}{2} \cdot \dfrac{1}{2} \cdot \dfrac{1}{2} \cdot \dfrac{1}{2} \cdot \dfrac{1}{2} = \dfrac{1}{32}$

EXAMPLE: $\sqrt{0.01} = 0.1$

because $(0.1)^2 = 0.1 \cdot 0.1 = 0.01$

Or, looking at this in fractional form,

$\sqrt{\dfrac{1}{100}} = \dfrac{1}{10}$ because $\left(\dfrac{1}{10}\right)^2 = \dfrac{1}{10} \cdot \dfrac{1}{10} = \dfrac{1}{100}$

EXAMPLE: $\sqrt{0.0004} = 0.02$

because $(0.02)^2 = 0.02 \cdot 0.02 = 0.0004$

EXAMPLE: $\sqrt{0.1}$ and $\sqrt{0.001}$ are both irrational.

However, $\sqrt{0.01}$ and $\sqrt{0.0001}$ are not, since $\sqrt{0.01} = 0.1$ and $\sqrt{0.0001} = 0.01$. This is exactly parallel to the fact that $\sqrt{10}$ and $\sqrt{1000}$ are irrational, whereas $\sqrt{100}$ and $\sqrt{10,000}$ are not.

EXAMPLE: $\sqrt[3]{0.001} = 0.1$, $\sqrt[3]{0.008} = 0.2$

because $(0.1)^3 = 0.1 \cdot 0.1 \cdot 0.1 = 0.001$

and

$(0.2)^3 = 0.2 \cdot 0.2 \cdot 0.2 = 0.008$

EXAMPLE: $\sqrt{1.21} = 1.1$

because $(1.1)^2 = 1.1 \cdot 1.1 = 1.21$

EXAMPLE: $\sqrt{9.61} = 3.1$

because $(3.1)^2 = 9.61$

6.2   THINGS YOU CAN AND CAN'T DO WITH RADICALS

It is extremely easy, and extremely wrong, to say that $\sqrt{a+b} = \sqrt{a} + \sqrt{b}$. The reason it is wrong is that if you replace a and b by numbers, $\sqrt{a+b}$ and $\sqrt{a} + \sqrt{b}$ almost always come out different, and therefore can't be written with an equals sign between them.

For example, consider $a = 9$, $b = 16$

Then

$$\sqrt{a + b} = \sqrt{9 + 16} = \sqrt{25} = 5$$

and

$$\sqrt{a} + \sqrt{b} = \sqrt{9} + \sqrt{16} = 3 + 4 = 7$$

which are definitely different

Therefore

$$\sqrt{9 + 16} \neq \sqrt{9} + \sqrt{16}$$

And in general:

$$\boxed{\sqrt{a + b} \neq \sqrt{a} + \sqrt{b}}$$

It is equally wrong to say that $\sqrt{a - b} = \sqrt{a} - \sqrt{b}$.
For example, try $a = 25$, $b = 9$

Then

$$\sqrt{a - b} = \sqrt{25 - 9} = \sqrt{16} = 4$$

and

$$\sqrt{a} - \sqrt{b} = \sqrt{25} - \sqrt{9} = 5 - 3 = 2$$

different

so

$$\sqrt{25 - 9} \neq \sqrt{25} - \sqrt{9}$$

In general:

$$\boxed{\sqrt{a - b} \neq \sqrt{a} - \sqrt{b}}$$

However, radicals are not completely badly behaved. For example it *is* true
that

$$\boxed{\sqrt{a \cdot b} = \sqrt{a} \cdot \sqrt{b}}$$

For example, if $a = 9$, $b = 4$

then

$$\sqrt{a \cdot b} = \sqrt{9 \cdot 4} = \sqrt{36} = 6$$

and

$$\sqrt{a} \cdot \sqrt{b} = \sqrt{9} \cdot \sqrt{4} = 3 \cdot 2 = 6$$

so

$$\sqrt{9 \cdot 4} = \sqrt{9} \cdot \sqrt{4}$$

For any positive values of a and b that you pick, you will always find that the values of $\sqrt{a \cdot b}$ and $\sqrt{a} \cdot \sqrt{b}$ come out equal.

To prove that $\sqrt{a \cdot b} = \sqrt{a} \cdot \sqrt{b}$ in general,   try squaring $\sqrt{a} \cdot \sqrt{b}$:

$$(\sqrt{a} \cdot \sqrt{b})^2 = \sqrt{a} \cdot \sqrt{b} \cdot \sqrt{a} \cdot \sqrt{b}$$

$$= \sqrt{a} \cdot \sqrt{a} \cdot \sqrt{b} \cdot \sqrt{b} \qquad \text{(because you can always change the order of multiplication)}$$

$$= a \cdot b \qquad \text{(because } \sqrt{a} \cdot \sqrt{a} = a, \text{ and } \sqrt{b} \cdot \sqrt{b} = b\text{)}$$

So

$$(\sqrt{a} \cdot \sqrt{b})^2 = a \cdot b$$

But remember that $\sqrt{a \cdot b}$ was defined as that positive number which when squared gives $a \cdot b$. Since $\sqrt{a} \cdot \sqrt{b}$ does just that, and is positive, it must be the same number as $\sqrt{a \cdot b}$, so

$$\sqrt{a \cdot b} = \sqrt{a} \cdot \sqrt{b}.$$

Also, it is true that

$$\boxed{\sqrt{\frac{a}{b}} = \frac{\sqrt{a}}{\sqrt{b}}}$$

For example, try $a = 81$, $b = 16$.

$$\sqrt{\frac{a}{b}} = \sqrt{\frac{81}{16}} = \frac{9}{4} \quad \text{because} \quad \left(\frac{9}{4}\right)^2 = \frac{9}{4} \cdot \frac{9}{4} = \frac{81}{16}$$

$$\frac{\sqrt{a}}{\sqrt{b}} = \frac{\sqrt{81}}{\sqrt{16}} = \frac{9}{4} \quad \text{because} \quad \sqrt{81} = 9, \sqrt{16} = 4$$

So

$$\sqrt{\frac{81}{16}} = \frac{\sqrt{81}}{\sqrt{16}}$$

For any values of a and b that you pick, $\sqrt{\frac{a}{b}}$ and $\frac{\sqrt{a}}{\sqrt{b}}$ will all always come out equal. This can be proved in general in the same way as $\sqrt{a \cdot b} = \sqrt{a} \cdot \sqrt{b}$.

6.3 SIMPLIFYING AND MANIPULATING RADICALS

Since most radicals are irrational, they cannot usually be simplified or reduced to lowest terms in the way that fractions can. However, there is a generally agreed way of putting a radical in "simplest" form. Let's look at the case of square roots. $\sqrt{2}$, for example, is in the simplest form, because there's really nothing you can do with it. $\sqrt{12}$ on the other hand, can be simplified:

$$\sqrt{12} = \sqrt{4 \cdot 3} = \sqrt{4} \cdot \sqrt{3} \quad \text{(since } \sqrt{a \cdot b} = \sqrt{a} \cdot \sqrt{b} \text{)}$$
$$= 2 \cdot \sqrt{3}$$

$2 \cdot \sqrt{3}$ is considered the simplest possible form of $\sqrt{12}$ because the number under the radical sign is as small as possible.

EXAMPLE: $\sqrt{20} = 2 \cdot \sqrt{5}$

because $\sqrt{20} = \sqrt{4 \cdot 5} = \sqrt{4} \cdot \sqrt{5} = 2 \cdot \sqrt{5}$

EXAMPLE: *Simplify $\sqrt{24}$.*

$\sqrt{24} = \sqrt{8 \cdot 3}$

which is not much help since neither 8 nor 3 are perfect squares, and so we can't go any further.

Trying again:

$\sqrt{24} = \sqrt{4 \cdot 6} = \sqrt{4} \cdot \sqrt{6} = 2 \cdot \sqrt{6}$

which is simpler.

General Method of Simplifying Square Roots of Integers

1. Factor the integer into prime factors.
2. Group together any perfect squares.
3. Take the square root of the perfect squares and leave the rest under the radical sign.

EXAMPLE: *Simplify* $\sqrt{72}$.

STEP 1 $72 = 2^3 \cdot 3^2$

STEP 2 $72 = \underbrace{2^2 \cdot 3^2} \cdot 2 \longleftarrow$ this is what is left over when we put all
the perfect squares at the front

perfect
squares

STEP 3 $\sqrt{72} = \sqrt{2^2 \cdot 3^2 \cdot 2}$

$= \sqrt{2^2} \cdot \sqrt{3^2} \cdot \sqrt{2}$

$= 2 \cdot 3 \cdot \sqrt{2}$

$= 6 \cdot \sqrt{2}$ (the simplest possible form of $\sqrt{72}$)

EXAMPLE: *Simplify* $\sqrt{81,000}$.

STEP 1 $81,000 = 2^3 \cdot 3^4 \cdot 5^3$

STEP 2 $81,000 = \underbrace{2^2 \cdot 3^2 \cdot 3^2 \cdot 5^2} \cdot \underbrace{2 \cdot 5}$

perfect   the leftovers
squares

STEP 3 $\sqrt{81,000} = \sqrt{2^2 \cdot 3^2 \cdot 3^2 \cdot 5^2 \cdot 2 \cdot 5}$

$= \sqrt{2^2} \cdot \sqrt{3^2} \cdot \sqrt{3^2} \cdot \sqrt{5^2} \cdot \sqrt{2 \cdot 5}$

$= 2 \cdot 3 \cdot 3 \cdot 5 \cdot \sqrt{2 \cdot 5}$

$= 90 \cdot \sqrt{10}$

This method can easily be generalized to simplifying cube and higher roots.

EXAMPLE: *Simplify* $\sqrt[3]{720}$.

STEP 1   Factor:

$720 = 2^4 \cdot 3^2 \cdot 5$

STEP 2   This time, group all the perfect cubes at the front:

$720 = 2^3 \cdot 2 \cdot 3^2 \cdot 5$

STEP 3   Take the cube root of all the perfect cubes:

$\sqrt[3]{720} = \sqrt[3]{2^3 \cdot 2 \cdot 3^2 \cdot 5}$

$= \sqrt[3]{2^3} \cdot \sqrt[3]{2 \cdot 3^2 \cdot 5}$

$= 2 \cdot \sqrt[3]{90}$ (the simplest possible form of $\sqrt[3]{720}$)

Simplifying Roots of Fractions

The method above can be used to simplify the radicals in the denominator or numerator of a fraction. Some people feel strongly that a fraction should never be left with a radical in the denominator, and therefore insist that it be moved by a process called *rationalizing the denominator*. Radicals in the denominator have never caused me any trouble, so I always leave them there, but in case you want to get them out, here is the *general method for rationalizing the denominator*. The trick in this method is to realize that any irrational root of any integer can be made into an integer by multiplying by some irrational number. For example, $\sqrt{2}$ can be multiplied by another $\sqrt{2}$ to give 2 and $\sqrt[3]{5}$ can be multiplied by $\sqrt[3]{5} \cdot \sqrt[3]{5}$ to give 5. Therefore, provided you can figure out what irrational number is needed, you just multiply your fraction top and bottom by it, and the radical in the denominator becomes an integer. Of course you pay a price for this tidying up in the denominator—you now have a radical in the numerator.

EXAMPLE: *Rationalize $\dfrac{1}{\sqrt{2}}$.*

If you multiply $\sqrt{2}$ by itself, you get 2, by definition. So try multiplying $\dfrac{1}{\sqrt{2}}$ top and bottom by $\sqrt{2}$. $^{\bullet}$

$$\frac{1}{\sqrt{2}} = \frac{1}{\sqrt{2}} \cdot \frac{\sqrt{2}}{\sqrt{2}} = \frac{\sqrt{2}}{\sqrt{2} \cdot \sqrt{2}} = \frac{\sqrt{2}}{2}$$

$\dfrac{\sqrt{2}}{2}$ now has no radicals in the denominator, and therefore is in some ways a more respectable fraction.

EXAMPLE: *Rationalize $\dfrac{1}{\sqrt{18}}$.*

If you simplify the bottom, you find

$$\frac{1}{\sqrt{18}} = \frac{1}{3 \cdot \sqrt{2}}$$

Therefore multiplying by $\sqrt{2}$ top and bottom should help:

$$\frac{1}{\sqrt{18}} = \frac{1}{3 \cdot \sqrt{2}} = \frac{1}{3 \cdot \sqrt{2}} \cdot \frac{\sqrt{2}}{\sqrt{2}} = \frac{\sqrt{2}}{3 \cdot \sqrt{2} \cdot \sqrt{2}} = \frac{\sqrt{2}}{3 \cdot 2} = \frac{\sqrt{2}}{6}$$

So

$$\frac{1}{\sqrt{18}} = \frac{\sqrt{2}}{6}$$

which is a more respectable fraction.

If you didn't think of simplifying the $\sqrt{18}$ first, you might have multiplied top and bottom by $\sqrt{18}$. This would have given you

$$\frac{1}{\sqrt{18}} = \frac{1}{\sqrt{18}} \cdot \frac{\sqrt{18}}{\sqrt{18}} = \frac{\sqrt{18}}{18}$$

which is another more respectable fraction.

Although it doesn't look like it, $\dfrac{\sqrt{18}}{18}$ is actually the same as the answer above. Here's why: Suppose we simplify $\dfrac{\sqrt{18}}{18}$, using $\sqrt{18} = 3 \cdot \sqrt{2}$.

Then

$$\frac{\sqrt{18}}{18} = \frac{3 \cdot \sqrt{2}}{18} = \frac{\sqrt{2}}{6}$$

EXAMPLE: *Rationalize $\dfrac{1}{\sqrt[3]{2}}$.*

This is a little different because of the cube root. However, remember that $\sqrt[3]{2}$ was defined so that:

$$(\sqrt[3]{2})^3 = \sqrt[3]{2} \cdot \sqrt[3]{2} \cdot \sqrt[3]{2} = 2$$

Therefore if you multiply $\sqrt[3]{2}$ by $\sqrt[3]{2} \cdot \sqrt[3]{2}$, or $(\sqrt[3]{2})^2$, you will get 2. So try multiplying $\dfrac{1}{\sqrt[3]{2}}$ top and bottom by $(\sqrt[3]{2})^2$:

$$\frac{1}{\sqrt[3]{2}} = \frac{1}{\sqrt[3]{2}} \cdot \frac{(\sqrt[3]{2})^2}{(\sqrt[3]{2})^2} = \frac{(\sqrt[3]{2})^2}{\sqrt[3]{2} \cdot \sqrt[3]{2} \cdot \sqrt[3]{2}} = \frac{(\sqrt[3]{2})^2}{2}$$

This last fraction certainly has a rational denominator, though it now has a quite irrational numerator.

You can also simplify a string of radicals that are added and subtracted. The idea is to simplify each of the radicals using the method at the beginning of this section, and then to add or subtract wherever possible. The trick is to remember that you can only—meaning ONLY—add or subtract if you have multiples of the SAME radical. For example, $4 \cdot \sqrt{7} + 2 \cdot \sqrt{7}$ means four $\sqrt{7}$'s plus two $\sqrt{7}$'s, making a total of six $\sqrt{7}$'s, so

$$4 \cdot \sqrt{7} + 2 \cdot \sqrt{7} = (4 + 2) \cdot \sqrt{7} = 6 \cdot \sqrt{7}$$

On the other hand, $4 \cdot \sqrt{7} + 2 \cdot \sqrt{5}$ is already as simple as it can be and so has to be left alone.

EXAMPLE: *Simplify* $\sqrt{75} + \sqrt{3} - \sqrt{12}$.

$\sqrt{3}$ is already as simple as possible, but

$$\sqrt{75} = \sqrt{5^2 \cdot 3} = 5 \cdot \sqrt{3}$$

and

$$\sqrt{12} = \sqrt{2^2 \cdot 3} = 2 \cdot \sqrt{3}$$

so

$$\sqrt{75} + \sqrt{3} - \sqrt{12} = 5 \cdot \sqrt{3} + \sqrt{3} - 2 \cdot \sqrt{3}$$

Now we have five $\sqrt{3}$'s, plus one $\sqrt{3}$, minus two $\sqrt{3}$'s which gives us four $\sqrt{3}$'s altogether. So

$$\sqrt{75} + \sqrt{3} - \sqrt{12} = 5 \cdot \sqrt{3} + \sqrt{3} - 2 \cdot \sqrt{3}$$
$$= (5 + 1 - 2) \cdot \sqrt{3}$$
$$= 4 \cdot \sqrt{3}$$

This problem shows the value of being able to simplify radicals. Had $\sqrt{75}$ and $\sqrt{12}$ not been expressed as multiples of $\sqrt{3}$, we could not have done the addition and subtraction that reduced this expression to one term.

EXAMPLE: *Simplify* $3 \cdot \sqrt{2} + 5 \cdot \sqrt{3} + \sqrt{3}$.

None of the radicals here can be simplified, so we just add:

$$3 \cdot \sqrt{2} + 5 \cdot \sqrt{3} + \sqrt{3} = 3 \cdot \sqrt{2} + (5 + 1) \cdot \sqrt{3}$$
$$= 3 \cdot \sqrt{2} + 6 \cdot \sqrt{3}$$

Now unfortunately there's no way we can add $3 \cdot \sqrt{2}$ and $6 \cdot \sqrt{3}$, and so $3 \cdot \sqrt{2} + 6 \cdot \sqrt{3}$ is the answer.

EXAMPLE: *Multiply out and simplify* $\sqrt{5} \cdot \left(\sqrt{5} - \dfrac{1}{\sqrt{5}} \right)$.

Multiplying out means using the distributive law:

$$a \cdot (b - c) = a \cdot b - a \cdot c$$

with a = $\sqrt{5}$, b = $\sqrt{5}$, c = $\dfrac{1}{\sqrt{5}}$. So

$$\sqrt{5} \cdot \left(\sqrt{5} - \frac{1}{\sqrt{5}} \right) = \sqrt{5} \cdot \sqrt{5} - \sqrt{5} \cdot \frac{1}{\sqrt{5}}$$
$$= 5 - 1$$
$$= 4$$

EXAMPLE: *Multiply out and simplify* $\sqrt{2} \cdot (\sqrt{8} + 3 \cdot \sqrt{18})$.

Using the distributive law we see that

$$\sqrt{2} \cdot (\sqrt{8} + 3 \cdot \sqrt{18}) = \sqrt{2} \cdot \sqrt{8} + \sqrt{2} \cdot 3 \cdot \sqrt{18}$$
$$= \sqrt{2 \cdot 8} + 3 \cdot \sqrt{2 \cdot 18}$$
$$= \sqrt{16} + 3 \cdot \sqrt{36}$$
$$= 4 + 3 \cdot 6$$
$$= 22$$

EXAMPLE: *Multiply out and simplify* $(2 + \sqrt{3}) \cdot (\sqrt{5} + \sqrt{3})$.

The distributive law can be used on this problem too
if you take $a = 2 + \sqrt{3}$ and $b = \sqrt{5}$ and $c = \sqrt{3}$. Then

$$a \cdot (b + c) = a \cdot b + a \cdot c$$

gives

$$(2 + \sqrt{3}) \cdot (\sqrt{5} + \sqrt{3}) = (2 + \sqrt{3}) \cdot \sqrt{5} + (2 + \sqrt{3}) \cdot \sqrt{3}$$

Now if you change the order of multiplication on the right to get

$$= \sqrt{5} \cdot (2 + \sqrt{3}) + \sqrt{3} \cdot (2 + \sqrt{3})$$

you can use the distributive law again

$$= \sqrt{5} \cdot 2 + \sqrt{5} \cdot \sqrt{3} + \sqrt{3} \cdot 2 + \sqrt{3} \cdot \sqrt{3}$$

and at last we can simplify slightly

$$= 2 \cdot \sqrt{5} + \sqrt{15} + 2 \cdot \sqrt{3} + 3$$

And that's all that can be done.

CHAPTER 6 EXERCISES

Simplify:

1. $\sqrt{45}$
2. $\sqrt{75}$
3. $\sqrt{\frac{1}{80}}$
4. $\sqrt{96}$
5. $\sqrt{84,000}$

6. $\dfrac{1}{\sqrt{270}}$
7. $\sqrt[3]{24}$
8. $\sqrt[3]{108}$
9. $\sqrt{4^3}$
10. $\sqrt[3]{-54}$

11. $\sqrt[4]{81}$
12. $\sqrt{1.21}$
13. $\sqrt{0.0144}$
14. $\sqrt[3]{0.027}$
15. $\sqrt[3]{\dfrac{1}{0.125}}$

Rationalize the denominator:

16. $\dfrac{1}{\sqrt{10}}$ 19. $\dfrac{1}{\sqrt[4]{27}}$ 22. $\dfrac{1}{\sqrt{24}}$

17. $\dfrac{1}{\sqrt{12}}$ 20. $\dfrac{1}{3\sqrt{3}}$ 23. $\dfrac{1}{\sqrt{96}}$

18. $\dfrac{1}{\sqrt[3]{4}}$ 21. $\dfrac{1}{\sqrt[3]{4^2}}$

Simplify:

24. $3 \cdot \sqrt{2} + \sqrt{2} \cdot (4 - 5\sqrt{2})$

25. $\dfrac{\sqrt{20}}{3} + \dfrac{5}{\sqrt{45}}$

26. $\sqrt{14}\left(\dfrac{1}{\sqrt{2}} - \dfrac{1}{\sqrt{7}}\right)$

27. $\sqrt{15} + \sqrt{20}$

28. $\sqrt{12} + \sqrt{75}$

29. $\sqrt{1 - \left(\dfrac{1}{2}\right)^2} \cdot \left(\sqrt{3} + \dfrac{1}{\sqrt{3}}\right)$

30. $\sqrt{\dfrac{1}{2}} + \sqrt{\dfrac{25}{2}} - \sqrt{\dfrac{2}{9}}$

31. $\sqrt{3} \cdot \sqrt{21} - \dfrac{\sqrt{21}}{\sqrt{3}}$

32. $(2 \cdot \sqrt{3} - 5 \cdot \sqrt{2}) \cdot (\sqrt{2} - 4 \cdot \sqrt{3})$

33. $(\sqrt{10} - \sqrt{5})^2$

34. $\sqrt[3]{\dfrac{27}{2}} - \dfrac{\sqrt[3]{12}}{\sqrt[3]{3}}$

35. $\sqrt{0.4 \cdot (0.4 + 0.5)}$

36. $\sqrt[3]{0.3 \cdot (0.04 + 0.05)}$

Are the following correct? Why or why not?

37. $\sqrt{7} = \sqrt{3} + \sqrt{4}$

38. $\sqrt{12} = \sqrt{3} \cdot \sqrt{4}$

39. $\sqrt[3]{8 + 27} = 2 + 3 = 5$

40. $2 \cdot \sqrt[3]{4} - 3 \cdot \sqrt[3]{4} = (2 - 3) \cdot (\sqrt[3]{4} - \sqrt[3]{4}) = 0$

41. $\sqrt[3]{4} \cdot \sqrt{2} = \sqrt[3]{8}$

42. $\sqrt{3 \cdot 4} = 3\sqrt{4} = 3 \cdot 2 = 6$

Fill in the blanks with "is greater than" or "is less than":

43. $\dfrac{1}{\sqrt{3}}$ _____ $\dfrac{1}{\sqrt{2}}$

44. $\dfrac{1}{\sqrt{3}}$ _____ $\dfrac{2}{3}$

45. $\dfrac{1}{\sqrt[3]{4}}$ _____ $\dfrac{\sqrt[3]{4^2}}{3}$

46. $\sqrt[3]{28}$ _____ $\sqrt[3]{8.1}$

47. $\dfrac{\sqrt[3]{20}}{\sqrt[2]{10}}$ _____ $\dfrac{\sqrt[2]{11}}{\sqrt[3]{24}}$

Simplify:

48. $\dfrac{2}{2 - \sqrt{3}}$ (Hint: Multiply by $\dfrac{2 + \sqrt{3}}{2 + \sqrt{3}}$)

49. $\dfrac{3 + \sqrt{5}}{3 - \sqrt{5}} - \dfrac{3 - \sqrt{5}}{3 + \sqrt{5}}$

50. $\dfrac{1}{\sqrt{5} + \sqrt{2}}$

51. $\dfrac{\sqrt{6} - \sqrt{2}}{2 \cdot \sqrt{2} - \sqrt{6}}$

52. Reflooring a bathroom took 1.21 square yards of rubber tiles. If the floor is square, can you stand in the center and put a hand on opposite walls at the same time? (Find the size of the bathroom; then estimate your reach.)

7 SCIENTIFIC NOTATION

7.1 DEFINITION OF SCIENTIFIC NOTATION AND CONVERSION TO DECIMAL NOTATION

The point of scientific notation is that it enables us to write down in compact form numbers that often occur in scientific calculations, but that are extremely unwieldy written in the usual form. For example, 30,000,000,000 centimeters per second is the speed of light; 0.000000000001 centimeters is the size of the nucleus of an atom; and 0.000000000000000000000000766 and 623,000,000,000,000,000,000,000 are both constants that occur frequently in chemical and physical calculations.

Having a more compact way to write these numbers is not just a matter of convenience; it is a matter of necessity. Imagine doing a long calculation involving such numbers—we would spend the whole time counting and writing zeros. Inevitably, the calculations would take forever and be full of mistakes.

Before introducing scientific notation, it is necessary to say something about powers of ten.

Powers of Ten

First, remember from Section 2.7 that

$$10^1 \quad \text{means} \quad 10$$

$$10^2 \quad \text{means} \quad 10 \cdot 10 = 100$$

$$10^3 \quad \text{means} \quad 10 \cdot 10 \cdot 10 = 1000$$

$$10^4 \quad \text{means} \quad 10 \cdot 10 \cdot 10 \cdot 10 = 10,000$$

and so on.

10^1, 10^2, 10^3, 10^4 are called *powers of ten;* the 1, 2, 3, 4 are called *exponents.* We will now derive the exponent rules and use them to define some other powers of ten.

The Exponent Rules

First, observe that:

$$10^2 \cdot 10^4 = (10 \cdot 10) \cdot (10 \cdot 10 \cdot 10 \cdot 10)$$
$$= 10 \cdot 10 \cdot 10 \cdot 10 \cdot 10 \cdot 10 = 10^6$$

therefore

$$10^2 \cdot 10^4 = 10^{2+4} = 10^6$$

So to multiply powers of ten, you must add the exponents.
    In general:

Exponent Rule 1

$10^m \cdot 10^n = 10^{m+n}$ where *m, n* are positive integers

---

If now we look at

$$\frac{10^5}{10^2} = \frac{10 \cdot 10 \cdot 10 \cdot \cancel{10} \cdot \cancel{10}}{\cancel{10} \cdot \cancel{10}} = 10 \cdot 10 \cdot 10 = 10^3$$

we find that

$$\frac{10^5}{10^2} = 10^{5-2} = 10^3$$

So, in order to divide powers of ten, we subtract the exponents.
    In general:

Exponent Rule 2

$\dfrac{10^m}{10^n} = 10^{m-n}$ where *m, n* are positive integers with *m* greater than *n*

Lastly, consider:

$$(10^3)^2 = 10^3 \cdot 10^3 = (10 \cdot 10 \cdot 10) \cdot (10 \cdot 10 \cdot 10) = 10^6$$

so

$$(10^3) = 10^{3 \cdot 2} = 10^6$$

Therefore to raise 10^m to a power, we multiply the exponents.
In general:

Exponent Rule 3

$(10^m)^n = 10^{m \cdot n}$ where m, n are positive integers

Definition of 10, 10^{-1}, 10^{-2} ,...

So far we have defined 10^m when m is a positive integer and discovered the three exponent rules. We are about to define 10^m when m is zero or a negative integer, and we would like to do so in such a way that these three rules remain true. There is absolutely no reason why these rules *have* to stay true when m and n are zero or negative, because the rules were never derived for those cases. However, they are immensely simple and useful rules, and we would certainly like them to apply for all values of m and n. Therefore, we shall purposely pick the one way of defining 10 raised to the zero or negative powers that does make these rules remain true.

Let us suppose that the second rule is true in the case where $m = n = 2$. Then

$$\frac{10^2}{10^2} = 10^{2-2} = 10^0$$

This tells us that if 10^0 is going to satisfy the second rule, it must equal

$$\frac{10^2}{10^2} \quad \text{or} \quad 1$$

so we are forced to define

$$10^0 = 1$$

Now let us suppose the second rule is true in the case $m = 0$, $n = 1$. Then

$$\frac{10^0}{10^1} = 10^{0-1} = 10^{-1}$$

So if 10^{-1} is to satisfy the second rule, 10^{-1} must equal

$$\frac{10^0}{10^1} \quad \text{or} \quad \frac{1}{10}$$

Therefore we are forced to define

$$10^{-1} = \frac{1}{10}$$

Similarly, 10^{-2} must be defined as

$$\frac{1}{10^2} = \frac{1}{100}$$

and 10^{-3} must be defined as

$$\frac{1}{10^3} = \frac{1}{1000}$$

and so on.

Summarizing:

Powers of Ten

Definitions: $10^1 = 10$

$10^2 = 10 \cdot 10 = 100$

$10^3 = 10 \cdot 10 \cdot 10 = 1000$

etc.

$10^0 = 1$

$10^{-1} = \dfrac{1}{10^1} = \dfrac{1}{10}$

$10^{-2} = \dfrac{1}{10^2} = \dfrac{1}{100}$

$10^{-3} = \dfrac{1}{10^3} = \dfrac{1}{1000}$

Exponent Rules:

1. $10^m \cdot 10^n = 10^{m+n}$

2. $\dfrac{10^m}{10^n} = 10^{m-n}$

3. $(10^m)^n = 10^{m \cdot n}$

And now back to scientific notation. You can already see that 10^4 is a more compact way of writing 10,000, and certainly 10^{10} is a more compact form of 10,000,000,000. Similarly, $10^{-8} = 0.00000001$ and 10^{-8} is more compact. Luckily, 30,000 can be written as $3 \cdot 10,000$, so the compact form of 10,000 can also be used to rewrite 30,000 as follows:

$$30,000 = 3 \cdot 10,000 = 3 \cdot 10^4$$

In the same way,

$$0.00000005 = 5 \cdot 0.00000001 = 5 \cdot 10^{-8}$$

and the speed of light, 30,000,000,000, can be written as $3 \cdot 10^{10}$, which is certainly a much more manageable form.

If you want to write 3,150,000 in a brief form using powers of ten, there are several possibilities. You could say

$$3,150,000 = 315 \cdot 10,000 = 315 \cdot 10^4$$

or

$$3,150,000 = 31.5 \cdot 100,000 = 31.5 \cdot 10^5$$

or

$$3,150,000 = 3.15 \cdot 1,000,000 = 3.15 \cdot 10^6 \qquad \text{(scientific notation)}$$

or

$$3,150,000 = 0.315 \cdot 10,000,000 = 0.315 \cdot 10^7$$

(To check this, multiply each of the numbers on the right and see that you do get 3,150,000.)

Any of these forms is more compact that the original number, but for the sake of consistency, things are usually put in the third form, which is called *scientific notation.*

A number is said to be in *scientific notation* when it is in the form

(number between 1 and 10) · power of 10

The number between 1 and 10 is always expressed as a decimal. It can be as small as 1 and up to but not actually equal to 10. You should notice that you may often find scientific notation written with a × between the decimal and the power of ten:

$$3.15 \times 10^6$$

We will stick to a dot, however.

EXAMPLE: *Put 51,623 in scientific notation.*

Here, the number between 1 and 10 must be 5.1623.

Now, what power of ten must you multiply 5.1623 by to get 51,623?

The decimal point in 51,623 is right after the 3, so you need to move the decimal point in 5.1623 four places to the right, which means that you must multiply by 10^4. Therefore,

$$51,623 = 5.1623 \cdot 10^4$$

EXAMPLE: *Put 0.015 in scientific notation.*

Here the number between 1 and 10 must be 1.5. The decimal point in 1.5 must be moved two places to the left to give 0.015. This means dividing 1.5 by 10^2, or multiplying by $\frac{1}{10^2} = 10^{-2}$. Therefore,

$$0.015 = 1.5 \cdot 10^{-2}$$

EXAMPLE: *Put 0.00000128 in scientific notation.*

Here the number between 1 and 10 must be 1.28, and the decimal point in 1.28 must be moved six places to the left to give 0.00000128. So you must multiply 1.28 by 10^{-6} to get 0.00000128. Therefore,

$$0.00000128 = 1.28 \cdot 10^{-6}$$

To Put a Number in Scientific Notation:

1. Write down the corresponding number between 1 and 10.
2. Count the number of places, say n, you must move the decimal point in the number between 1 and 10 to give the original number.
3. If the decimal point has to be moved to the right, multiply the number between 1 and 10 by 10^n.
4. If the decimal point has to be moved to the left, multiply the number between 1 and 10 by 10^{-n}.

EXAMPLE: *Put 80,516 in scientific notation.*

$$80,516 = 8.0516 \cdot 10^4$$

EXAMPLE: *Put 0.0751 in scientific notation.*

$$0.0751 = 7.51 \cdot 10^{-2}$$

EXAMPLE: *Put $3.52 \cdot 10^6$ in decimal notation.*

$$3.52 \cdot 10^6 = 3,520,000$$

EXAMPLE: *Put $8.1 \cdot 10^{-9}$ in decimal notation.*

$$8.1 \cdot 10^{-9} = 8.1 \cdot \left(\frac{1}{10^9}\right) = \frac{8.1}{10^9} = 0.0000000081$$

(The last two examples are done as shown in the section on multiplying and dividing decimals by powers of ten.)

7.2 MULTIPLYING AND DIVIDING IN SCIENTIFIC NOTATION

Next we will consider operations in scientific notation. As with fractions, addition and subtraction are much messier than multiplication and division, so we will start with multiplication.

In multiplication and division in scientific notation, the idea is to multiply or divide the numbers between 1 and 10 in the way that you usually do decimals, and to do the powers of ten separately, using the exponent laws.

EXAMPLE: *Evaluate $(3.1 \cdot 10^4) \cdot (2 \cdot 10^2)$.*

$$
\begin{aligned}
(3.1 \cdot 10^4) \cdot (2 \cdot 10^2) &= (3.1 \cdot 2) \cdot 10^4 \cdot 10^2 \\
&= 6.2 \cdot 10^{4+2} \\
&= 6.2 \cdot 10^6
\end{aligned}
$$

Note: The 6.2 comes from multiplying the 3.1 and the 2; the 10^6 comes from multiplying the 10^4 and the 10^2.

EXAMPLE: *Evaluate $\dfrac{4.2 \cdot 10^5}{2.1 \cdot 10^2}$.*

$$\frac{4.2 \cdot 10^5}{2.1 \cdot 10^2} = \frac{4.2}{2.1} \cdot \frac{10^5}{10^2} = 2 \cdot 10^{5-2} = 2 \cdot 10^3$$

Note: The 2 comes from dividing 2.1 into 4.2; the 10^3 from dividing 10^2 into 10^5.

EXAMPLE: *Find $(6.1 \cdot 10^8) \cdot (3.2 \cdot 10^{-3})$.*

$$
\begin{aligned}
(6.1 \cdot 10^8) \cdot (3.2 \cdot 10^{-3}) &= 6.1 \cdot 3.2 \cdot 10^8 \cdot 10^{-3} \\
&= 19.52 \cdot 10^{8-3} \\
&= 19.52 \cdot 10^5
\end{aligned}
$$

Notice that the answer here is not in scientific notation, because the decimal 19.52 is not between 1 and 10. If you want the answer in scientific notation, this can easily be fixed by writing 19.52 in scientific notation:

$$19.52 = 1.952 \cdot 10^1$$

So the answer is

$$19.52 \cdot 10^5 = 1.952 \cdot 10^1 \cdot 10^5$$
$$= 1.952 \cdot 10^6$$

EXAMPLE: *Find* $\dfrac{(1.2 \cdot 10^{-5}) \cdot (1.5 \cdot 10^3)}{(9 \cdot 10^{-9})}$.

$$\frac{(1.2 \cdot 10^{-5}) \cdot (1.5 \cdot 10^3)}{(9 \cdot 10^{-9})} = \frac{1.2 \cdot 1.5}{9} \cdot \frac{10^{-5} \cdot 10^3}{10^{-9}}$$

$$= 0.2 \cdot 10^{-5+3-(-9)}$$

$$= 0.2 \cdot 10^{-5+3+9}$$

$$= 0.2 \cdot 10^7$$

Again, this is not in scientific notation at the moment. But we can fix it by saying:

$$0.2 = 2 \cdot 10^{-1}$$

and so

$$0.2 \cdot 10^7 = 2 \cdot 10^{-1} \cdot 10^7$$
$$= 2 \cdot 10^6$$

7.3   ADDING AND SUBTRACTING IN SCIENTIFIC NOTATION

Suppose we had to add $2.1 \cdot 10^3 + 3.5 \cdot 10^5$. Just as with fractions, where people have a tendency to add the tops and the bottoms separately, so here people have a tendency to add the decimal part and the exponents separately.

Thus, for example, someone might say: $2.1 \cdot 10^3 + 3.5 \cdot 10^5 = 5.6 \cdot 10^8$ However, this is just as wrong as adding the tops and bottoms of fractions. The best way to see that this method does not work is to write the numbers out in full. This gives you:

$$2.1 \cdot 10^3 + 3.5 \cdot 10^5 = 2{,}100 + 350{,}000$$

$$= 352{,}100$$

So the right answer, 352,100, is *much* smaller than and very different from $5.6 \cdot 10^8$.

So, how *do* you add or subtract numbers in scientific notation? Obviously you could write each of the numbers out in full and add or subtract the way you usually do. But that would take away the point of having scientific notation in the first place, which was *not* to have to write the numbers out in full.

Alternatively (and this is the clever way), if you rearrange the numbers so that all the powers of ten are the same, then you can add or subtract the decimal parts and leave the power of ten alone.

EXAMPLE: *Evaluate $2.1 \cdot 10^3 + 3.5 \cdot 10^5$.*

You can write $3.5 \cdot 10^5$ as $350 \cdot 10^3$ (because both equal 350,000). So

$$2.1 \cdot 10^3 + 3.5 \cdot 10^5 = 2.1 \cdot 10^3 + 350 \cdot 10^3$$

Now, if you have 2.1 of something plus 350 of something (in this case 10^3), you surely have 352.1 of that thing altogether. Therefore,

$$2.1 \cdot 10^3 + 350 \cdot 10^3 = (2.1 + 350) \cdot 10^3$$
$$= 352.1 \cdot 10^3$$
$$= 3.521 \cdot 10^2 \cdot 10^3$$
$$= 3.521 \cdot 10^5$$

so

$$2.1 \cdot 10^3 + 3.5 \cdot 10^5 = 3.521 \cdot 10^5$$

Alternatively, you could have written everything in terms of 10^5. To write

$$2.1 \cdot 10^3 \quad \text{as} \quad (\text{something}) \cdot 10^5$$

means making the power of ten bigger by a factor of 10^2. The decimal in front must therefore be decreased by a factor of 10^2 to keep the entire number the same size:

10^2 times larger

$$2.1 \cdot 10^3 = 0.021 \cdot 10^5$$

10^2 times smaller

Then

$$2.1 \cdot 10^3 + 3.5 \cdot 10^5 = 0.021 \cdot 10^5 + 3.5 \cdot 10^5$$

Again, 0.021 of something plus 3.5 of something (in this case 10^5) is clearly 3.521 of that thing. Therefore,

$$0.021 \cdot 10^5 + 3.5 \cdot 10^5 = (0.021 + 3.5) \cdot 10^5$$
$$= 3.521 \cdot 10^5$$

Thus,

$$2.1 \cdot 10^3 + 3.5 \cdot 10^5 = 3.521 \cdot 10^5$$

EXAMPLE: *Evaluate $9.76 \cdot 10^9 - 7.5 \cdot 10^8$.*

$$9.76 \cdot 10^9 - 7.5 \cdot 10^8 = 9.76 \cdot 10^9 - 0.75 \cdot 10^9$$
$$= (9.76 - 0.75) \cdot 10^9$$
$$= 9.01 \cdot 10^9$$

The first step of the problem used this manipulation:

$$\left(\begin{array}{c} \text{10 times larger} \\ 7.5 \cdot 10^8 = 0.75 \cdot 10^9 \\ \text{10 times smaller} \end{array} \right)$$

EXAMPLE: *Evaluate $4.21 \cdot 10^{25} - 1.85 \cdot 10^{26}$.*

$$4.21 \cdot 10^{25} - 1.85 \cdot 10^{26} = 0.421 \cdot 10^{26} - 1.85 \cdot 10^{26}$$
$$= (0.421 - 1.85) \cdot 10^{26}$$
$$= -1.429 \cdot 10^{26}$$

The first step of the problem used this manipulation:

$$\left(\begin{array}{c} \text{10 times larger} \\ 4.21 \cdot 10^{25} = 0.421 \cdot 10^{26} \\ \text{10 times smaller} \end{array} \right)$$

You can go through exactly the same process in the case when you have negative exponents.

EXAMPLE: *Evaluate $7.2 \cdot 10^{-9} + 1.3 \cdot 10^{-10}$.*

First let's convert everything into multiples of 10^{-10}. You have to remember that

$$10^{-9} = \frac{1}{10^9} \quad \text{and} \quad 10^{-10} = \frac{1}{10^{10}}$$

so 10^{-10} is smaller than 10^{-9}. Thus, writing

$$7.2 \cdot 10^{-9} \quad \text{as} \quad (\text{something}) \cdot 10^{-10}$$

means making the power of ten 10 times smaller, and so the decimal must be made 10 times larger:

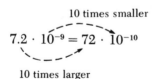

Going on with the problem:

$$7.2 \cdot 10^{-9} + 1.3 \cdot 10^{-10} = 72 \cdot 10^{-10} + 1.3 \cdot 10^{-10}$$
$$= 73.3 \cdot 10^{-10}$$
$$= 7.33 \cdot 10^{1} \cdot 10^{-10}$$
$$= 7.33 \cdot 10^{-9}$$

This process is clearly very like that of adding fractions, in which you have to get the denominators equal and then you can add the numerators, leaving the denominators alone. Here, you must make the powers of ten equal and then you can add the decimal parts, leaving the powers of ten alone. This similarity is not an accident; the reason for it should be clear if the previous example is rewritten as a fraction problem:

$$7.2 \cdot 10^{-9} + 1.3 \cdot 10^{-10} = 7.2 \cdot \frac{1}{10^{9}} + 1.3 \cdot \frac{1}{10^{10}}$$
$$= \frac{7.2}{10^{9}} + \frac{1.3}{10^{10}}$$

The L.C.D. for these fractions is 10^{10}, so the first fraction must be multiplied, top and bottom, by 10 (because $10^{1} \cdot 10^{9} = 10^{10}$). Therefore,

$$7.2 \cdot 10^{-9} + 1.3 \cdot 10^{-10} = \frac{72}{10^{10}} + \frac{1.3}{10^{10}}$$
$$= \frac{73.3}{10^{10}} \quad \text{since the denominators are now the same, we can add the numerators}$$
$$= 73.3 \cdot \frac{1}{10^{10}}$$
$$= 73.3 \cdot 10^{-10}$$
$$= 7.33 \cdot 10^{-9}$$

Comparing this calculation with the one above, you can see that getting the powers of ten equal (in this case to 10^{-10}) corresponds exactly to getting the denominators equal (in this case, to 10^{10}).

EXAMPLE: *Evaluate $4.05 \cdot 10^{-19} - 10^{-21}$.*

Realizing that 10^{-21} is the same as $1.0 \cdot 10^{-21}$, we can rewrite everything in terms of 10^{-19} as follows:

10^{2} times larger

$$10^{-21} = 1.0 \cdot 10^{-21} = 0.01 \cdot 10^{-19}$$

10^{2} times smaller

so

$$4.05 \cdot 10^{-19} - 10^{-21} = 4.05 \cdot 10^{-19} - 0.01 \cdot 10^{-19}$$
$$= 4.04 \cdot 10^{-19}$$

EXAMPLE: *Evaluate $1.2 \cdot 10^{-19} \cdot (5.2 \cdot 10^{10} + 4 \cdot 10^{9})$.*

The order of operations tells us that we must work out what is in the parentheses first:

$$1.2 \cdot 10^{-19} \cdot (5.2 \cdot 10^{10} + 4 \cdot 10^{9}) = 1.2 \cdot 10^{-19} \cdot (5.2 \cdot 10^{10} + 0.4 \cdot 10^{10})$$
$$= 1.2 \cdot 10^{-19} \cdot (5.6 \cdot 10^{10})$$

And then multiply:

$$= (1.2 \cdot 5.6) \cdot (10^{-19} \cdot 10^{10})$$
$$= 6.72 \cdot 10^{-9}$$

EXAMPLE: *Evaluate*

$$\frac{7 \cdot 10^{7} - 2.8 \cdot 10^{6}}{10^{-7} + 1.24 \cdot 10^{-6} - 1.02 \cdot 10^{-6}}$$

This means divide the whole of the bottom into the whole of the top, so we work out the top and bottom separately and then divide:

$$\frac{7 \cdot 10^{7} - 2.8 \cdot 10^{6}}{10^{-7} + 1.24 \cdot 10^{-6} - 1.02 \cdot 10^{-6}}$$

$$= \frac{7 \cdot 10^{7} - 0.28 \cdot 10^{7}}{1 \cdot 10^{-7} + 12.4 \cdot 10^{-7} - 10.2 \cdot 10^{-7}}$$

$$= \frac{6.72 \cdot 10^{7}}{3.2 \cdot 10^{-7}}$$

$$= \frac{6.72}{3.2} \cdot 10^{7-(-7)}$$

$$= 2.1 \cdot 10^{14}$$

CHAPTER 7 EXERCISES

Write in scientific notation:

1. 1,567,000	5. 0.023	9. 11%
2. 0.92	6. 3.12	10. 1.7%
3. $\dfrac{0.7}{10}$	7. 0.00065	11. 0.000000038
	8. 31.750	12. 1,300,000,000
4. $\dfrac{0.0032}{8}$		

Express in scientific notation, and then as a percentage:

13. 0.00706	15. 2.23	17. $\frac{1}{2}$
14. 0.230	16. 0.00000678	18. $\frac{1}{4}$

Express in scientific notation, and then as a decimal:

19. 0.0270% 22. 2.060% 25. $\frac{1}{3}$
20. 321% 23. 101.3% 26. $\frac{1}{5}$
21. 0.0003% 24. 98.09%

Evaluate, writing the answer in scientific notation:

27. $\dfrac{4 \cdot 10^2}{2 \cdot 10^{-2}}$

35. $4.7 \cdot 10^{-9} - 1.7 \cdot 10^{-9}$

36. $2.2 \cdot 10^{-12} - 1.2 \cdot 10^{-13}$

37. $\sqrt{4 \cdot 10^{-4}}$

28. $\dfrac{5 \cdot 32 \cdot 10^{-2}}{0.2}$

38. $\dfrac{0.9}{10^{-3}} \cdot \dfrac{1}{3}$

29. $\dfrac{(2.1 \cdot 10^{-2}) \cdot (3 \cdot 10^{-5})}{7 \cdot 10^{-4}}$

39. $\dfrac{4.2}{10^{13}} + 0.1 \cdot 10^{-11}$

30. $2.1 \cdot 10^{-9} - 17 \cdot 10^{-11}$

40. $\dfrac{1.2}{10^{-4}} - \dfrac{0.3}{10^{-5}}$

31. $\dfrac{5.9 \cdot 10^{99}}{10^{-100}}$

41. $\sqrt{10^{-2} \cdot (311 - 310)}$

32. $\frac{1}{4} \cdot (4 \cdot 10^{-3}) \cdot 10^5$
33. $0.00036 + 4.2 \cdot 10^{-4}$

42. $(0.6) \cdot \left(\dfrac{3}{10^8}\right)$

34. $3.1 \cdot 10^{-9} + 1.2 \cdot 10^{-10}$

43. $\frac{1}{4} \cdot (2.1 \cdot 10^{-21} - 17 \cdot 10^{-22})$

Express your answers in scientific notation:

44. (a) What is the volume of a spherical raindrop of radius $1.5 \cdot 10^{-3}$ meter?

(b) What is the weight of this raindrop if the density of water is 10^6 grams/cubic meter (i.e., if 1 cubic meter of water weighs 10^6 grams)?

45. In 1960, the total electric energy generated in the United States was 842 billion kilowatt-hours, and the population was 179.3 million. Calculate the energy generated per capita (i.e. per person.)

46. How many spherical ball bearings of diameter 10^{-1} inches can you fit in a circular track (see Figure 7.1) of radius 0.7 inch? (Take $\pi = \frac{22}{7}$ here.)

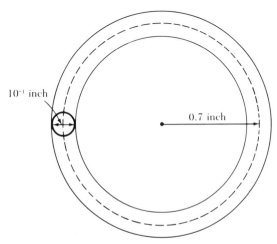

FIG. 7.1

47. If light travels at $3 \cdot 10^{10}$ centimeters per second, find:
    (a) The magnitude of a light year, in centimeters. (A light year is the distance traveled by light in one year.)
    (b) The number of atoms of carbon, placed end to end in a line, that a ray of light can pass by in 1 hour. Assume that each atom is a sphere of diameter 1.5 angstrom (Note: 10^{-8} centimeter = 1 angstrom, the atomic unit of length.)

Scientists use the concept of significant figures to indicate the accuracy of various measurements. We will define the number of significant figures in a number to be the number of digits in front of the power of ten when the number is written in scientific notation. Thus $12 = 1.2 \cdot 10^1$ has two significant figures, as does $0.012 = 1.2 \cdot 10^{-2}$. Notice that 3.0 implies greater accuracy than 3; we say that 3.0 has two significant figures whereas 3 has one. How many significant figures are in each of the following numbers?

48. 13.2   52. $1.0 \cdot 10^2$ 56. 0.0126
49. 0.06   53. $1 \cdot 10^2$ 57. 11%
50. 1.045   54. 0.0302   58. 10^8
51. $1.00 \cdot 10^2$ 55. 1475

8 ALGEBRA: VARIABLES AND EXPRESSIONS

8.1 ALGEBRA IS FOUNDED ON ARITHMETIC

Since algebra is only arithmetic in fancy dress, all that you need to do algebra successfully is to understand arithmetic reasonably well and to keep your head. The operations done in algebra are exactly those of arithmetic. The difference between the two subjects is that in arithmetic one works with specific numbers, be they fractions, decimals, or worse, but always with specific numbers; in algebra one achieves more generality by working with letters that stand for any or many numbers. But these letters stand for exactly the same old numbers we've been dealing with all along, and so they operate by exactly the same old rules.

8.2 VARIABLES AND EXPRESSIONS

In algebra letters are used to stand for numbers. You have already seen one example of this in the use of π for the ratio of the circumference of a circle to its diameter. π is written instead of the number because the number is irrational and cannot be written down exactly in either fractional or decimal form. Thus the symbol π has the advantage of being both accurate (it means precisely the number you get by dividing the diameter of a circle into the circumference) and being briefer than the two most common approximations to π, namely $\frac{22}{7}$ and 3.14.

Most letters that one uses in algebra, however, are different from π because we do not know exactly what they stand for. Sometimes we find out during a problem, but more often we do not, and we must treat the letter as though it were any possible number. The results that we get, therefore, are true for any possible number, rather than for a specific case. It is this that makes algebra so much more powerful than arithmetic.

Letters that stand for numbers are called *variables* if they may stand for any one of a group of numbers, and are called *constants* if you happen to know that they stand for one fixed number (like π).

Letters can be combined just as numbers can: $a + b$ means the sum of the number represented by a and the number represented by b; ab, which is written instead of $a \cdot b$, means the product of the number represented by a and the number represented by b; and $\frac{a}{b}$ means the number represented by a divided by the number represented by b ($a \div b$ is seldom used).

Numbers and letters can also be combined: $2 + p$ means 2 plus the number represented by p; $\frac{15x}{7}$ means 15 times the number represented by x divided by 7; $3(x + 5)$ means 3 times the number you get by adding x and 5.

Convention About Multiplication

In algebra, ab is written to mean a times b. So if you see no sign between between two letters or between a letter and a number, then a multiplication sign is understood. You'll be glad to know, however, that this convention doesn't apply to two numbers—24 still means twenty-four and not $2 \cdot 4$. The product of two numbers is still expressed by putting a dot between them.

Any combination of letters and numbers connected by symbols such as, $+$, $-$, and so on, is called an *expression*, or an *algebraic expression*. For example,

$$x + y \qquad \frac{2x + y^2}{z + 3} \qquad \frac{ax - \sqrt{py + cz}}{t^3} \qquad 5x^2 + 4x - 3$$

are expressions.

If somehow you know the values of the variables in an expression, substituting the values of those variables will give you the number that the expression represents. This is called *evaluating the expression.*

EXAMPLE: *Evaluate* $xy + \frac{z^2}{2}$ *when* $x = 2$, $y = -3$, *and* $z = 4$.

Substituting:

$$xy + \frac{z^2}{2} = 2(-3) + \frac{4^2}{2} = -6 + \frac{16}{2} = -6 + 8 = 2$$

EXAMPLE: *Evaluate* $\dfrac{2(2t^2 + 3)}{2t + 1}$ *when* $t = 3$.

Remember the order of operations. When $t = 3$;

$$\frac{2(2t^2 + 3)}{2t + 1} = \frac{2(2 \cdot 3^2 + 3)}{(2 \cdot 3 + 1)}$$

Work with parentheses first. Inside them, do squares first, then multiplication.

$$= \frac{2(2 \cdot 9 + 3)}{(2 \cdot 3 + 1)}$$

$$= \frac{2(18 + 3)}{(6 + 1)}$$

Now do addition to finish inside parentheses.

$$= \frac{2(\overset{3}{\cancel{21}})}{\underset{1}{\cancel{7}}}$$

Cancel to simplify.

$$= 2 \cdot 3$$

$$= 6$$

In an expression consisting of a number of smaller expressions connected by $+$ or $-$ signs, each of the smaller parts is called a *term*. For example,

$$xy^2 + z - \frac{(a + 1)}{b}$$

has terms xy^2, z and $\dfrac{(a + 1)}{b}$

EXAMPLE: $x^2y\sqrt{z + 5}$ has only one term, namely itself.

$\left(\dfrac{a - b}{c - d}\right) x + \left(\dfrac{p + q}{r + s}\right)$ has two terms, namely

$\left(\dfrac{a - b}{c - d}\right) x$ and $\left(\dfrac{p + q}{r + s}\right)$

In an expression such as

$$2a^2b + c^2 - ab^2 + 3a^2b$$

$2a^2b$ and $3a^2b$ are said to be *like terms* because they contain the same combinations of the same variables. $2a^2b$ and c^2 are *unlike* terms. $2a^2b$ and $-ab^2$ are also unlike terms because the first contains an a^2 and a b while the second has an a and a b^2. Similarly, $y + y^2 + y^2x$ has three unlike terms.

An expression that is the sum (or difference) of terms, as above, but in which each of the terms is the product of numbers and of variables raised to positive powers, is called a *polynomial*. For example, these are polynomials:

$$x^2y + 5x^2 + \frac{a^2}{12} \qquad 7t^5 - 2t^2 + \frac{t}{3} + 8$$

and these are not:

$$\frac{2}{x^2} + \frac{3}{x} \qquad 3\sqrt{x} + 5 \qquad \frac{1}{2x^2}$$

We shall meet mostly polynomials in one variable (i.e., involving one letter) such as

$$7t^5 + t^3 - 2t^2 + \frac{t}{3} + 8$$

The number in front of each term is called the *coefficient* of that term. For example, in

$$7t^5 + t^3 - 2t^2 + \frac{t}{3} + 8$$

the coefficient of the t^5 term is 7; the coefficient of the t^3 term is 1 (because $t^3 = 1 \cdot t^3$), the coefficient of the t^2 term is -2 (notice the sign comes too), the coefficient of the t term is $\frac{1}{3}$ (because $\frac{t}{3} = \frac{1}{3} \cdot t$), and the constant term (so called because it has no variable) is 8. Extending this idea, we say that in

$$ax^2 - \frac{c}{d}x$$

the coefficient of x^2 is a and the coefficient of x is $-\frac{c}{d}$.

In a polynomial of one variable, the highest power to which the variable is raised is called the *degree* of the polynomial. For example,

$$6x^2 + 12x^5 - 9 \qquad \text{has degree 5}$$

$$16x^{14} + 1 \qquad \text{has degree 14}$$

$$x + 5 \qquad \text{has degree 1}$$

CHAPTER 8 EXERCISES

For each of the following, state how many terms are in each expression and give the degree of each expression that is a polynomial in one variable.

1. $x + 12$

2. $7p^2 + 8p + 1$

3. 41

4. $64z^3 + \dfrac{1}{z}$

5. $18f^6 - 2f^5 + 6f^3 + \dfrac{19f}{x}$

6. $84r^2 + r$

7. Evaluate the expressions in Problems 1–6 for $p = 2$, $z = \frac{1}{4}$, $f = 0$, $r = -1$, $x = 2$, $y = -5$.

Evaluate the following expressions:

8.	$x^3 + 2x^2 - 5x$	when $x = 2$
9.	$3x + 7$	when $x = 5$
10.	$3(x + 7)$	when $x = 5$
11.	$\dfrac{2y + 0.8}{3y + 0.6}$	when $y = 0.5$
12.	$(2p)^2 - (2p^2)$	when $p = -3$
13.	$\sqrt{a(a + 0.5)}$	when $a = 0.4$
14.	$10a^2 + 2ax - x + 5$	when $a = 5$, $x = -10$
15.	$10x^2 - 5xy + y^2x - 2$	when $x = -2$, $y = 3$
16.	$\dfrac{(a + b)c + d}{-bc + 2a}$	when $a = 1$, $b = -2$, $c = 3$, $d = -1$
17.	$x^2yz - 4xyz^2$	when $x = 1$, $y = 2$, $z = 3$

How many terms are there in each of the following expressions, and what are they?

18. $\sqrt{a^2 + 5} + a^2 + 5$

19. $\left(\dfrac{p + q}{p - q}\right) t + p$

20. $\dfrac{2z^2 + 3z^3 + 4z^4}{2 + z}$

21. $3a(x^2 + 1) + \dfrac{b}{x^2 + 1} + (a + b + 2)\dfrac{x^2 + 1}{x^2 + 2}$

22. $\left(\dfrac{p + q}{p - q} + q\right) p - q\,(p + q^2) + \dfrac{p^2 + q^2 - pq}{p - q}$

23. In $14z^4 - 3z^3 - z^2 - \dfrac{z}{2} - 5$, what are the coefficients of z^4, z^3, z^2, z, and what is the constant term?

24. In $2(a + b)^3 - 3(a + b)^2 + (a + b)$, what are the coefficients of $(a + b)^3$, $(a + b)^2$ and $(a + b)$?

25. If a and b are constants and x is a variable, what are the coefficients of x^3, x^2, x in $ax^3 + (a + b)x^2 + \dfrac{x}{b}$?

Which of the following are like terms?

26. $2Bp$, pQ, pBQ, $\dfrac{4pQ}{21}$

27. a^2b, $2ab^2$, $-3ba^2$, $5aba$, $(2ab)b$

28. $(b + a)^2$, $2(a + b)^3$, $3(a + b)^2$

29. $x(x + 2)$; $\dfrac{x}{x + 2}$; $\dfrac{(x + 2)x}{3}$

30. $3axy$; ay^2; $\dfrac{yx}{2}$ (*a* constant)

9   POLYNOMIALS

9.1   ADDING AND SUBTRACTING POLYNOMIALS

There are two things to remember when adding and subtracting polynomials:
1. Watch your signs!
2. You can only combine like terms.

The reason for the second is that you can only combine similar things. You can't combine $2x$ and $3x^2$ for the same reason that you can't combine 2 peanuts and 3 oranges, so the sum $2x + 3x^2$ cannot be reduced to one term.

The possibility of combining like terms enables us to simplify some expressions.

EXAMPLE: *Simplify $2a^2b + 3a^2b - ab^2$.*

$$2a^2b + 3a^2b - ab^2 = (2 + 3)a^2b - ab^2 = 5a^2b - ab^2$$

This cannot be simplified any further, because $5a^2b$ and ab^2 are unlike terms.

EXAMPLE: *Simplify $2x^2 + 5x^3 - 3x^2 + x$.*

$$2x^2 + 5x^3 - 3x^2 + x = 5x^3 - x^2 + x$$

This can go no further, since x^3, x^2, and x are unlike terms.

EXAMPLE: *Add $2x^2 + 3a$ to $5x^2 + 6a$.*

$$2x^2 + 3a + 5x^2 + 6a = 7x^2 + 9a$$

EXAMPLE: *Add $4x^2 - 2x + 3$ to $-3x^2 - 4x - 1$.*

$4x^2 - 2x + 3 + (-3x^2 - 4x - 1)$

$\quad = 4x^2 - 2x + 3 - 3x^2 - 4x - 1$ (because $+(-a) = -a$)

$\quad = x^2 - 6x + 2$

EXAMPLE: *Add $a + a^3$ to $2a^2$.*

The expression $a + a^3 + 2a^2$ cannot be simplified, so this is the answer.

EXAMPLE: *Subtract $2x + 3$ from $5x + 9$.*

This means write down $5x + 9$ first, and subtract $2x + 3$ from it:

$5x + 9 - (2x + 3)$

This is where you have to watch the signs: the $-$ applies to the whole of the second polynomial (not just the first term), so it is important *not* to drop the parentheses (or you will have $5x + 9 - 2x + 3$, in which the $-$ sign does apply just to the first term). Then you use the distributive law to "distribute the minus sign":

$-(a + b) = -a - b$

So

$5x + 9 - (2x + 3) = 5x + 9 - 2x - 3$

$\quad\quad = 3x + 6$

Note: If you had forgotten to use the distributive law, and had just dropped the parentheses, you would have had $5x + 9 - 2x + 3 = 3x + 12$, which is definitely not the right answer.

EXAMPLE: *Subtract $4 - 2x^2 - 3x$ from $3x - 3x^2 - 9$.*

$3x - 3x^2 - 9 - (4 - 2x^2 - 3x)$

$\quad = 3x - 3x^2 - 9 - 4 + 2x^2 + 3x$ (because $-(-2x^2) = 2x^2$ and $-(-3x) = 3x$)

$\quad = 6x - x^2 - 13$

PROBLEM SET   9.1

Simplify by combining like terms:

1. $5t + 3t$
2. $4x^4 - 2x^3 - 5x^4$
3. $3xy - 4xy - 6xy$
4. $a + b - 2a + 3b - 2$
5. $3as^2 + 4s^2 - 4as^2 + 5s^2$
6. $a^2b + 2ab^2 + 3a^2b$
7. $3p^2t^3 + 2p^2t^3 - t^3p^2$
8. $2r^2ts - 5s^2rt - r^2st$
9. $3x^2 - xy - 5yx + x^2 + y^2$
10. $5xyz + 3xz - 2x^2zy + 3zxy^2$

Perform the indicated operation:

11. $5x^2 + 2x - 4 - (x^3 - 5x - 2)$
12. $3c + b + (4c - b)$
13. $7t^2 - rt - 10r^2 + (5rt - 3r^2)$
14. $x^2 + x^3 - 3x + (4 - 5x^2 + 3x^3)$
15. $10 - 8x^2 - 5x - (3x^2 + 5)$
16. $y^4 + 2y^3 - 1 + (3y^4 + 2y^3 - y^2 + 5y + 1)$
17. $-7m + 12m^2 + 3 - (2m^2 - m - 6)$
18. $5x^2 + 4 - (10 - 2x^2) - 3x^2 + 1 + x$
19. $a^2 - 3ab + b^2 + 2c - (4ab + b^2) + (2a^2 + b^2 - 3c)$
20. $2x^2y^2 + 8x^2y + 4xy + (x^2y^2 - 3x^2y + 7xy^2) - (x^2y + 6xy^2)$

9.2   MULTIPLYING POLYNOMIALS

To multiply two polynomials, use the distributive law:

$$a(b + c) = ab + ac$$

which holds when a, b, and c, are algebraic expressions. You may have to use the law several times, but first we will do the simplest kind of example, where you only have to use the law once.

EXAMPLE: $3x^2(y + tz^2)$.

This is of the form $a(b + c)$ (with $a = 3x^2$, $b = y$, $c = tz^2$), so use

a $(b + c)$ $=$ ab + ac to get

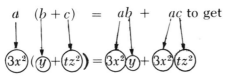

Going from $3x^2(y + tz^2)$ to $3x^2y + 3x^2tz^2$ is called *multiplying out*

$3x^2(y + tz^2)$. Any number of terms in the parentheses can be treated the same way.

EXAMPLE: *Multiply out $2x^2(x^3 - 3x^2 + 4x)$.*

$2x^2(x^3 - 3x^2 + 4x)$

$= 2x^2 \cdot x^3 - 2x^2 \cdot 3x^2 + 2x^2 \cdot 4x$

$= 2x^2 \cdot x^3 - 2 \cdot 3 \cdot x^2 \cdot x^2 + 2 \cdot 4 \cdot x^2 \cdot x$ (changing the order of multiplication)

$= 2x^5 - 6x^4 + 8x^3$ (since $x^2 \cdot x^3 = x^5$, $x^2 \cdot x^2 = x^4$, $x^2 \cdot x = x^3$)

In order to multiply out a product in which each factor has two terms, you will need to use the distributive law twice.

EXAMPLE: *Multiply out $(5x + y)(2t + s)$.*

First think of $5x + y$ as one number, and identify it with the a in the distributive law. Then, letting $b = 2t$ and $c = s$,

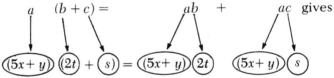

Rewrite each of the products on the right-hand side in the opposite order, and you will see that you can use the distributive law again:

$(5x + y)(2t + s) = 2t(5x + y) + s(5x + y)$

$= 2t5x + 2ty + s5x + sy$

$= 10tx + 2ty + 5sx + sy$

Convention says that the numbers are written on the left-hand side of each term, so $s5x$ is rewritten as $5sx$. Otherwise, this cannot be simplified any further.

EXAMPLE: *Multiply out $(2x - 3)(x + 5)$.*

$(2x - 3)(x + 5) = (2x - 3)x + (2x - 3)5$

$= x(2x - 3) + 5(2x - 3)$ (switching order of multiplication)

$= 2x^2 - 3x + 10x - 15$ (using distributive law again)

$= 2x^2 + 7x - 15$ ($-3x$ and $10x$ are like terms and so can be combined)

EXAMPLE: *Multiply out $(x - 2)(x + 2)$.*

$$(x - 2)(x + 2) = x(x - 2) + 2(x - 2)$$
$$= x^2 - 2x + 2x - 4$$
$$= x^2 - 4$$

Notice (for future reference) that $x^2 - 4 = x^2 - 2^2$ (since $4 = 2^2$), so $x^2 - 4$ is the *difference of two squares*. We have shown that $(x - 2)(x + 2) = x^2 - 2^2$, which tells us that the difference of two squares can be rewritten as a product, because $x^2 - 2^2$ is the product of $(x - 2)$ and $(x + 2)$. This has no use whatever on this page, but it is of the greatest importance a few pages hence!

EXAMPLE: *Expand $(x + y)^2$. (Expand means multiply out.)*

$$(x + y)^2 = (x + y)(x + y) = x(x + y) + y(x + y)$$
$$= x^2 + xy + yx + y^2$$
$$= x^2 + 2xy + y^2 \qquad \text{since } yx = xy$$

Notice (again for future reference) that we have shown that something of the form $x^2 + 2xy + y^2$ can be rewritten as a square, namely $(x + y)^2$. The form $x^2 + 2xy + y^2$ is said to be a *perfect square*.

Also, please note that $(x + y)^2$ *is not the same as* $x^2 + y^2$. There is a geometrical way of seeing the difference between these two quantities. It goes as follows: x^2 represents the area of a square of side x, y^2 represents the area of a square of side y, and $(x + y)^2$ represents the area of a square of side $(x + y)$. Draw a square of side $(x + y)$, as in Figure 9.1.

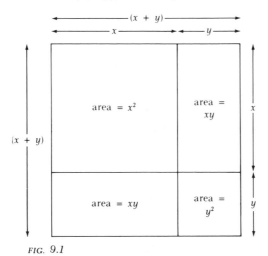

FIG. *9.1*

Area of top left square $= x^2$

Area of bottom right square $= y^2$

Area of each rectangle $= xy$

Area of whole figure $= (x + y)^2$

Area of top left and bottom right squares $= x^2 + y^2$

So $(x + y)^2$ and $x^2 + y^2$ are clearly different in general. Also, since the area of the whole figure is the sum of the areas of the smaller pieces that compose it, $(x + y)^2 = x^2 + y^2 + 2xy$.

But back to multiplying out.

EXAMPLE: *Expand $(p + q + r)^2$.*

$$(p + q + r)^2 = (p + q + r)\,(p + q + r)$$

Think of this as one quantity for the moment.

$$= (p + q + r)p + (p + q + r)q + (p + q + r)r$$

Switch the order of multiplication and use the distributive law again:

$$= p(p + q + r) + q(p + q + r) + r(p + q + r)$$

$$= p^2 + pq + pr + qp + q^2 + qr + rp + rq + r^2$$

And collect terms:

$$= p^2 + q^2 + r^2 + 2pq + 2qr + 2pr$$

This method can be used to multiply out any number of polynomials, with any number of terms. (Of course it may be very tedious, but it can always be done.)

EXAMPLE: *Multiply out $(x + 1)\,(x + 2)(x + 3)$.*

Multiply two of these together first, and then multiply by the last one.

$$(x + 1)(x + 2)(x + 3) = (x + 1)\,[x(x + 2) + 3(x + 2)]$$

Leave this alone for the moment.

$$= (x + 1) \, [x^2 + 2x + 3x + 6]$$
$$= (x + 1) \, [x^2 + 5x + 6]$$

Now start with the $(x + 1)$:

$$= x^2(x + 1) + 5x(x + 1) + 6(x + 1)$$
$$= x^3 + x^2 + 5x^2 + 5x + 6x + 6$$
$$= x^3 + 6x^2 + 11x + 6$$

In practice this process of multiplying out is shortened considerably. Let us look again at how we multiplied $(2x - 3)(x + 5)$ out to get $2x^2 + 7x - 15$. What we did amounted to multiplying everything in the first parentheses by everything in the second (with due respect to the signs) and then combining like terms. So you can think of it like this:

$$(2x - 3)(x + 5) = 2x \cdot x + 2x \cdot 5 - 3 \cdot x - 3 \cdot 5$$
$$= 2x^2 + 10x - 3x - 15$$
$$= 2x^2 + 7x - 15 \dots \text{as above}$$

This is the method most people use when they're doing it in their heads.

PROBLEM SET   9.2

Multiply out:

1. $2a^2(a^3 - 2a)$
2. $4x(5 - x - 7x^2)$
3. $(3ab + a^2bc) \, 2bc$
4. $(r + 3s) \, (r - s)$
5. $(2y - 3) \, (y + 4)$
6. $(x - 2) \, (x + 2)$
7. $(a + s)^2$
8. $(x - 3)^2$
9. $(a + b + c)^2$
10. $(x - \frac{1}{2}y) \, (x + \frac{1}{2}y)$

11. $(2x + 7) \, (3x - 1)$
12. $(ab + 5) \, (3 + 2a)$
13. $(p^2 + q^2) \, (r + st)$
14. $(3x + 2y) \, (2x - y)$
15. $(x + 2)^3$
16. $(x - y) \, (x^2 + xy + y^2)$
17. $(x + 2) \, (x + 3) \, (x - 4)$
18. $\left(\dfrac{x}{3} + \dfrac{3}{x} \right)^2$

19. $(2b + 3) \, (-4 + 3b) \, (b - 5)$
20. $(x - y) \, (x - y) + (2x + y) \, (2x + y)$
21. $(r - 5) \, (r + 2) + (r + 7) \, (r - 4)$
22. $-(a - b + 2c) \, (a - b + 2c) + (2a + b) \, (2a - b)$
23. $-3r^2sp^2(-rs + 6sp - 2r^3sp^{10})$
24. $(m - 2) \, (m + 7) - 2(m - 9) \, (m + 5)$
25. $(y^2 + 4y - 3) \, (y - 1) + (2y^2 + y - 4) \, (y^2 + y)$

9.3   DIVIDING POLYNOMIALS

Two polynomials in the same variable can be divided into one another by a method exactly like the numerical division that you may have learned in school. First, a numerical example:

$$
\begin{array}{r}
49 \\
3{\overline{)\,148}} \\
\end{array}
$$

multiply $(4 \cdot 3)$— — — →12 ¦
                             bring down

after subtracting— — — →28
multiply $(9 \cdot 3)$ — — — →27

                    1----→remainder

This shows that 3 goes into 148 forty-nine times with 1 left over, so

$$\frac{148}{3} = 49 + \frac{1}{3}$$

148 is called the *dividend*, 3 is the *divisor*, 49 is the *quotient*, and 1 is the *remainder*.

Here's how the method works for polynomials. Suppose we want to divide $x + 1$ into $2x^2 + 5x + 3$. The degree of the divisor, $x + 1$, must be less than the degree of the dividend, $2x^2 + 5x + 3$, and both polynomials must be written with their powers of x in descending order.

The problem should be set up like a numerical division problem:

$$
\begin{array}{r}
\text{quotient} \\[-2pt]
\text{divisor} \overline{)\,\text{dividend}} \\
x + 1 {\overline{)\,2x^2 + 5x + 3}}
\end{array}
$$

The first stage is to take the *highest term* (meaning the term that has the highest power of x in it) in the divisor—in this case x—and divide it into the highest term of the dividend, here $2x^2$. The result, which in this case is $\frac{2x^2}{x} = 2x$, is the first term of the quotient, and is written above the corresponding term of the dividend. So now we have

$$
\begin{array}{r}
2x \\
x + 1{\overline{)\,2x^2 + 5x + 3}}
\end{array}
$$

The divisor is now multiplied by the first term of the quotient ($2x$), and the result—here $2x^2 + 2x$—is written under the corresponding terms of the dividend and subtracted from them:

$$\begin{array}{r}
2x \\
x + 1 \overline{)\, 2x^2 + 5x + 3\,} \\
\text{subtract} \dashrightarrow 2x^2 + 2x \\
\hline
0 \quad + 3x
\end{array}$$

Of course the $2x$ was chosen in such a way that the highest terms (those with the largest powers of x in them) always cancel out in the subtraction. The next term of the divisor, in this case 3, is brought down and added to the result of the subtraction. The process then starts over again, using whatever is on the bottom line (here $3x + 3$) instead of the dividend. Since $\dfrac{3x}{x} = 3$, the last term in the quotient is 3.

$$\begin{array}{r}
2x + 3 \\
x + 1 \overline{)\, 2x^2 + 5x + 3\,} \\
\text{multiply } [2x(x + 1)] \text{---} \dashrightarrow 2x^2 + 2x \quad \downarrow \qquad \text{brought down} \\
\text{after subtracting---} \dashrightarrow 3x + 3 \\
\text{multiply } [3(x + 1)] \text{---} \dashrightarrow 3x + 3 \\
\hline
0 \dashrightarrow \text{remainder}
\end{array}$$

The fact that the last subtraction leaves 0 means that there is no remainder; if it had not been 0, the remainder would have appeared instead of the 0. So $x + 1$ goes into $2x^2 + 5x + 3$ exactly $2x + 3$ times, with nothing left over. Hence

$$\frac{2x^2 + 5x + 3}{x + 1} = 2x + 3$$

EXAMPLE: *Divide $x^3 + 2x + 1$ into $2x^2 + 3x$.*

This cannot be done because the degree of the divisor is greater than the degree of the dividend. If you try to do it, you run into trouble when you divide the highest terms of each and get

$$\frac{2x^2}{x^3} = \frac{2}{x}$$

which is not a possible term in a polynomial.

EXAMPLE: *Divide $x^2 + 2x - 1$ into $x^3 - 4x^2 + 3$.*

First notice that the dividend, $x^3 - 4x^2 + 3$, has no x term. In order to keep a separate column for every power of x, it is useful to think of it as $x^3 - 4x^2 + 0x + 3$. So this is how the problem looks:

$$
\begin{array}{r}
x - 6 \\
x^2 + 2x - 1 \overline{) x^3 - 4x^2 + 0x + 3}
\end{array}
$$

multiplied ⟶ $x^3 + 2x^2 - x$

after subtracting ⟶ $0 - 6x^2 + \boxed{x} + 3$ $\;(0x - (-x) = 0x + x = x)$

multiplied ⟶ $-6x^2 - 12x + 6$

$\boxed{13x} - 3$ $\;(x - (-12x) = x + 12x = 13x)$

Since x^2 cannot be divided into $13x$, this is as far as we can go. The $13x - 3$ is the remainder.

So when we divide $x^2 + 2x - 1$ into $x^3 - 4x^2 + 3$, we get $x - 6$ and a remainder of $13x - 3$. Just as in the numerical example, the remainder is written as a fraction:

$$\frac{x^3 - 4x^2 + 3}{x^2 + 2x - 1} = x - 6 + \frac{13x - 3}{x^2 + 2x - 1}$$

PROBLEM SET   9.3

Perform the division:

1. $\dfrac{xy + x}{x}$

2. $\dfrac{ab^2 + a^2b}{ab}$

3. $\dfrac{Q + Qst}{Q}$

4. $\dfrac{x^5 + 2x^4 - x^3}{x^2}$

5. $\dfrac{15x^2 - 45x}{15x}$

6. $\dfrac{x^3 - 7x^2 + 10x}{-x}$

7. $\dfrac{30a^4 - 36a^3 + 3a^2}{3a^2}$

8. $\dfrac{7x^2 - 9x + 2}{7x - 2}$

9. $\dfrac{1 - 6x + 9x^2}{1 - 3x}$

10. $\dfrac{4x^2 - 7x + 3}{x - 1}$

Divide and give the remainder:

11. Divide $x^2 - 5x + 2$ by $x - 3$.

12. Divide $x^2 + 1$ by $x - 1$.

13. Divide $7y^2 + 2y - 63$ by $y + 3$.

14. Divide $x^3 - 4x^2 + 7x + 4$ by $x + 2$.

15. Divide $y^3 - 3y^2 - 6$ by $y + 4$.

16. Divide $3 + 7y^2 - 22y$ by $y - 3$.

17. Divide $2x^2 + 11x + 12$ by $4 + x$.
18. Divide $2x^2 + 9x + 4$ by $2x + 1$.
19. Divide $3x^2 + 5x - 15$ by $x - 2$.
20. Divide $f^3 + f^2 - 12f + 15$ by $f - 3$.
21. Divide $x^2 - x - 2$ into $x^3 + 5x^2 - 7x + 2$.
22. Divide $a + b$ into $a^3 + b^3$.
23. Divide $a - b$ into $a^3 - b^3$.
24. Divide $y + 1$ into $2y^3 - y^2 - 2y + 1$.
25. Divide $2x + 3y$ into $12x^2 + 8xy - 15y^2$.

Divide the following:

26. $\dfrac{11x^3 + 3x^4 + 7x^2 - 9}{x + 3}$

27. $\dfrac{2a^3 - 9a^2b + 7ab^2 + 6b^3}{a - 3b}$

28. $\dfrac{8t^3 + 15 - 34t + 4t^2}{5 + 2t}$

29. $\dfrac{-16ab + 15b^2 + 4a^2}{2a - 3b}$

30. $\dfrac{x^4 + 4x^3y + 4xy^3 + y^4 + 6x^2y^2}{x + y}$

9.4   ORDER OF OPERATIONS, THE DISTRIBUTIVE LAW, AND SIMPLIFYING POLYNOMIALS

The *order of operations* is the same for algebraic as for numerical expressions. It is

1. Parentheses, innermost first
2. Exponents
3. Multiplication and division
4. Addition and subtraction

Also, as we saw earlier in this chapter, the *distributive law* for numbers,

$$a(b + c) = ab + ac$$

works just as well when a, b, and c are expressions.

These rules can be used to simplify some of the nastiest looking polynomials.

EXAMPLE: *Simplify* $5(a - b) - 3(a + 2b)$.

Using the distributive law to remove each of these sets of parentheses will allow you to combine the like terms:

$$5(a - b) - 3(a + 2b) = 5a - 5b - 3a - 6b \quad \text{(by the distributive law)}$$
$$= 2a - 11b \quad \text{(combining } 5a - 3a \text{ and } -5b - 6b)$$

EXAMPLE: *Simplify* $3[4q - (q - 2)]$.

Here you must start with the inner parentheses first and work outwards. The innermost parentheses, $(q - 2)$, cannot be simplified any further since we don't know what q is. Therefore the next set of parentheses to consider is

$$[4q - (q - 2)]$$

This can be simplified by using the distributive law to remove the round parentheses, and then by collecting like terms (as in the example above). However, if one works on different parts of the expression separately and then attempts to put the pieces together to make up the original expression, it is almost impossible not to lose something (usually a minus sign) on the way. Therefore it is very important to *write out the whole problem at every step*.

$$3[4q - (q - 2)] = 3[4q - q + 2] \quad \text{(using the distributive law on the parentheses)}$$

$$= 3[3q + 2] \quad \text{(collecting } q\text{'s)}$$

$$= 9q + 6 \quad \text{(distributive law on the brackets)}$$

EXAMPLE: *Simplify* $3y - 2y [4 - (3 - y)]$.

The order of operations applies here too, and this means that you must not subtract the $2y$ from the $3y$ first, although that might look a reasonable thing to do. Instead, the order of operations tells you to do the innermost parentheses first and the subtraction last. The inner parentheses, $(3 - y)$, cannot be simplified. If however the inner parentheses are removed by the distributive law, what's in the brackets can be simplified. Again, it is *much* safer to write out the whole problem every time:

$$3y - 2y[4 - (3 - y)] = 3y - 2y[4 - 3 + y] \quad \text{(distributive law on parentheses)}$$

$$= 3y - 2y[1 + y] \quad \text{(collecting terms)}$$

$$= 3y - 2y - 2y^2 \quad \text{(distributive law on brackets)}$$

$$= y - 2y^2 \quad \text{(collecting terms)}$$

EXAMPLE: *Simplify* $(3y - 2y) [4 - (3 - y)]$.

In this case you *are* meant to subtract the $2y$ from the $3y$ right away because they are in a set of parentheses, and parentheses are first in the order of operations.

$$(3y - 2y)\,[4 - (3 - y)] = y\,[4 - (3 - y)] \qquad \text{(combining } 3y - 2y)$$

$$= y\,[4 - 3 + y\,]$$

$$= y\,[1 + y\,]$$

these steps are the same as those in the problem above

$$= y + y^2 \qquad \text{(distributive law on brackets)}$$

EXAMPLE: *Simplify* $-2\Big(-(b + c) - \{2 - [b - (b - c)]\}\Big).$

This one is just to give you an idea how messy these problems can get! Actually one seldom comes across expressions like this in "real" problems, or even in math books. The reason they are included is that it is crucial that you:

1. Develop a sixth sense for minus signs and parentheses. You have to be able to manipulate them without a thought.

2. Realize that complicated-looking problems work by exactly the same set of rules as simple ones. Oddly enough, the most frequent reason that people can't do a complicated problem is not that such problems are really any harder than the easy ones, but that when faced with something complicated, they sometimes get scared and lose their heads. They then start using some completely wrong method, and one they know is wrong when they try it on a simple example.

Back to the problem! Notice that the different kinds of parentheses enable you to distinguish one set from another, and that there should obviously be as many left-hand as right-hand ends. Again, the whole problem *must* be written out at every step.

$$-2\Big(-(b + c) - \{2 - [b - (b - c)]\}\Big)$$

$$= -2\Big(-(b + c) - \{2 - [b - b + c]\}\Big) \qquad \text{(using distributive law on inner ( ))}$$

$$= -2\Big(-(b + c) - \{2 - [c]\}\Big) \qquad \text{(collecting terms inside [ ])}$$

$$= -2\Big(-(b + c) - \{2 - c\}\Big) \qquad \text{(dropping unnecessary [ ] around } c)$$

$$= -2\left(-b - c - 2 + c\right)$$ (using distributive law to remove ( ) and { } so that $\left(\ \right)$ can be simplified)

$$= -2\left(-b - 2\right)$$ (combining terms inside $\left(\ \right)$)

$$= 2b + 4$$ (using distributive law on $\left(\ \right)$)

Therefore

$$-2\left(-(\dot{b} + c) - \{2 - [b - (b - c)]\}\right) = 2b + 4$$

So we certainly succeeded in simplifying the expression—and, I hope you will agree, we managed this by *exactly* the same method as we used in the other problems.

PROBLEM SET   9.4

Simplify the following expressions:

1. $3 - [r - (2 - r)]$
2. $2x(x - 2) - x(2x - 5)$
3. $2(x - 3) + (3x - 1)$
4. $-a[2(a - 1) + 4a]$
5. $2p - 3[1 - (4 + p)]$
6. $-[5 - 4(1 - x) - (1 + x)]$
7. $4(5 - d) - 3[5 - 2(d - 2)]$
8. $x - [x - (x - y)]$
9. $(1.1p)^2 - 1.1p^2$

10. $\dfrac{-6y^2}{q} - \left(-\dfrac{6y^2}{q}\right)$
11. $\frac{1}{2}(6c - 8d) - \frac{2}{3}(9d - 30c)$
12. $-2(x - 3) - 3(3x - 1)$
13. $(a^2 - r^2z) - (r^2z - a^2)$
14. $3x - 5[(4x + 1) - (3 - x)]$
15. $8a - (a + 3c) - 2(3a - c)$
16. $-(q - 8) - 2[2 - 2(q - 1)]$

17. $3[a - 2(a - 2)] + 2[3a - 2(2a - 2)]$
18. $-4[x - (7x - 1)] + 2[(x - 1) + 2(x + 1)]$
19. $0.1z + (0.32)t + 0.2(t + z)$
20. $7a - a\{4 - 2a[a - (1 + a) + 1] + 1\} + 1 - a$
21. $6k - 2\{k + 4 - 2[2k - 1 - 2(1 - k)]\}$
22. $4m - 2n - \{m - [n + m - 2(n - m + 1)]\}$
23. $b - \{b - 5[2 - (b - 2) - 1 - b] - 1\}$
24. $-2a(5a - 1) - a(a - \{5 - a - [1 - a(3 - 2)]\})$

25. By changing the form of the expression

$$x(y + 1) - x(y - 1) - 2y$$

show that it is divisible by $(x - y)$.

9.5   FACTORING:
           WHAT IT MEANS

Factoring polynomials is the opposite of multiplying them out. For example, in Section 9.2 we showed that

$$(2x - 3)(x + 5) = 2x^2 + 7x - 15$$

To get the expression on the right, we multiplied out the product on the left. If, however, we had started with $2x^2 + 7x - 15$ and then had rewritten it as $(2x - 3)(x + 5)$, we would have *factored* $2x^2 + 7x - 15$. Alternatively, $(2x - 3)(x + 5)$ is said to be the *factored form* of $2x^2 + 7x - 15$.

To factor an expression therefore means to write it as a product of two or more expressions. This is exactly the same idea as in arithmetic where a number is said to be factored when it is written as the product of two or more integers. For example, 12 is said to be factored when it is written as $12 = 3 \cdot 4$ or as $12 = 3 \cdot 2 \cdot 2$. Alternatively, $3 \cdot 4$ or $3 \cdot 2 \cdot 2$ is a factored form of 12. So $(2x - 3)(x + 5)$ is said to be factored, because it is the product of two quantities, $(2x - 3)$ and $(x + 5)$, i.e., because it is in the form:

(something) *times* (something)

$$(2x - 3) \cdot (x + 5)$$

[Note that the parentheses around the $(2x - 3)$ and the $(x + 5)$ mean that they are each to be treated as *one* number, or *one* quantity]. Similarly,

$$5(a + b)(2a - b)$$

is factored because it is in the form

(something) *times* (something) *times* (something)

$$5 \cdot (a + b) \cdot (2a - b)$$

On the other hand,

$$(a + b)(2a - b) + 5$$

is *not* factored. Although the first part of the expression is written as a product $[(a + b)(2a - b)$ is the product of $(a + b)$ and $(2a - b)]$, in order to get the whole expression, you must *add* the 5. The problem is that instead of being in the form

(something) times (something) times (something)

$(a + b)(2a - b) + 5$ is the form

(something) times (something) plus (something)

$$(a + b) \cdot (2a - b) + 5$$

Now, not all expressions can be factored. If you pick an expression at random, the chances are that it can't be. However, many of the ones you meet in math books do turn out to be factorable—they were chosen that way to make the problems come out neatly.

An expression that cannot be factored is called *prime*. An expression that has been factored into several factors, none of which can be factored further, is said to be factored into *prime factors*, or *fully factored*.

9.6   FACTORING: COMMON FACTORS

Now we get to the business of how one actually does factor things, and unfortunately it is not as straightforward as multiplying out. However, all factoring is based on one principle, that of *taking out a common factor*, which means using the distributive law backwards. If you have an expression that has two (or more) terms, each of which is the product of several factors, and if every term contains the same factor, then this factor is called a *common factor* (because it is shared by all the terms). For example, in

$$2ps + 5pt^2$$

The first term is $2 \cdot p \cdot s$, the second is $5 \cdot p \cdot t \cdot t$, and so each term has p as a factor. The p is therefore the common factor in these two terms.

Since the distributive law can be written as

$$ab + ac = a(b + c)$$

you can think of the law as saying that whenever all the terms in an expression have a common factor, in this case a, you can rewrite the expression as

(common factor) times (the rest of the expression with the common
factor removed from each term)

Therefore the example above can be rewritten:

$$2ps + 5pt^2 = p(2s + 5t^2)$$
$$ab + ac = a(b + c)$$

The fact that

$$2ps + 5pt^2 = p(2s + 5t^2)$$

can be checked by multiplying out $p(2s + 5t^2)$ using the distributive law.

The process of rewriting $2ps + 5pt^2$ as $p(2s + 5t^2)$ is called *taking out the common factor.* Since $p(2s + 5t^2)$ is a factored form of $2ps + 5pt^2$, taking out a common factor is a method of factoring certain expressions.

EXAMPLE: *Factor $2xy + 4z^2$.*

The common factor here is 2, because the second term can be written as $2 \cdot 2 \cdot z^2$. So

$$2xy + 4z^2 = 2xy + 2 \cdot 2z^2$$
$$= 2(xy + 2z^2)$$

EXAMPLE: *Factor $-4x - 8y$.*

The common factor is -4. Therefore

$$-4x - 8y = -4(x + 2y)$$

You have to watch the signs very carefully. Notice that when you take out the -4 the second term inside the parentheses must be $+2y$, because $(-4)(+2y) = -8y$.

EXAMPLE: *Factor $-2x + 6y$.*

The common factor is -2. Therefore

$$-2x + 6y = -2(x - 3y)$$

Again, you have to watch the signs carefully in order to get the second term right. We want to end up with $+6y$ when we multiply out, and we are taking out a -2. Since minus $\cdot$ minus $=$ plus, the second term inside the parentheses had better have a $-$ sign. Check: $(-2)(-3y) = +6y$.

You can also factor this by taking out a 2, giving

$$-2x + 6y = 2(-x + 3y)$$

so it is clearly possible to factor an expression in more than one way.

EXAMPLE: *Factor $5p^2q + 10pq^2 - 5pqr$.*

$$5p^2q = 5 \cdot p \cdot p \cdot q$$
$$10pq^2 = 2 \cdot 5 \cdot p \cdot q \cdot q$$
$$-5pqr = -5 \cdot p \cdot q \cdot r$$

There is a factor of $5pq$ in every term, and this is the common factor. Therefore

$$5p^2q + 10pq^2 - 5pqr = 5pq \cdot p + 5pq \cdot 2q - 5pq \cdot r$$
$$= 5pq(p + 2q - r)$$

EXAMPLE: *Factor $2x^2 + 6x^3 + 3x^4$.*

The common factor here is x^2, since

$$2x^2 + 6x^3 + 3x^4 = 2 \cdot x^2 + 6x \cdot x^2 + 3x^2 \cdot x^2$$

Therefore

$$2x^2 + 6x^3 + 3x^4 = x^2(2 + 6x + 3x^2)$$

EXAMPLE: *Factor $4x^2(a - b) - x(a - b)^2 + 15(a - b)^3$.*

There is no numerical factor nor any power of x that occurs in every term here. However, an $(a - b)$ does occur in every term, and this is the common factor. Therefore,

$$4x^2(a - b) - x(a - b)^2 + 15(a - b)^3$$
$$= 4x^2 \cdot (a - b) - x(a - b) \cdot (a - b) + 15(a - b)^2 \cdot (a - b)$$
$$= (a - b) \, [4x^2 - x(a - b) + 15(a - b)^2]$$

Sometimes it is possible to take out two common factors in sucession.

EXAMPLE: *Factor $ax + ay + 2x + 2y$.*

The first two terms of this expression have a common factor of a, and the second two have one of 2. Suppose we factor each pair of terms:

$$ax + ay + 2x + 2y = a(x + y) + 2(x + y)$$

[Note: This by itself *does not* constitute factoring the entire expression because of the plus sign between the $a(x + y)$ and the $2(x + y)$.]

Now we notice (because this was a specially constructed expression!) that the terms on the right have a common factor of $(x + y)$, so we can write:

$$ax + ay + 2x + 2y = a(x + y) + 2(x + y)$$
$$= (x + y) \, (a + 2)$$

And now the expression *is* factored, because it is written as the product of two expressions, namely $(x + y)$ and $(a + 2)$. So this rather unlikely looking method of factoring four terms in pairs can enable you to factor the whole expression. But this only works if it is possible to take out a second common factor.

EXAMPLE: *Factor $xy^2 + 2x^2y + 3ay + 6ax$.*

Looking at the terms in pairs, we see that the first pair has a common factor of xy, and the second pair of $3a$. Therefore,

$$xy^2 + 2x^2y + 3ay + 6ax = xy(y + 2x) + 3a(y + 2x)$$

Luckily, there's another common factor, $(y + 2x)$, so we can go on:

$$= (y + 2x)\ (xy + 3a)$$

EXAMPLE: *Factor: $x^4 + 3x^3 + 6x + 18$.*

Grouping pairwise:

$$x^4 + 3x^3 + 6x + 18 = (x^4 + 3x^3) + (6x + 18) = x^3(x + 3) + 6(x + 3)$$
$$= (x + 3)\ (x^3 + 6)$$

To check this or any other factoring problem, multiply out and you should get the original expression back:

$$(x + 3)\ (x^3 + 6) = x \cdot x^3 + x \cdot 6 + 3 \cdot x^3 + 3 \cdot 6$$
$$= x^4 + 6x + 3x^3 + 18 \dots \text{OK!}$$

EXAMPLE: *Factor $2x^3 - 5x^2 - 4x + 10$.*

First factor pairwise:

$$2x^3 - 5x^2 - 4x + 10 = x^2(2x - 5) - 2(2x - 5)$$

Note: You have to watch signs very carefully. When you take out a factor of -2 from the last two terms, you must put a minus in front of the 5 inside the parentheses. To check this, multiply out $-2(2x - 5)$; you should get the $-4x + 10$ back.

Anyway, we again have a common factor, this time $(2x - 5)$. So we can go on:

$$2x^3 - 5x^2 - 4x + 10 = x^2(2x - 5) - 2(2x - 5)$$
$$= (2x - 5)(x^2 - 2)$$

EXAMPLE: *Factor $x^5 + x^3 + x^2 + 1$.*

The first pair of terms has a common factor of x^3, but the second pair seems to have nothing in common—not surprisingly since 1 has no factors besides itself. So we have

$$x^5 + x^3 + x^2 + 1 = x^3(x^2 + 1) + x^2 + 1$$

This doesn't look too helpful until you notice that the first pair of terms has $x^2 + 1$ as a factor, and $x^2 + 1$ *is* the second pair of terms. Suppose, then, that we take out a factor of 1 from the second pair of terms (you can always do that, since anything is 1 times itself):

$$x^5 + x^3 + x^2 + 1 = x^3(x^2 + 1) + x^2 + 1$$
$$= x^3(x^2 + 1) + 1(x^2 + 1)$$

Usually taking out a factor of 1 doesn't help much, but here it shows us that $(x^2 + 1)$ is a common factor, and so you can go on to get:

$$x^5 + x^3 + x^2 + 1 = x^3(x^2 + 1) + 1(x^2 + 1)$$
$$= (x^2 + 1)\,(x^3 + 1)   \text{ which is factored.}$$

EXAMPLE: *Factor $5x^4 - 3x^3 - 5x + 3$.*

The first two terms have a common factor of x^3, the second two seem to have none, so we are left with

$$5x^4 - 3x^3 - 5x + 3 = x^3(5x - 3) - 5x + 3$$

The only ray of hope for a second common factor is that the $(5x - 3)$ that came out of the first two terms is exactly the last pair of terms—except that the signs are wrong.

Suppose we take out a -1 from the last two terms. This is also something you can always do (provided you keep the signs straight) and sometimes it's surprisingly helpful. Here it makes things "look right":

$$5x^4 - 3x^3 - 5x + 3 = x^3(5x - 3) - 5x + 3$$
$$= x^3(5x - 3) - 1(5x - 3)   (-1(5x - 3) = -5x + 3)$$

Since there's now a common factor of $(5x - 3)$ we can go on, getting:

$$= (5x - 3)\,(x^3 - 1)$$

Later sections will actually enable you to factor this further, but as far as we're concerned at the moment, it's done.

You may have the feeling that this method of grouping terms in pairs ("pairwise factoring") must be very useless in practice, because the original expression has to be so special for it to work. What happens when we don't get a second common factor after taking out the first pair? For example, remember $ax + ay + 2x + 2y$, which factored like this:

$$ax + ay + 2x + 2y = a(x + y) + 2(x + y) = (x + y)(a + 2)$$

Now $ax + ay + 2x + 2y$ doesn't have to be changed by much to ensure that $x + y$ is *not* a common factor after you have taken out the first pair of common factors. For example, suppose we were unfortunate enough to be given $ax + ay + 2x + 4y$ instead. As before, we could take an a out of the first pair of terms, and a 2 out of the second pair:

$$ax + ay + 2x + 4y = a(x + y) + 2(x + 2y)$$

But since $(x + y)$ and $(x + 2y)$ are different, this time there is nothing more that we can do. So $ax + ay + 2x + 4y$ cannot be factored, at least not by this method.

Of course, the reason I have included so much about this "pairwise" method is that it *is* useful. So useful, in fact, that it gets the next section to itself.

PROBLEM SET 9.6

Name the largest factor common to:
1. $36x^2ya^3$ and xy^2
2. $12cd^3f^4$ and $15c^3d^2f^9$
3. $4ax$ and $8x$
4. $25b^2$ and $10b^3$
5. $12x^4yz^2$ and $32x^3y^4z^4$
6. $55L^2T^2$ and $990LT^2$ and $45L^2T$

Factor a (-1) out of:
7. $-7xz - y$
8. $-5 + 2z^2$
9. $-x - 1 + z$
10. $a - b + (c - d)$

Factor:
11. $x^2 + 10x$
12. $abc + dbc$
13. $3ab - 6ac$
14. $2\pi r^2 + 2\pi rh$
15. $15bx + 20ax$
16. $2x^2 + 5x$
17. $6a^5 + 2a^3$
18. $3x^2y + 3y^2x$
19. $12RT^2 - 3QT$
20. $\dfrac{Ba}{2} + \dfrac{ba}{2}$
21. $\dfrac{L^3T}{\pi} - \dfrac{9LT^3}{\pi}$
22. $2r^3 - r^2 + 4r$
23. $(x + y)^2 - 3(x + y)$
24. $6ax^2 - 3a^2x + 3ax$
25. $11zy^2 + 22t^2zy + 11atzy$
26. $2x^2y^2z + z^2yx + y^2z^2$
27. $3a^2b + 9ab^3c + a^2c27$
28. $r^4s^2t^2 + rs^2t^2 - rs^4t^4$
29. $a^3bc^3d^2 + 4acd - 3abd^2$
30. $15p^4q^6r^4 - 20p^3q^9r^7 + 10r^6p$

31. Show, by factoring the left-hand side, that the following is an identity (i.e., it is true for all values of n):

$$n(n + 1)(n + 2) - n^2(n + 1) = 2n(n + 1)$$

Use the pairwise factoring method to factor the following:

32. $a^2 + ac + ab + bc$ 35. $2xy - 4x - y + 2$

33. $9ax - 9bx - 2ar + 2br$ 36. $a^3 - a^2b - b + a$

34. $mn + mk + ln + lk$ 37. $3a(x - y) - x + y$

38. $(a - 2b)^2 - 3a + 6b$

39. $x(x + 1) + 2(x + 1) + x(x - 3) + 2(x - 3)$

40. $5st^2 + 20st - 3.5p^{1/2}t - 14p^{1/2}$

9.7   FACTORING: SYSTEMATIC METHOD FOR QUADRATIC POLYNOMIALS

The kind of polynomials that we most often want to factor are *quadratics*, meaning those of degree 2. For example,

$$2x^2 + 5x - 3$$

or

$$p^2 - 9$$

or

$$5 + 8y - y^2$$

are all quadratics. They occur frequently and so you may be glad to know that there is a method for factoring them that *always* works, provided the expression is factorable.

Let's investigate quadratics in detail. Since factoring is multiplying out done backwards, it is useful to look first at what happens when we multiply out something like $(x + 2)(x + 3)$:

$$(x + 2)(x + 3) = x(x + 2) + 3(x + 2)$$

$$= x^2 + 2x + 3x + 6$$

$$= x^2 + 5x + 6$$

If you start at the $x^2 + 2x + 3x + 6$ and read backwards, you will see that it looks just like the "pairwise factoring" of the last section:

$$x^2 + 2x + 3x + 6 = x(x+2) + 3(x+2)$$
$$= (x+2)\,(x+3)$$

So, supposing you wanted to factor

$$x^2 + 5x + 6$$

and supposing someone told you to rewrite it as

$$x^2 + 2x + 3x + 6$$

then you could go ahead and factor by the "pairwise" method.
Let's do some more examples, to make sure the one above wasn't a fluke.

EXAMPLE: *Factor $x^2 + 7x + 12$ by first writing it as $x^2 + 3x + 4x + 12$.*

$$x^2 + 7x + 12 = x^2 + 3x + 4x + 12$$
$$= x(x+3) + 4(x+3)$$
$$= (x+3)\,(x+4)$$

EXAMPLE: *Factor $x^2 + 4x - 5$ by first writing it as $x^2 - x + 5x - 5$.*

$$x^2 + 4x - 5 = x^2 - x + 5x - 5$$
$$= x(x-1) + 5(x-1)$$
$$= (x-1)\,(x+5)$$

This method obviously works, but there's one *big* flaw. You have to know how to rewrite the expression properly, because if you split up the x term wrong, you won't get the second common factor. For example, if instead of rewriting

$$x^2 + 7x + 12 \quad \text{as} \quad x^2 + 3x + 4x + 12$$

we had rewritten it as $x^2 + x + 6x + 12$

We would have got:

$$x^2 + 7x + 12 = x^2 + x + 6x + 12 = x(x+1) + 6(x+2)$$

and then we would have been stuck, because $(x+1)$ and $(x+2)$ are different. So obviously, unless we can find a way of splitting up the x term correctly, this method won't do us much good. Fortunately however there is a method for splitting the x term.

Method for Factoring a Quadratic:

First split the *x* term

Suppose we want to factor $ax^2 + bx + c$ (*a* is the coefficient of x^2, *b* is the coefficient of *x*, and *c* is the constant).

1. Multiply *a* and *c* (remember the signs).
2. Write down *b* (remember the sign).
3. Look for a pair of numbers that when multiplied together make *ac* and when added together make *b*.
   This is done by:
   (*i*) Writing down all the pairs of numbers (with signs) that multiply together to give *ac*.
   (*ii*) Going through the pairs to check which add to give *b*.
4. Rewrite the *x* term as the sum of two terms with these numbers as coefficients.

Then factor pairwise

Note: If the variable is not called *x*, then "*x* term" should be taken to mean that term in the quadratic containing only the first power of the variable.

Why this method works will be left until the end of this section. First we'll show that it does work by using it on something we know. Suppose we see how it tells us to split the *x* term of

$$x^2 + 7x + 12$$

Here $a = 1$, $b = 7$, $c = 12$. So

$$ac = 12$$

$$b = 7$$

Factors of 12 are:

12	and	1
−12	and	−1
6	and	2
−6	and	−2
3	and	4
−3	and	−4

These factors added:

$$12 + 1 = 13$$
$$(-12) + (-1) = -13$$
$$6 + 2 = 8$$
$$(-6) + (-2) = -8$$
$$3 + 4 = 7$$
$$(-3) + (-4) = -7$$

So 3 and 4 is the right pair. This means that $7x$ must be replaced by $3x + 4x$ (or $4x + 3x$; it makes no difference) for the pairwise factoring to work—which is the way we did in fact split the *x* term. You might check that the method applied to

$$x^2 + 4x - 5$$

does tell you to split $4x$ into $-x + 5x$ or $5x - x$. Now a new one.

EXAMPLE: *Factor $x^2 - 6x - 27$.*

First we must split the x term. Here $a = 1$, $b = -6$, $c = -27$. Therefore

$$ac = -27$$
$$b = -6$$

Factors of -27 are :	These factors added:
-27 and 1	$(-27) + 1 = -26$
27 and -1	$27 + (-1) =\ \ 26$
-9 and 3	$(-9) + 3 =\ \ -6$
9 and -3	$9 + (-3) =\ \ \ \ 6$

So -9 and 3 is the required pair, and we should rewrite $-6x$ as $-9x + 3x$. Therefore

$$x^2 - 6x - 27 = x^2 - 9x + 3x - 27$$
$$= x(x - 9) + 3(x - 9)$$
$$= (x - 9)(x + 3).$$

Therefore $x^2 - 6x - 27$ factors into $(x - 9)\,(x + 3)$. So that you can see that it makes no difference whether you replace $-6x$ by $-9x + 3x$ or by $3x - 9x$, let's do it the other way as well:

$$x^2 - 6x - 27 = x^2 + 3x - 9x - 27$$
$$= x(x + 3) - 9(x + 3)$$
$$= (x + 3)(x - 9)$$

So the factors come out in the opposite order, but you get exactly the same answer.

EXAMPLE: *Factor $2x^2 + 7x + 6$.*

Here $a = 2$, $b = 7$, $c = 6$. The fact that a is no longer 1 makes no difference: the method works just the same.

$$ac = 2 \cdot 6 = 12$$
$$b = 7$$

Factors of 12 are 12 and 1, -12 and -1, 6 and 2, -6 and -2, 4 and 3, -4 and -3. If we are looking for a pair that adds to 7, we can ignore the pairs of negative numbers, because their sum is negative. Now,

$$12 + 1 = 13$$
$$6 + 2 = 8$$
$$4 + 3 = 7$$

So the pair we want is 4 and 3. Therefore

$$2x^2 + 7x + 6 = 2x^2 + 4x + 3x + 6$$
$$= 2x(x + 2) + 3(x + 2)$$
$$= (x + 2)(2x + 3)$$

Now a more complicated one.

EXAMPLE: *Factor $3x^2 + 25x + 28$.*

Here $a = 3$, $b = 25$, $c = 28$, so

$ac = 84$ and $b = 25$.

In order to work out all the pairs of numbers whose product is 84, it is helpful to have the prime factors of 84.

$$84 = 12 \cdot 7 = 4 \cdot 3 \cdot 7 = 2 \cdot 2 \cdot 3 \cdot 7$$

If two numbers are going to have a product of 84, then between them they must have all the prime factors of 84, and no other factors except possibly a 1. Therefore, in order to find two such numbers, we can just divide these factors into two groups. For example,

$2 \cdot 2$ and $3 \cdot 7$ i.e.   4 and 21

have all the prime factors of 84 between them, and their product is 84 (check that $4 \cdot 21 = 84$).

If now we want to find all possible pairs of numbers whose product is 84, then what we must do is think up all the possible ways of dividing the prime factors of 84 into two groups. Here goes:

First, we could put just a 1 in the first group, and everything else in the second:

1 and $2 \cdot 2 \cdot 3 \cdot 7$ i.e.   1 and 84

Don't forget that the same numbers made negative will work too, so another possible pair is

$$-1 \text{ and } -84$$

Or, we could have one of the factors in the first group, and three in the second:

2 and $2 \cdot 3 \cdot 7$	i.e.	2 and 42
3 and $2 \cdot 2 \cdot 7$		3 and 28
7 and $2 \cdot 2 \cdot 3$		7 and 12

And, of course, the corresponding negative pairs:

$$-2 \text{ and } -42$$
$$-3 \text{ and } -28$$
$$-7 \text{ and } -12$$

Or, we could put two factors in the first group and two in the second:

$2 \cdot 2$ and $3 \cdot 7$	i.e.	4 and 21
$2 \cdot 3$ and $2 \cdot 7$		6 and 14
$2 \cdot 7$ and $2 \cdot 3$		these give the same factors as
$3 \cdot 7$ and $2 \cdot 2$		the previous two, only in the opposite order

And the corresponding negatives:

$$-4 \text{ and } -21$$
$$-6 \text{ and } -14.$$

If we start putting three factors in the first group and one in the second, we'll get only numbers that we've had before, but in the opposite order (for example, $2 \cdot 2 \cdot 3$ and 7 gives 12 and 7 again). Therefore we now have all the possible pairs of numbers whose products are 84.

And now we have to check which of these pairs adds to 25. Since it couldn't be one of the negative pairs, we only need to check the positive ones:

$1 + 84 = 85$	$3 + 28 = 31$	$4 + 21 = 25$
$2 + 42 = 44$	$7 + 12 = 19$	$6 + 14 = 20$

So *the pair we're looking for is 4 and 21.*

Now (at long last!) we can go ahead and factor $3x^2 + 25x + 28$:

$$3x^2 + 25x + 28 = 3x^2 + 4x + 21x + 28$$
$$= x(3x + 4) + 7(3x + 4)$$
$$= (3x + 4)(x + 7)$$

CHECK: It's *always* a good idea to check your answers. For factoring problems this is particularly easy: multiply out your answer and you should end up with whatever you started with. In this case:

$$(3x + 4)(x + 7) = 3x^2 + 21x + 4x + 28$$
$$= 3x^2 + 25x + 28$$

So it looks as though we factored right!

Now for a really complicated one!

EXAMPLE: *Factor $4x^2 + 44x - 75$.*

Here $a = 4$, $b = 44$, $c = -75$, so $ac = -300$ and $b = 44$.
To find the pairs of numbers whose products are -300, we are going to need the prime factors of 300:

$$300 = 3 \cdot 100 = 3 \cdot 2 \cdot 50 = 3 \cdot 2 \cdot 2 \cdot 25 = 3 \cdot 2 \cdot 2 \cdot 5 \cdot 5$$

Now to find the numbers whose product is -300 and whose sum is 44. First, notice that in order to multiply to -300, one number must be positive and one negative. Second, in order to add to 44, the larger number must be positive. Third, remember that all the prime factors of 300 must be divided between the two numbers we are looking for (or they won't multiply to -300). So all we have to do is divide these factors into two groups, and fix the signs by putting a negative in front of the smaller number in each pair. Therfore, out of all the possible pairs whose products are -300, the ones we might be interested in are

-1 and $2 \cdot 2 \cdot 3 \cdot 5 \cdot 5$ i.e.	-1 and 300
-2 and $2 \cdot 3 \cdot 5 \cdot 5$	-2 and 150
-3 and $2 \cdot 2 \cdot 5 \cdot 5$	-3 and 100
-5 and $2 \cdot 2 \cdot 3 \cdot 5$	-5 and 60
$-2 \cdot 2$ and $3 \cdot 5 \cdot 5$	-4 and 75
$-2 \cdot 3$ and $2 \cdot 5 \cdot 5$	-6 and 50

$$-2 \cdot 5 \text{ and } 2 \cdot 3 \cdot 5 \quad \text{i.e.} \quad -10 \text{ and } 30$$

$$-3 \cdot 5 \text{ and } 2 \cdot 2 \cdot 5 \qquad -15 \text{ and } 20$$

$$5 \cdot 5 \text{ and } -2 \cdot 2 \cdot 3 \qquad 25 \text{ and } -12$$

Now all we have to do is check which pair adds to 44. Looking down the list, it is obviously −6 and 50. Therefore

$$4x^2 + 44x - 75 = 4x^2 - 6x + 50x - 75$$
$$= 2x(2x - 3) + 25(2x - 3)$$
$$= (2x - 3)(2x + 25)$$

CHECK: Multiply out again; and hope you get the original expression:

$$(2x - 3)(2x + 25) = 4x^2 + 50x - 6x - 75$$
$$= 4x^2 + 44x - 75 \qquad \text{So we did it right!}$$

Difficulties with this Method

1. It's long and tedious. The last examples certainly were, but that was partly because we wrote out everything in such detail. In practice, you can skip some of the steps and you don't have to write down the reasons and so it's quicker. It is also possible to factor things by guesswork and trial and error. Doing it that way is perfectly fine—provided that you always check your answer. However, it is sometimes helpful to have a systematic method to fall back on!

2. You may not get a common factor at the second stage. This means that you must have made a mistake in splitting up the x term.

3. You may not be able to find a pair of numbers that multiply to make ac and add to make b. In this case the expression cannot be factored by any method. For example, try to factor

$$x^2 + 7x - 12$$

Here $a = 1$, $b = 7$, $c = -12$, so

$$ac = -12$$
$$b = 7$$

Factors of −12 are

$$-12 \text{ and } \quad 1 \text{ which add to } -11$$
$$12 \text{ and } -1 \text{ which add to } \quad 11$$

6 and -2 which add to $\quad 4$

-6 and $\quad 2$ which add to $\quad -4$

4 and -3 which add to $\quad 1$

-4 and $\quad 3$ which add to $\quad -1$

Since none of these pairs add to 7, we can conclude that

$$x^2 + 7x - 12$$

cannot be factored, and so is called *prime*.

Why the Systematic Method Works

Suppose we are trying to factor a typical quadratic, such as $ax^2 + bx + c$, where a, b, c are integers. If it factors, it will break down into something like $(mx + n)(px + q)$, where m, n, p, q are integers. But $(mx + n)(px + q) = mpx^2 + (mq + np)x + nq$, and if this is to be equal to $ax^2 + bx + c$, corresponding coefficients must be equal. Therefore

$$a = mp \qquad b = (mq + nq) \qquad c = nq$$

When we split up bx, we look for a pair of numbers that when multiplied make $ac = mpnq$ and when added make $b = mq + np$. The numbers mq and np do exactly that, so bx is split up into $mqx + npx$. Then the expression can be factored "pairwise" as follows:

$$ax^2 + bx + c = mpx^2 + mqx + npx + nq$$

$$= mx(px + q) + n(px + q)$$

$$= (mx + n)(px + q)$$

PROBLEM SET 9.7

Factor using the systematic method. (Note: They can't all be factored!)

1. $6x^2 + 13x + 5$
2. $5r^2 - 130r + 800$
3. $z^2 + 9z + 14$
4. $2y^2 - 17y + 21$
5. $x^2 - 4x - 12$
6. $9 - x^2$
7. $t^2 + 10 - 7t$
8. $6x^2 - x - 12$
9. $-3 - r + 2r^2$
10. $4x^2 + 5 - 21x$
11. $x^2 + 3x - 6$
12. $50 - 20s - 16s^2$
13. $10b^2 - 43b + 45$
14. $66 + 4y^2 + 35y$
15. $-16 + 10x - x^2$
16. $c^2 + 1 + c$
17. $5 + 14a - 3a^2$
18. $6t^2 + 6 + 13t$
19. $6x^2 + 16x - 6$
20. $-m^2 + 12 + m$
21. $3k^2 + 1 - 4k$
22. $2p^2 + 4p + 5$
23. $a^2 - 99a - 100$
24. $8d + 15 + d^2$
25. $2x^2 + 7x - 15$

**9.8   FACTORING:
             STANDARD FORMS**

Factoring is a terribly useful process, but if you always do it by the systematic method it can be very slow. Consequently, it is worth memorizing the factors of a few expressions that come up really often. One whose factors you should be able to pick out immediately is a *perfect square*, which is developed below.

Suppose we factor

$$x^2 + 2x + 1$$

Here $a = 1, b = 2, c = 1$, so $ac = 1$ and the possible pairs are 1 and 1 or -1 and -1. The first pair adds to 2, so

$$x^2 + 2x + 1 = x^2 + x + x + 1$$
$$= x(x + 1) + 1(x + 1)$$
$$= (x + 1)(x + 1)$$
$$= (x + 1)^2$$

Therefore $(x^2 + 2x + 1) = (x + 1)^2$—a perfect square.

Now suppose we try

$$x^2 + 4x + 4$$

where $a = 1, b = 4, c = 4$, so $ac = 4$ and the pair we need is 2 and 2.

$$x^2 + 4x + 4 = x^2 + 2x + 2x + 4$$
$$= x(x + 2) + 2(x + 2)$$
$$= (x + 2)(x + 2)$$
$$= (x + 2)^2$$

Therefore $x^2 + 4x + 4 = (x + 2)^2$—again a perfect square.

By the same method, you can show that

$$x^2 + 6x + 9 = (x + 3)^2$$
$$x^2 + 8x + 16 = (x + 4)^2$$
$$x^2 + 10x + 25 = (x + 5)^2 \quad \text{and so on.}$$

Similarly,

$$x^2 - 2x + 1 = (x - 1)^2$$
$$x^2 - 4x + 4 = (x - 2)^2$$
$$x^2 - 6x + 9 = (x - 3)^2 \quad \text{and so on.}$$

You should be able to see a pattern emerging, namely:

> Anything that looks like
>
> $$x^2 + 2ax + a^2$$
>
> is a *perfect square* because
>
> $$x^2 + 2ax + a^2 = (x + a)^2$$
>
> (a is a constant, either positive or negative).

Now, quadratics like $x^2 + 2ax + a^2$ aren't the only kind of perfect square, but they are certainly the most common. Any others you happen to come across can always be factored by the systematic method if you don't recognize them as perfect squares.

The second standard form that you should definitely be able to recognize is the *difference of squares*. This is an expression of the form.

$$(\text{something})^2 - (\text{something})^2$$

For example, $x^2 - 2^2 = x^2 - 4$, $x^2 - a^2$, $(2x)^2 - 3^2 = 4x^2 - 9$.

We will factor some differences of squares and see what pattern emerges:

EXAMPLE: *Factor $x^2 - 4$.*

Think of this as $x^2 + 0x - 4$, and follow the usual method with $a = 1$, $b = 0$, $c = -4$. You will find that the x term, $0 \cdot x$, splits into $2x - 2x$.

$x^2 - 4 = x^2 + 2x - 2x - 4$

$\qquad = x(x + 2) - 2(x + 2)$

Therefore

$x^2 - 4 = (x + 2)(x - 2)$

EXAMPLE: *Factor $x^2 - 1$.*

$x^2 - 1 = x^2 - 1^2$, so this a difference of squares also. Now

$x^2 - 1 = x^2 + x - x - 1$

$\qquad = x(x + 1) - 1(x + 1)$ (you have to remember to factor out a -1 here)

Therefore

$x^2 - 1 = (x + 1)(x - 1)$

EXAMPLE: *Factor $4x^2 - 9$.*

Thinking of this as $4x^2 + 0x - 9$ leads you to split the x term into $6x - 6x$:

$$4x^2 - 9 = 4x^2 + 6x - 6x - 9$$
$$= 2x(2x + 3) - 3(2x + 3)$$

Therefore

$$4x^2 - 9 = (2x + 3)(2x - 3)$$

By similar reasoning,

$$x^2 - 9 = (x + 3)(x - 3)$$
$$x^2 - 16 = (x + 4)(x - 4)$$
$$9x^2 - 4 = (3x + 2)(3x - 2)$$
$$16x^2 - 9 = (4x + 3)(4x - 3) \qquad \text{and so on.}$$

Anything that looks like:
$$a^2x^2 - b^2$$
is a *difference of squares*, and factors as follows:
$$a^2x^2 - b^2 = (ax + b)(ax - b)$$
(where a, b are any numbers).

PROBLEM SET   9.8

Factor:

1. $b^2 - a^2$
2. $9 - r^2$
3. $2x^2 - 8$
4. $x^2 + 4x + 4$
5. $9 + d^2 + 6d$
6. $4t^2 - 4t + 1$
7. $9p^2 - 4$
8. $25a^2 - 49b^2$
9. $16 + y^2 + 8y$
10. $4d^2 - 12d + 9$

11. $8 - 8d^2$
12. $x^2 - 14x + 49$
13. $a^2b^2 - 16$
14. $ax^2 - a$
15. $75 + 12y^2 - 60y$
16. $(x + y)^2 - z^2$
17. $56t + 16t^2 + 49$
18. $x^2 + a^2 + 2ax$
19. $(x^2 + 9)^2 - 36x^2$
20. $k^2(a + 2) - 9a - 18$

9.9 FACTORING: EXAMPLES

Many of the quadratics that come up in problems are ones that can be done directly by using the systematic method (Section 9.7) or by recognizing them as standard forms (Section 9.8). A great many more can be made to fit into these forms if you look at them right—meaning that you can actually do a wide variety of problems by what you already know.

EXAMPLE: *Factor $x^2 - 8xy + 12y^2$.*

This can be done by the systematic method:

$$x^2 - 8xy + 12y^2 = x^2 - 6xy - 2xy + 12y^2$$
$$= x(x - 6y) - 2y(x - 6y)$$

Therefore

$$x^2 - 8xy + 12y^2 = (x - 6y)(x - 2y)$$

[Compare: $x^2 - 8x + 12 = (x - 6)(x - 2)$.]

EXAMPLE: *Factor $a^4 - 9$.*

The trick here is to see that $(a^2)^2 = a^2 \cdot a^2 = a^4$, so that

$$a^4 - 9 = (a^2)^2 - 9$$

This tells us that $a^4 - 9$ is really the difference of squares, and remembering that

$$x^2 - 9 = (x + 3)(x - 3)$$

gives

$$a^4 - 9 = (a^2 + 3)(a^2 - 3)$$

EXAMPLE: *Factor $x^4 - 10x^2 + 25$.*

This is a perfect square with x^2 replacing x. Therefore

$$x^4 - 10x^2 + 25 = (x^2)^2 - 10(x^2) + 25$$
$$= (x^2 - 5)^2$$

[Compare: $x^2 - 10x + 25 = (x - 5)^2$.]

EXAMPLE: *Factor $ax^2 - 16a$.*

There is a common factor of a in these terms. Take this out first:

$$ax^2 - 16a = a(x^2 - 16)$$

Now $a(x^2 - 16)$ is factored, but it is not yet fully factored because the $x^2 - 16$ is a difference of squares and can be factored too. Therefore,

$$ax^2 - 16a = a(x^2 - 16)$$

$$= a(x + 4)\,(x - 4) \qquad \text{This is now fully factored.}$$

EXAMPLE: *Factor $2\pi c x^4 - 50\pi c^3$.*

This is just like the last one, only a little more complicated. Taking out the common factor, $2\pi c$, leaves $x^4 - 25c^2$, which is a difference of squares since $x^4 - 25c^2 = (x^2)^2 - (5c)^2$. So

$$2\pi c x^4 - 50\pi c^3 = 2\pi c(x^4 - 25c^2) \qquad \text{(not yet fully factored)}$$

$$= 2\pi c(x^2 + 5c)(x^2 - 5c) \qquad \text{(fully factored)}$$

EXAMPLE: *Factor $2p^2 + 2p - 60$.*

First take out the common factor of 2, and then use the systematic method:

$$2p^2 + 2p - 60 = 2(p^2 + p - 30)$$

$$= 2(p^2 + 6p - 5p - 30)$$

$$= 2[p(p + 6) - 5(p + 6)]$$

$$= 2(p + 6)(p - 5)$$

If we forgot to take out the 2 at the beginning, we could have used the systematic method right away. The price of such forgetfulness, however, is having to deal with larger coefficients:

$$2p^2 + 2p - 60 = 2p^2 + 12p - 10p - 60$$

$$= 2p(p + 6) - 10(p + 6)$$

$$= (p + 6)(2p - 10) \qquad \underleftarrow{} \; \begin{array}{l}\text{common factor of 2}\\\text{comes from here}\end{array}$$

$$= 2(p + 6)(p - 5) \qquad \text{as before.}$$

It is *always* a good idea to look for and take out any common factors first, because it makes the coefficients you are left with so much smaller and easier to work with.

EXAMPLE: *Factor $12ax^2 + 42ax - 90a$.*

There is a common factor of $6a$ and we had better take it out first.

Otherwise the size of the coefficients will make the systematic method unbearable.

$$12ax^2 + 42ax - 90a = 6a(2x^2 + 7x - 15)$$
$$= 6a(2x^2 + 10x - 3x - 15)$$
$$= 6a\,[2x(x + 5) - 3(x + 5)]$$
$$= 6a(x + 5)(2x - 3)$$

EXAMPLE: *Factor* $2k^2m^2 + 7km - 15$

We factored $2x^2 + 7x - 15$ into $(x + 5)(2x - 3)$ in the example above; this one is the same except with km replacing x:

$$2k^2m^2 + 7km - 15 = 2(km)^2 + 7(km) - 15$$
$$= (km + 5)(2km - 3)$$

EXAMPLE: *Factor* $2(a - b)^2 + 7(a - b) - 15.$

Here the x in $2x^2 + 7x - 15$ is replaced by $(a - b)$. Therefore,

$$2(a - b)^2 + 7(a - b) - 15 = [(a - b) + 5]\,[2(a - b) - 3]$$

EXAMPLE: *Factor* $(a + b)^2 - c^2.$

This is a difference of squares, and therefore

$$(a + b)^2 - c^2 = [(a + b) + c]\,[(a + b) - c]$$
$$= (a + b + c)(a + b - c)$$

EXAMPLE: *Factor* $5q^2 + 5 + 26q.$

The trick here is to realize that the terms are not written in the usual order. The term that needs splitting is the $26q$, *not* the 5. Suppose we rewrite the expression as

$$5q^2 + 26q + 5$$

Now we can use the systematic method to get

$$5q^2 + 26q + 5 = (5q + 1)(q + 5)$$

EXAMPLE: *Factor* $4x^2 + x + 3x^3.$

The order of the terms is playing tricks here too. There is an x term in the middle all right, but there is an x^3 term instead of a constant term. The problem turns out to be that there is common factor of x, which must be factored out before we can go on:

$$4x^2 + x + 3x^3 = x(4x + 1 + 3x^2)$$

It's now clear that the expression in parentheses is not in the usual order. Realizing that the term that needs splitting is the **4x**, we get:

$$4x^2 + x + 3x^3 = x(4x + 1 + 3x^2)$$

$$= x(3x^2 + 4x + 1)$$

$$= x(3x + 1)(x + 1)$$

PROBLEM SET   9.9

Factor if possible using the methods presented so far:

1. $x^2 + 3xy + 2y^2$
2. $2b^2 - 2bt - 4t^2$
3. $x^4 - 4$
4. $a^2 - 4ab - 12b^2$
5. $p^3 - p^3r^2$
6. $x^3y - 9xy^3$
7. $-2 + 2x^2 + 3x$
8. $-8xy + x^2 + 16y^2$
9. $1 - 17nt - 12n^2t^2$
10. $r^2 + s^2 + 2rs - t^2$
11. $12(c - d)^2 - 5(c - d) - 2$
12. $-6 + 3L^2K^2 + 7LK$
13. $(a + b)^2 + 2(a + b) + 1$
14. $4cx + x^2 - d^2 + 4c^2$
15. $a^2 + b^2 + c^2$
16. $56df + 16d^2 + 49f^2$
17. $r^2 - s^2 + 6xs - 9x^2$
18. $4a^2 + 8ab^2 + 4b^4$
19. $x^3 - x$
20. $\dfrac{x^2}{4} + x + 1$
21. $49 - 4y^2$
22. $3m^2 + \frac{1}{4} + 2m$
23. $a^3 - 6a^2 + a^4$
24. $-12(c + d) + (c + d)^2 + 20$
25. $49xy + 2x^2 + 21y^2$
26. $-4 + r^2 + 2rs + s^2$
27. $6a^2 - 15b^2 + ab$
28. $(m + n)^2 + 11m + 24 + 11n$
29. $-20bc + 25c^2 - 100x^2 + 4b^2$
30. $2x^2 - xy - 10y^2$

31. $(t + 7)^2 - (s + 4)^2$
32. $r^2s^2 + 8r^2s + 16r^2$
33. $2h^4 - 2$
34. $\pi^2(p - q)^2 - 12\pi(p - q) + 20$
35. $x^3 + 3x^2 + 2x$
36. $-y^2 + a^2 + b^2 + 2ab$
37. $z^2 - x^2 + w^2$
38. $x^2 - 4y^2 - 3x + 6y$
39. $\frac{4}{3}x^2 + \frac{10}{3}x + \frac{4}{3}$
40. $4x^3y - 9xy^3$
41. $a^2x^2 - 2abxy + b^2y^2$
42. $\dfrac{4K^2}{9} - \dfrac{4KB}{3} + B^2$
43. $2(s - a)^2 - s + a - 3$
44. $(x^2 + 2xy + y^2) + (x + y) - 6$
45. $12x^2 - 10y^2 + 19xy$
46. $9a^2 - 12a + 4 - 4b^2$
47. $5\pi t^2 + 20\pi^2 t + 20\pi^3$
48. $-20a^2b + 25b^2 + 4a^4$
49. $a^4 - b^4$
50. $12(c - r)^2 + 5r - 2 - 5c$
51. $p^2 + 2pq + q^2 + 2p + 2q + 1$
52. $ax^2 + 2\sqrt{a}\,\sqrt{b}\,xy + by^2$
53. $x - 3\sqrt{x} + 2$
54. $6x - \sqrt{x} - 12$
55. $20 + 20\dfrac{p}{q} + 5\dfrac{p^2}{q^2}$
56. $(x + y)(x - y) - (c^2 + 2cy)$
57. $\dfrac{ur^2}{2\pi} + \dfrac{urs}{\pi} + \dfrac{us^2}{2\pi}$

9.10 FACTORING: HIGHER-DEGREE POLYNOMIALS

Factoring polynomials of arbitrary degree is rather more haphazard than factoring quadratics, as there is no systematic method for polynomials of degree higher than two. However, there are a few good ways of approaching a general factoring problem, and that's what this section is for.

1. Don't forget what you know about quadratics: Some polynomials of higher degree can be factored by the same methods.

Some of the examples of Section 9.9 are polynomials of degree greater than 2, so you have already had a chance to see how the systematic method can be persuaded to help factor things other than quadratics. Typical examples are as follows:

EXAMPLE: *Factor $8x^5 + 18x^4 - 5x^3$.*

First notice that there is a common factor of x^3:

$$8x^5 + 18x^4 - 5x^3 = x^3(8x^2 + 18x - 5)$$

Now $8x^2 + 18x - 5$ can be factored by the systematic method, giving

$$8x^5 + 18x^4 - 5x^3 = x^3(8x^2 + 18x - 5)$$
$$= x^3(2x + 5)(4x - 1)$$

EXAMPLE: *Factor $4x^4 - 5x^2 + 1$.*

The trick here is to notice that since $(x^2)^2 = x^4$, the polynomial $4x^4 - 5x^2 + 1$ can be made to look like a quadratic with x^2 replacing x:

$$4x^4 - 5x^2 + 1 = 4(x^2)^2 - 5(x^2) + 1$$

This can be factored into $(4x^2 - 1)(x^2 - 1)$ by the systematic method.

But since $(4x^2 - 1)$ and $(x^2 - 1)$ are both differences of squares, they can be factored also:

$$4x^4 - 5x^2 + 1 = (4x^2 - 1)(x^2 - 1)$$
$$= (2x + 1)(2x - 1)(x + 1)(x - 1)$$

EXAMPLE: *Factor $x^6 - 2x^3 + 1$.*

You will find that this can be done by a similar method to the one above. It can be made to look like a quadratic with x^3 replacing x.

2. If you have one factor, long division will give you the other.

EXAMPLE: *Factor $x^3 - 2x^2 - 5x + 6$, if one factor is $(x - 1)$.*

Knowing that $(x - 1)$ is a factor of $x^3 - 2x^2 - 5x + 6$ means that $(x - 1)$ divides exactly into $x^3 - 2x^2 - 5x + 6$. Doing the long division:

$$
\begin{array}{r}
x^2 - x - 6 \\
x - 1 \overline{)\, x^3 - 2x^2 - 5x + 6} \\
x^3 - x^2 \\
\hline
-x^2 - 5x \\
-x^2 + x \\
\hline
-6x + 6 \\
-6x + 6 \\
\hline
0
\end{array}
$$

The remainder comes out to be zero because $(x - 1)$ is a factor of $x^3 - 2x^2 - 5x + 6$.

Therefore,

$$\frac{x^3 - 2x^2 - 5x + 6}{x - 1} = (x^2 - x - 6)$$

So

$$x^3 - 2x^2 - 5x + 6 = (x - 1)(x^2 - x - 6)$$

Now $x^2 - x - 6$ can itself be factored:

$$x^2 - x - 6 = (x - 3)(x + 2)$$

So the full factorization of $x^3 - 2x^2 - 5x + 6$ is

$$x^3 - 2x^2 - 5x + 6 = (x - 1)(x^2 - x - 6)$$
$$= (x - 1)(x - 3)(x + 2)$$

EXAMPLE: *Factor $x^4 + x^3 + 2x^2 + x + 1$, given that $x^2 + x + 1$ is a factor.*

Divide $x^2 + x + 1$ into $x^4 + x^3 + 2x^2 + x + 1$:

$$
\begin{array}{r}
x^2 + 1 \\
x^2 + x + 1 \overline{)\, x^4 + x^3 + 2x^2 + x + 1} \\
x^4 + x^3 + x^2 \\
\hline
x^2 + x + 1 \\
x^2 + x + 1 \\
\hline
0
\end{array}
$$

This tells you that

$$\frac{x^4 + x^3 + 2x^2 + x + 1}{x^2 + x + 1} = x^2 + 1$$

Therefore

$$x^4 + x^3 + 2x^2 + x + 1 = (x^2 + x + 1)(x^2 + 1)$$

Any attempt to factor either $(x^2 + 1)$ or $(x^2 + x + 1)$ further will turn out to be fruitless, because you won't be able to split the x term. Therefore you have to conclude that this is as far as $x^4 + x^3 + 2x^2 + x + 1$ can be factored.

EXAMPLE: *Decide whether $x^2 - 3$ or $x^2 + 3$ is a factor of $x^3 - 2x^2 + 3x - 6$ and find the other factor.*

The point of this problem is to realize that if something is a factor of $x^3 - 2x^2 + 3x - 6$ then it must divide exactly into $x^3 - 2x^2 + 3x - 6$, which means a zero remainder. The way to find out whether $x^2 - 3$ or $x^2 + 3$ is a factor, then, is to divide both of them into $x^3 - 2x^2 + 3x - 6$ and see which of them gives a remainder of zero. Here goes:

$$
\begin{array}{r}
x - 2 \\
x^2 - 3 \overline{)\ x^3 - 2x^2 + 3x - 6} \\
\underline{x^3 \qquad\ -3x} \\
-2x^2 + 6x - 6 \\
\underline{-2x^2 \qquad +6} \\
6x - 12 \qquad \text{(remainder is not zero)}
\end{array}
$$

Therefore $x^2 - 3$ is not a factor of $x^3 - 2x^2 + 3x - 6$.

$$
\begin{array}{r}
x - 2 \\
x^2 + 3 \overline{)\ x^3 - 2x^2 + 3x - 6} \\
\underline{x^3 \qquad\ +3x} \\
-2x^2 \qquad -6 \\
\underline{-2x^2 \qquad -6} \\
0 \qquad \text{(remainder is zero)}
\end{array}
$$

This makes it plain that $x^2 + 3$ is a factor of $x^3 - 2x^2 + 3x - 6$, and that $x - 2$ is the other one. Therefore

$$x^3 - 2x^2 + 3x - 6 = (x^2 + 3)(x - 2)$$

Since $x^2 + 3$ doesn't factor, this is as far as it goes.

3. Guessing linear factors of a polynomial p.

Now that you know how to check whether or not something is a factor (by division), here is a method of deciding what it might be reasonable to check.

It will not give you every possible factor, only those which are linear, meaning those with an x term, but no x^2, x^3, or higher-order terms. The idea is this:

If $(x - k)$ is a factor of p, then k must divide exactly into p's constant term.

If $(lx - k)$ is a factor of p, then l must divide exactly into the coefficent of p's highest term and k must divide exactly into the constant term.

(l and k can be positive or negative, but both must be integers.)

To see how this works, look at what happens when you multiply out

$$(x - 5)\ (x - 3)\quad \text{to get } x^2 - 8x + 15.$$

The constant term, 15, is the product of the -5 and the -3, and therefore 5 and 3 divide exactly into 15. Similarly,

$$x^3 - 10x^2 + 31x - 30 \text{ is the product of } (x - 5),\ (x - 3),\ \text{and } (x - 2).$$

The constant term, -30, arises as the product of -5, -3, and -2. Therefore 5, 3, and 2 all divide -30.

Here's how you can use these facts. Suppose you wanted to find the factors of $x^3 - 2x^2 - x + 2$. You might as well start by seeing if it has any factors of the form $(x - k)$. If it does, k will have to divide 2, so k will have to be ± 2 or ± 1 (meaning $+2$ or -2 or $+1$ or -1). Therefore, possible factors are

$$(x - 2)\ \text{ or }\ [x - (-2)]\ \text{ or }\ (x - 1)\ \text{ or }\ [x - (-1)],\ \text{i.e.}$$

$$(x - 2)\ \text{ or } (x + 2) \text{ or } (x - 1) \text{ or } (x + 1).$$

As above, you can tell which, if any, of these actually are factors by dividing each one into $x^3 - 2x^2 - x + 2$. Those that give zero remainders are factors. You can check that $(x - 2)$, $(x - 1)$, and $(x + 1)$ all give zero remainders, and so all three are factors. Now $x^3 - 2x^2 - x + 2$ couldn't have any more factors besides these three, or the highest term wouldn't be x^3. So it looks like

$$x^3 - 2x^2 - x + 2 \text{ factors into } (x - 2)\ (x - 1)\ (x + 1).$$

To check, multiply out the product on the right, which tells you for sure that

$$x^3 - 2x^2 - x + 2 = (x - 2)\ (x - 1)\ (x + 1).$$

There are three things worth noticing about this:

1. You need not consider factors of the form $(x + k)$, because they came automatically from $(x - k)$ when k is negative. [For example, see how we got the $(x + 1)$.] But *you must remember to look for positive and negative values for k.*

2. In the example above we did not need to consider factors of the form $(lx - k)$. This was because the highest degree term, x^3, had coefficient 1. When the coefficient of the highest degree term is not 1—but only then—you do have to bother with factors of the form $(lx - k)$.

3. If none of the factors you are testing gives a zero remainder upon division, then none of them actually is a factor. This is perfectly possible—it just means that the expression has no factors of the form $(x - k)$ or $(lx - k)$.

EXAMPLE: *Factor $x^3 + 1$.*

If $(x - k)$ is a factor, k must divide 1 and so must be $+1$ or -1. Therefore the possible factors are

$$(x - 1) \quad \text{and} \quad [x - (-1)] = (x + 1)$$

Division tells you that $(x - 1)$ is not a factor (it gives a remainder of 2) but that $(x + 1)$ is. The division also tells you that

$$\frac{x^3 + 1}{x + 1} = x^2 - x + 1$$

So $x^3 + 1$ factors into

$$x^3 + 1 = (x + 1)(x^2 - x + 1)$$

Now $x^2 - x + 1$ won't factor any further (as unsuccessful attempts to split the x term will tell you), and so that's it for $x^3 + 1$.

EXAMPLE: *Factor $3x^3 + 4x^2 - 5x - 2$.*

Since the coefficent of x^3 is 3, we are going to have to consider factors of the form $(lx - k)$, where l divides 3 and k divides -2. Now l must be ± 3 or ± 1, and k must be ± 2 or ± 1, and so the possible factors are

$$(3x - 2), (3x + 2), (3x - 1), (3x + 1)$$

$$(-3x - 2), (-3x + 2), (-3x - 1), (-3x + 1)$$

$$(x - 2), (x + 2), (x - 1), (x + 1)$$

$$(-x - 2), (-x + 2), (-x - 1), (-x + 1)$$

Now this looks like a horribly large number of possible factors to check until you realize that each one really appears twice. Notice that $(-3x - 2) = (-1)(3x + 2)$ and $(-3x + 2) = (-1)(3x - 2)$ and so on, so everything in the second row is (-1) times something in the first. Similarly, everything in the fourth row is (-1) times something in the third. Now if $(3x + 2)$ is a factor of some polynomial, $(-1)(3x + 2)$ will be a factor too. Therefore we need only check the first and third rows. Notice that in the first and third rows l is 3 or 1; in the second and fourth, l is -3 or -1. On the other hand, k takes both signs in each row. Therefore, in practice, the best way to get just the factors you want to check is to *take l positive and k both positive and negative.*

Doing the division tells us that $(3x + 1)$, $(x + 2)$, and $(x - 1)$ are factors. Common sense tells us there can't be any other factors if the highest term is to be $3x^3$. Or, if your common sense deserts you, you can multiply these three factors together to see that you don't need any others to get $3x^3 + 4x^2 - 5x - 2$. And even if it doesn't desert you, you should multiply them together just to check. In any case, we have

$$3x^3 + 4x^2 - 5x - 2 = (3x + 1)\,(x + 2)\,(x - 1)$$

EXAMPLE: *Factor $6 - 11x + 6x^2 - x^3$.*

It's probably a good idea to write the terms of this polynomial in the "usual" order,

$$-x^3 + 6x^2 - 11x + 6$$

before you begin. Since the coefficient of x^3 is -1 and the constant term is 6, l must be ±1 and k must be $\pm1, \pm2$, or ±3. Taking l to be positive means that the possible factors are

$$(x - 3),\ (x + 3),\ (x - 2),\ (x + 2),\ (x - 1),\ (x + 1)$$

Much long division tells us that $(x - 3)$, $(x - 2)$, and $(x - 1)$ are factors; $(x + 3)$, $(x + 2)$, and $(x + 1)$ are not. Now it's important to realize that you can't immediately conclude that $-x^3 + 6x^2 - 11x + 6$ factors into $(x - 3)\,(x - 2)\,(x - 1)$. In fact, if you multiply out this product you find:

$$(x - 3)\,(x - 2)\,(x - 1) = x^3 - 6x^2 + 11x - 6$$

which is our polynomial with all the signs changed. Hence

$$(-1)\,(x - 3)\,(x - 2)\,(x - 1) = (-1)\,(x^3 - 6x^2 + 11x - 6)$$
$$= -x^3 + 6x^2 - 11x + 6$$

Therefore

$$-x^3 + 6x^2 - 11x + 6 = (-1)\,(x - 3)\,(x - 2)\,(x - 1)$$

EXAMPLE: *Factor $x^4 + 3x^2 + 2$.*

The factors to be tested are

$$(x - 2),\ (x + 2),\ (x - 1),\ (x + 1)$$

Division gives you remainders of 30, 30, 6, and 6, which tells us that, unfortunately, none of these are factors and so $x^4 + 3x^2 + 2$ doesn't have any factors of the form $(x - k)$. Then we're stuck, not knowing if this thing doesn't factor at all, or whether it just doesn't factor by this method, until we realize that we can use the method at the very start of this section to see that

$$x^4 + 3x^2 + 2 = (x^2 + 1)(x^2 + 2)$$

And now, to prove we know enough to factor just about anything that factors, here's a really awful one:

EXAMPLE: *Factor $9x^4 - 18x^3 + 12x^2 - 6x + 3$.*

The first moral of this problem is that it pays to be lazy. We could start right off by factoring the 9 and the 3 and inventing a *very* large number of possible factors to check. However, before doing that, look at the coefficients for a moment. Notice that we can first take out a factor of 3:

$$9x^4 - 18x^3 + 12x^2 - 6x + 3 = 3(3x^4 - 6x^3 + 4x^2 - 2x + 1)$$

Now all we have to do is factor $3x^4 - 6x^3 + 4x^2 - 2x + 1$, which isn't nearly as bad as what we started with because the coefficients are smaller.

Possible factors of the form $(lx - k)$, with l positive and k positive or negative, are

$$(3x - 1), (3x + 1), (x - 1), (x + 1)$$

Division tells you that only $(x - 1)$ is a factor, and that

$$\frac{3x^4 - 6x^3 + 4x^2 - 2x + 1}{x - 1} = 3x^3 - 3x^2 + x - 1$$

Hence:

$$9x^4 - 18x^3 + 12x^2 - 6x + 3 = 3(x - 1)(3x^3 - 3x^2 + x - 1)$$

The next thing is to see what we can do with $3x^3 - 3x^2 + x - 1$. The possible factors are the same as before (work them out and see), and division shows that only $(x - 1)$ is a factor. (Actually you could have guessed that, because anything that is a factor of $3x^3 - 3x^2 + x - 1$ would have to be a factor of $3x^4 - 6x^3 + 4x^2 - 2x + 1$ too.) Now

$$\frac{3x^3 - 3x^2 + x - 1}{x - 1} = 3x^2 + 1$$

So

$$9x^4 - 18x^3 + 12x^2 - 6x + 3 = 3(3x^4 - 6x^3 + 4x^2 - 2x + 1)$$
$$= 3(x - 1)(3x^3 - 3x^2 + x - 1)$$
$$= 3(x - 1)(x - 1)(3x^2 + 1)$$
$$= 3(x - 1)^2(3x^2 + 1)$$

and, since $(3x^2 + 1)$ doesn't factor further, we are done.

4. Checking possible linear factors of a polynomial p.

You may feel that the one thing that is really wrong with the above method is all the long division that has to be done to check possible factors. Fortunately for everybody, there's a much neater way of doing the checking, and that's what we'll do now. Let me state the rule first and then illustrate it with an example.

If $(x - k)$ is a factor of p, then substituting $x = k$ into p gives zero.
If $(x - k)$ is not a factor of p, then substituting $x = k$ does not give zero.

If $(lx - k)$ is a factor of p, then substituting $x = \dfrac{k}{l}$ into p gives zero.

If $(lx - k)$ is not a factor of p, then substituting $x = \dfrac{k}{l}$ into p does not give zero.

Why does this work? Suppose we take the polynomial p to be $x^2 - 4x + 3$, with factors of $(x - 3)$ and $(x - 1)$. Now imagine k to be 3 and substitute $x = 3$ into p.

$$x^2 - 4x + 3 = 3^2 - 4 \cdot 3 + 3 = 9 - 12 + 3 = 0$$

You do get zero, corresponding to the fact that $(x - 3)$ is a factor. The reason that you're bound to get zero comes from looking at the factored form of p:

$$x^2 - 4x + 3 = (x - 1)(x - 3)$$

Substituting $x = 3$ gives

$$x^2 - 4x + 3 = (x - 1)(x - 3)$$
$$3^2 - 4 \cdot 3 + 3 = (3 - 1)(3 - 3)$$
$$9 - 12 + 3 = (2)(0)$$
$$0 = 0$$

On the other hand, if we substitute something other than $x = 1$ or $x = 3$—say we substituted $x = 5$—then neither of the factors on the right is zero, and so the right-hand side is not zero. This means that substituting $x = 5$ into p does not give zero—which is just as it should be, since $(x - 5)$ is not a factor.

$$5^2 - 4 \cdot 5 + 3 = (5 - 1)(5 - 3)$$
$$25 - 20 + 3 = \qquad (4)(2)$$
$$8 = 8$$

From the above examples you can see that if a polynomial has $(x - k)$ as a factor, then substituting $x = k$ will give you zero; if it does not have $(x - k)$ as a factor then it will not.

The same reasoning applies to polynomials having factors such as $(3x - 2)$ or $(4x - 5)$, i.e., factors of the form $(lx - k)$. The only value for x that makes the factor $(3x - 2)$ zero is $x = \frac{2}{3}$. Substituting $x = \frac{2}{3}$ will make the factor, and therefore the whole polynomial, zero. If there is no factor of $(3x - 2)$ in p, then substituting $x = \frac{2}{3}$ will make none of the factors zero and hence the polynomial nonzero.

Note: It's important to realize that the sign of the number you substitute and the sign of k in the factor are *opposite*. Also notice that you can think of factors such as $(x + 1)$ as $[x - (-1)]$, i.e., of the form $(x - k)$, with $k = -1$. This means that if $(x + 1)$ is a factor of p, then substituting $x = -1$ will give zero, and if $(x + 1)$ is not a factor, then substituting $x = -1$ won't give zero. Similarly for $(x + 2)$, $(x + 3)$, and so on.

Now to use this. The reason it's useful is that it provides a way of checking whether or not something is a factor of a given polynomial, and a much easier way than the long division of the last section.

EXAMPLE: *Factor $x^3 + 1$.*

When we did this earlier, we knew the possible factors were $(x - 1)$ and $(x + 1)$. Instead of dividing by $(x - 1)$ we substitute $x = 1$:

$p = x^3 + 1 = 1^3 + 1 = 2 \neq 0$

So $(x - 1)$ is not a factor of p. Now substitute $x = -1$:

$p = (-1)^3 + 1 = -1 + 1 = 0$

So $(x + 1)$ is a factor of p.

We do have to do long division to get the other factor, $(x^2 - x + 1)$, and, as before,

$x^3 + 1 = (x + 1)(x^2 - x + 1)$

EXAMPLE: *Factor $3x^3 + 4x^2 - 5x - 2$.*

When we did this problem earlier we had to check

$(3x - 2), (3x + 2), (3x - 1), (3x + 1)$

$(x - 2), (x + 2), (x - 1), (x + 1)$

which is a real bore by division but not so bad by substitution. The following calculations show that $(3x + 1), (x + 2)$, and $(x - 1)$ are the factors:

$$x = -\tfrac{1}{3}: \quad p = 3(-\tfrac{1}{3})^3 + 4(-\tfrac{1}{3})^2 - 5(-\tfrac{1}{3}) - 2$$
$$= -\tfrac{1}{9} + \tfrac{4}{9} + \tfrac{5}{3} - 2 = 0$$

$$x = -2: \quad p = 3(-2)^3 + 4(-2)^2 - 5(-2) - 2$$
$$= -24 + 16 + 10 - 2 = 0$$

$$x = 1: \quad p = 3(1)^3 + 4(1)^2 - 5(1) - 2 = 3 + 4 - 5 - 2 = 0$$

And, for example, this calculation shows that $(x + 1)$ is not a factor:

$$x = -1: \quad p = 3(-1)^3 + 4(-1)^2 - 5(-1) - 2$$
$$= -3 + 4 + 5 - 2 = 4$$

You can check the others in the same way.

EXAMPLE: *Show why $x^2 + 1$ doesn't factor.*

If $x^2 + 1$ factored, it would have to be into two factors of the form $(x - k)$ (there's no other way to get a quadratic). Suppose we're checking whether some particular $(x - k)$ is indeed a factor. That means substituting $x = k$ into $x^2 + 1$, giving

$k^2 + 1$

But no matter what k is, k^2 is never negative. Therefore $k^2 + 1$ is always at least 1, and in any case is never zero. So $(x - k)$ is not a factor of $x^2 + 1$ for any value of k, and therefore $x^2 + 1$ has no factors at all.

EXAMPLE: *Factor $x^3 - 3x^2 + 3x - 1$.*

Let's suppose that $x^3 - 3x^2 + 3x - 1$ has a factor of the form $(x - k)$ (and of course it might not have such a factor—indeed it might not factor at all). Then the two possibilities are $(x - 1)$ and $(x + 1)$, which we will check:

Substituting $x = 1$:

$$p = 1^3 - 3 \cdot 1^2 + 3 \cdot 1 - 1 = 1 - 3 + 3 - 1 = 0$$

Substituting $x = -1$:

$$p = (-1)^3 - 3(-1)^2 + 3(-1) - 1 = -1 - 3 - 3 - 1 = -8$$

Therefore $(x - 1)$ is a factor and $(x + 1)$ is not. Division tells us that

$$x^3 - 3x^2 + 3x - 1 = (x - 1)(x^2 - 2x + 1)$$

You might recognize $x^2 - 2x + 1$ as a perfect square, or you might use the systematic method to factor it. Or you might keep going in the same vein, realizing that the only possible factors of $x^2 - 2x + 1$ are $(x - 1)$ and $(x + 1)$, and checking them. In any case you should end up with

$$x^3 - 3x^2 + 3x - 1 = (x - 1)(x^2 - 2x + 1)$$
$$= (x - 1)(x - 1)^2$$

So

$$x^3 - 3x^2 + 3x - 1 = (x - 1)^3 \qquad \text{(a perfect cube)}$$

5. The number of factors you should expect.

It would be nice, when factoring a polynomial, to have some idea how many factors to expect. Unfortunately, it is not possible to say exactly how many factors a polynomial will have, but there are limits on the number of factors containing x's.

For example;

First-degree:	$2x + 2$	$= 2(x + 1)$	1 factor containing x's
Second-degree:	$x^2 + 3x + 2$	$= (x + 1)(x + 2)$	2 factors containing x's
Third-degree:	$x^3 - 3x^2 + 3x - 1 = (x - 1)^3$		3 factors containing x's

In each of the examples the number of factors containing x's is equal to the degree of the polynomial. There may be fewer factors containing x's than the degree of the polynomial:

Second-degree:	$2x^2 + 2 = 2(x^2 + 1)$	1 factor containing x's
Third-degree:	$x^3 - 1 = (x - 1)(x^2 + x + 1)$	2 factors containing x's

However, there cannot be more factors containing x's than the degree of the polynomial. For example, multiplying out:

$$(x - 1)(x + 3)(x - 5) \qquad \text{(3 factors)}$$

will give you an x^3 term, and so this must be a cubic (i.e., third-degree). Multiplying out

$$x(x^2 + 1)\,(x + 8) \qquad \text{(3 factors)}$$

will produce an x^4 term, and so this must be a fourth-degree polynomial.

In general, if you have a certain number of factors each containing an x, when you multiply them out each factor will contribute at least one x to the term of highest degree. Therefore the degree of that term, and hence the degree of the polynomial, must be at least as great as the number of factors.

Summary of Method for Factoring General Polynomials

1. Take out any common factors.
2. See if the polynomial can be factored by methods for quadratics. If not:
3. Line up all the possible factors of the form $(x - k)$ or $(lx - k)$, where

    k is a factor (positive or negative) of the constant term

    l is a positive factor of the coefficient of the highest power of x

4. Check which of these are factors by substituting $x = k$, or $x = \dfrac{k}{l}$, into p and seeing which give zero.
5. Find other factors by division.
6. Continue the process on other factors.
7. Check your answer by multiplying out.

PROBLEM SET 9.10

Factor the following expressions if possible:

1. $p^2q^4 - 64$
2. $-x^4 - 2x^3 + x^5$
3. $x^3y - 9xy^3$
4. $4 + 4a^2 + a^4$
5. $f^4 - 2f^2g^2 + g^4$
6. $3r^4 - 10 - r^2$
7. $2t - 4t^2 + 8t^3$
8. $64x^4 - \dfrac{9}{y^2}$
9. $a^3 - a^2b - abc - a^2c$
10. $t^2 + t^3 - 17t + 15$
11. $x^3 + 6x^2 - 30 - x$
12. $8ax^8 - 2a$
13. $2 - y^4 - y^2$
14. $144m - 24m^2 + m^3$
15. $-x^4 + 2xy - 3x^3 + 6y$
16. $\pi^6 - 12\pi^3 + 20$
17. $\dfrac{16a^2b^2}{c^6} - 9d^4$
18. $12 - s^4 + s^2$
19. $x^3 + xy^2 - yx^2 - y^3$
20. $12 - 4t - t^3 + 3t^2$

21. $3 + 8x^4 - 10x^2$
22. $30t + 8t^3 + 4t^2 + 15$
23. $-8m^2 + \dfrac{2m^4}{r^2}$

24. $a^3 - 6a^2 + a^4$
25. $(x + 2)^3 - 30 + 6(x + 2)^2 - (x + 2)$
    (Hint: Use Problem 11.)

Factor the following expressions. Expressions in the right column are factors of the given expression.

26. $x^3 - 6x^2 + 11x - 6$ $(x - 2)$
27. $12 + 2y^3 - 8y - 3y^2$ $(2y - 3)$
28. $-24 + 10x + 3x^2 - x^3$ $(x + 3)$
29. $d^3 - 14d - d^2 + 24$ $(d + 4)$
30. $4x^5 + 17x^3 - 3 + 13x - 19x^2 - 16x^4$ $[(x^2 + 1)(2x - 1)]$

Use Problem 23 in Section 9.3 to factor the following:

31. $t^3 - 27$
32. $y^3 - 8$
33. $54b^3 - 16$
34. $x^6 - 1$ (Hint: Use difference of squares, and Problems 22 and 23
35. $a^6 - b^6$ in Section 9.3.)

Factor the following expressions completely:

36. $y^4 - 4y^3 + 13y^2 - 36y + 36$
37. $15x^2t + 75xt^2 + x^3 + 125t^3$
38. $16x^3 - x - 15$
39. $11x + 16x^3 - 15$
40. $-5a^2b - 8b^3 + 7ab^2 + 6a^3$
41. $-36r^2s + 54rs^2 - 27s^3 + 8r^3$
42. $a^3 - 8 + 2a(a - 2)$
43. $2(x + 2)^2 (x - 3) + 3(x + 2) (x - 3)^2$
44. $(m + n)^3 + (m - n)^3$
45. $14a^{n+2} - 17a^{n+1}b - 6a^n b^2$, where n is a positive integer

9.11   COMPLETING THE SQUARE

Completing the square is a method of rewriting quadratic polynomials. It turns out to be useful in a number of unexpected places, including the solution of quadratic equations (see Section 14.3), maxima and minima (this section), as well as in calculus. So although at the moment completing the square may look like a superfluous trick, it is being included here because it will turn out to be useful later, and consequently is worth getting straight.

Completing the square is a way of turning expressions such as

$$x^2 + 2x \quad \text{or} \quad 2x^2 - 3x$$

into perfect squares. For example, if you look at

$$x^2 + 2x$$

and remember that

$$x^2 + 2x + 1$$

is a perfect square, because

$$x^2 + 2x + 1 = (x + 1)^2$$

you see that if we add 1 to $x^2 + 2x$ it becomes a perfect square. Adding 1 to $x^2 + 2x$ to get $(x + 1)^2$ is called completing the square on $x^2 + 2x$.
  Therefore:

Completing the square means adding whatever *constant* is necessary to an expression of the form

$$ax^2 + bx \qquad (a \text{ and } b \text{ are any numbers and } a \neq 0)$$

to make it a perfect square. Notice that since we are adding something to an expression, we are changing its value.

Case 1: Expressions Where the Coefficient of x^2 is 1.

Let us do some examples to see what pattern emerges. To complete the square on:

$x^2 + 2x$	add 1	because $x^2 + 2x + 1 = (x + 1)^2$
$x^2 + 4x$	add 4	because $x^2 + 4x + 4 = (x + 2)^2$
$x^2 + 6x$	add 9	because $x^2 + 6x + 9 = (x + 3)^2$
$x^2 - 2x$	add 1	because $x^2 - 2x + 1 = (x - 1)^2$
$x^2 - 4x$	add 4	because $x^2 - 4x + 4 = (x - 2)^2$
$x^2 - 6x$	add 9	because $x^2 - 6x + 9 = (x - 3)^2$

Notice that in every example the number added was

$$\left(\frac{\text{coefficient of } x}{2}\right)^2$$

For example, in $x^2 + 2x$, the coefficient of x is 2 and we added

$$\left(\frac{2}{2}\right)^2 = 1^2 = 1$$

In $x^2 - 6x$ the coefficient of x is -6 and we added

$$\left(\frac{-6}{2}\right)^2 = (-3)^2 = 9$$

Why is this? We are trying to make our expression into a perfect square, which means something of the form $(x + k)^2$, where k is a constant. Now if you multiply out $(x + k)^2$, you get

$$x^2 + 2kx + k^2$$

We start with the x^2 and x terms of this expression, namely,

$$x^2 + 2kx$$

and we have to find what number to add to this to make it a perfect square; in other words we have to find k^2. But in order to get k^2 from $x^2 + 2kx$ we have to:

1. Take the coefficient of the x term (this is $2k$).
2. Divide by 2 (giving k).
3. Square (giving k^2).

This amounts to calculating

$$\left(\frac{\text{coefficient of } x}{2}\right)^2$$

In other words:

To complete the square on an expression whose x^2 coefficient is 1, you must first add

$$\left(\frac{\text{coefficient of } x}{2}\right)^2$$

and then rewrite the expression as a perfect square.

EXAMPLE: *Complete the square on $x^2 + 10x$.*

The coefficient of x is 10, so we must add

$$\left(\frac{10}{2}\right)^2 = 5^2 = 25$$

Now we write $x^2 + 10x + 25$ as a perfect square as follows:

$$x^2 + 10x + 25 = x^2 + 2 \cdot 5x + (5)^2 = (x + 5)^2$$

EXAMPLE: *Complete the square on $x^2 - 16x$.*

The negative sign here makes no difference. The coefficient of x is -16, so we must add

$$\left(\frac{-16}{2}\right)^2 = (-8)^2 = 64$$

Now we write $x^2 - 16x + 64$ as a perfect square:

$$x^2 - 16x + 64 = x^2 - 2 \cdot (8)x + (8)^2 = (x - 8)^2$$

In all the examples we have done so far, the coefficient of x has been even. The same method works when the coefficient is odd; the answer merely becomes messier.

EXAMPLE: *Complete the square on $x^2 + 3x$.*

The coefficient of x is 3, so we must add

$$\left(\frac{3}{2}\right)^2 = \frac{9}{4}$$

Now we write $x^2 + 3x + \frac{9}{4}$ as a perfect square:

$$x^2 + 3x + \frac{9}{4} = x^2 + 2 \cdot \frac{3}{2}x + \left(\frac{3}{2}\right)^2 = \left(x + \frac{3}{2}\right)^2$$

EXAMPLE: *Complete the square on $x^2 - x$.*

The coefficient of x is -1, so we must add

$$\left(-\frac{1}{2}\right)^2 = \frac{1}{4}$$

Now we write $x^2 - x + \frac{1}{4}$ as a perfect square:

$$x^2 - x + \frac{1}{4} = x^2 - 2 \cdot \frac{1}{2}x + \left(\frac{1}{2}\right)^2 = \left(x - \frac{1}{2}\right)^2$$

The coefficient of x does not even need to be an integer.

EXAMPLE: *Complete the square on $x^2 + \frac{3}{2}x$.*

The coefficient of x is $\frac{3}{2}$, so we must add

$$\left(\frac{\frac{3}{2}}{2}\right)^2 = \left(\frac{3}{4}\right)^2 = \frac{9}{16}$$

Now we write $x^2 + \frac{3}{2}x + \frac{9}{16}$ as a perfect square:

$$x^2 + \frac{3}{2}x + \frac{9}{16} = x^2 + 2 \cdot \frac{3}{4}x + \left(\frac{3}{4}\right)^2 = \left(x + \frac{3}{4}\right)^2$$

EXAMPLE: *Complete the square on $x^2 + bx$.*

The coefficient of x is b, so we must add

$$\left(\frac{b}{2}\right)^2 = \frac{b^2}{4}$$

Now we write $x^2 + bx + \dfrac{b^2}{4}$ as a perfect square:

$$x^2 + bx + \frac{b^2}{4} = x^2 + 2 \cdot \frac{b}{2}x + \left(\frac{b}{2}\right)^2 = \left(x + \frac{b}{2}\right)^2$$

Case 2: Expressions in Which x^2 Can Have Any Coefficient. Suppose we want to convert

$$2x^2 + 8x$$

to the form

(constant) · (perfect square).

We will start by factoring out the 2 and will then make the expression inside the parentheses into a perfect square. Factoring out the 2 gives:

$$2x^2 + 8x = 2(x^2 + 4x)$$

and so we have to make $x^2 + 4x$ into a perfect square. This means adding 4 to the $x^2 + 4x$ inside the parentheses:

$$2(x^2 + 4x + 4) = 2(x + 2)^2$$

But since

$$\underbrace{2x^2 + 8x}_{\text{original expression}} + 8 = 2(x^2 + 4x + 4) = 2(x + 2)^2$$

adding 4 inside the parentheses is the same as adding 8 to the original expression. *Adding 8 to $2x^2 + 8x$ to get $2(x + 2)^2$ is called completing the square on $2x^2 + 8x$.* This is actually a slight abuse of notation, since we are converting $2x^2 + 8x$ to the form (constant) · (perfect square), rather than to a perfect square alone.

The example above shows that:

To complete the square on an expression whose x^2 coefficient is not 1, you must first factor out the coefficient of x^2, then complete the square on the expression inside the parentheses.

EXAMPLE: *Complete the square on $2x^2 - 6x$.*

Factor out the 2:

$$2x^2 - 6x = 2(x^2 - 3x)$$

Completing the square on $x^2 - 3x$ means adding

$$\left(\frac{3}{2}\right)^2 = \frac{9}{4}$$

inside the parentheses:

$$2(x^2 - 3x + \tfrac{9}{4}) = 2(x - \tfrac{3}{2})^2$$

This is equivalent to adding $\frac{9}{2}$ to the original expression, since

$$\underbrace{2x^2 - 6x}_{} + \frac{9}{2} = 2\left(x^2 - 3x + \tfrac{9}{4}\right) = 2\left(x - \tfrac{3}{2}\right)^2$$

EXAMPLE: *Complete the square on $3x^2 + 4x$.*

Factor out the 3 (note: unfortunately, this cannot be done without introducing fractions)

$$3x^2 + 4x = 3(x^2 + \tfrac{4}{3}x)$$

Completing the square on $x^2 + \tfrac{4}{3}x$ involves adding

$$\left(\frac{\tfrac{4}{3}}{2}\right)^2 = \left(\frac{2}{3}\right)^2 = \frac{4}{9}$$

inside the parentheses:

$$3(x^2 + \tfrac{4}{3}x + \tfrac{4}{9}) = 3(x + \tfrac{2}{3})^2$$

Multiplying out shows this is equivalent to adding $\tfrac{4}{3}$ to the original expression:

$$3(x^2 + \tfrac{4}{3}x + \tfrac{4}{9}) = \underbrace{3x^2 + 4x}_{} + \frac{4}{3}$$

Hence

$$3x^2 + 4x + \tfrac{4}{3} = 3(x + \tfrac{2}{3})^2$$

EXAMPLE: *Complete the square on $-5x^2 + x$.*

Factor out the -5:

$$-5x^2 + x = -5(x^2 - \tfrac{1}{5}x)$$

We must add

$$\left(\frac{-\tfrac{1}{5}}{2}\right)^2 = \left(\frac{-1}{10}\right)^2 = \frac{1}{100}$$

inside the parentheses:

$$-5(x^2 - \tfrac{1}{5}x + \tfrac{1}{100}) = -5(x - \tfrac{1}{10})^2$$

This is equivalent to adding

$$-5 \cdot \tfrac{1}{100} = -\tfrac{1}{20}$$

to the original expression, so

$$-5x^2 + x - \tfrac{1}{20} = -5(x^2 - \tfrac{1}{5}x + \tfrac{1}{100}) = -5(x - \tfrac{1}{10})^2$$

EXAMPLE: *Complete the square on $ax^2 + bx$.*

Factoring out the a:

$$ax^2 + bx = a\left(x^2 + \frac{b}{a}x\right)$$

Completing the square on $x^2 + \dfrac{b}{a}x$ means adding

$$\left(\frac{\frac{b}{a}}{2}\right)^2 = \left(\frac{b}{2a}\right)^2 = \frac{b^2}{4a^2}$$

inside the parentheses:

$$a\left(x^2 + \frac{b}{a}x + \frac{b^2}{4a^2}\right) = a\left(x^2 + 2 \cdot \frac{b}{2a}x + \left(\frac{b}{2a}\right)^2\right) = a\left(x + \frac{b}{2a}\right)^2$$

This is equivalent to adding $\dfrac{b^2}{4a}$ to the original expression, since multiplying out the left side gives

$$a\left(x^2 + \frac{b}{a}x + \frac{b^2}{4a^2}\right) = ax^2 + bx + \frac{b^2}{4a}$$

Hence

$$ax^2 + bx + \frac{b^2}{4a} = a\left(x + \frac{b}{2a}\right)^2$$

Use of Completing the Square for Maxima and Minima

Completing the square can be used to find the maximum or minimum values of certain expressions (meaning the largest or smallest values taken on by these expressions). The basic idea is this: Suppose you have an expression such as

$$x^2 + 5$$

As x varies, x^2 is always positive or zero. Whenever x^2 is greater then zero,

$x^2 + 5$ is greater then 5. The least value that $x^2 + 5$ can have is 5, and that occurs when x is zero.

EXAMPLE: *What is the largest, or maximum, value acquired by $-(x-1)^2 + 6$?*

As x varies, $(x-1)^2$ is always positive or zero, and therefore $-(x-1)^2$ is always negative or zero. So the greatest $-(x-1)^2$ can be is zero (when $x = 1$).

Therefore the greatest value acquired by $-(x-1)^2 + 6$ is 6 (and this occurs when $x = 1$).

EXAMPLE: *Find the minimum value of $x^2 + 4x - 3$.*

Looking at the previous example, you can see that what made it "do-able" was that all the x's were contained in a perfect square, and the perfect square was always either positive or zero. If we could get $x^2 + 4x - 3$ in the form

(perfect square) + (some constant)

we could apply the same reasoning. But $x^2 + 4x - 3$ can be put in this form by completing the square on $x^2 + 4x$. To do this we must add 4, because

$$x^2 + 4x + 4 = (x + 2)^2$$

But we cannot just add 4 to our original expression, or we will change its value (and hence the problem). Therefore, if we complete the square on $x^2 + 4x$ by adding 4, we must subtract that 4 also:

$$x^2 + 4x - 3 = x^2 + 4x + \overbrace{4 - 4}^{\text{zero}} - 3 = (x + 2)^2 - 7$$

Now $(x + 2)^2$ is always positive except when $x = -2$, and then it is zero. Therefore the least value taken on by

$$(x + 2)^2 - 7$$

is -7, which occurs when $x = -2$.

The same idea can be used in the following example.

EXAMPLE: *A ball is thrown vertically upwards from the ground. If t represents the time since it was thrown, then the ball's height above the ground at time t is given by the formula*

height $= -16t^2 + 32t$

Find the maximum height above the ground reached by the ball, and the time the ball gets there.

Here we must maximize $-16t^2 + 32t$. As in the last example, we must complete the square on $-16t^2 + 32t$ while preserving the value of the expression. To do this, we first find out what we must add to $-16t^2 + 32t$ to complete the square.

$$-16t^2 + 32t = -16(t^2 - 2t)$$

To complete the square, we add 1 inside the parentheses, which amounts to subtracting 16 from the original expression:

$$-16t^2 + 32t - 16 = -16(t^2 - 2t + 1) = -16(t - 1)^2$$

But if we still want our expression to represent the height of the ball above the ground, we must preserve its orginal value. This means that if we subtract 16 from it, we must also add 16:

$$\text{height} = -16t^2 + 32t = -16t^2 + 32t - 16 + 16$$
$$= -16(t^2 - 2t + 1) + 16$$
$$= -16(t - 1)^2 + 16$$

Now $-16(t - 1)^2$ is always negative, except at $t = 1$, when it is zero. Therefore the maximum value taken on by

$$-16t^2 + 32t = -16(t - 1)^2 + 16$$

is 16, and it occurs when $t = 1$.

Therefore the ball reaches a maximum height of 16 when $t = 1$.

PROBLEM SET   9.11

Complete the square on the following:

1. $y^2 + 4y$	6. $s^6 - 6s^3$	11. $r^2 + 6r$
2. $y^2 - 8y$	7. $2y^2 - 4y$	12. $-4p + p^2$
3. $x^2 - 3x$	8. $7a^2 - a^4$	13. $3x^2 - x$
4. $a^2 + 7a$	9. $b - 2b^2$	14. $12a + a^2$
5. $3r^2 + 6r$	10. $d^2 - \frac{2}{3}d$	

15. $x^2 + ax$ (where a is any constant)
16. $ax^2 + bx$ (where a and b are constants)
17. Find the minimum value of $x^2 + 6x + 2$.
18. Find the maximum value of $7y - 2y^2 + 7$.
19. Find the minimum value of $t^2 - 9 + 4t$.
20. Find the maximum value of $6p - 5p^2 + 1$.
21. Find the minimum value of $\dfrac{r}{2} + \dfrac{r^2}{4} - \dfrac{1}{4}$.

22. Find the maximum value of $-\frac{2}{3}y - y^2 + 1$.
23. Find the maximum value of $1 - 2t^2 - 4t$.
24. Find the minimum value of $2a^2 - 3a + 3$.
25. Find the minimum value of $ax^2 + bx + c$ assuming that a is a positive constant and b and c are any constants.

Find the maximum and minimum values obtained by the expressions below as the variable ranges between -10 and 10. Note that either the maximum or the minimum (or possibly both) will occur at 10 or -10.

26. $z^2 - 2z - 1$
27. $x^2 + 14x + 5$
28. $-13p + 4 + 3p^2$
29. $t^2 - \frac{4}{3}t - 5$
30. $3 + 2r^2 - 9r$

31. $3s^2 + 2 + 6s$
32. $-15 + 7d + 2d^2$
33. $m^2 + \frac{9}{2}m + \frac{7}{2}$
34. $12 + x^2 + 25x$
35. $3l - 10 + l^2$

36. The population of June flies on Crystal Lake beginning on June 1 is given by the formula

$$p = 30t - t^2$$

where time, t, is measured in days. What is the maximum population of flies on the lake in the month of June?

37. During a small depression in 1958, the Dow-Jones index of the Stock Market showed a dip with the equation

$$D.J. = 528 - 20t + t^2$$

Time, t, is measured in days after November 3. On what date was the market its lowest? What was the Dow-Jones index that day?

38. As a plane flies from Atlanta to Augusta, its height in feet, h, above the ground as a function of its distance, d, from Atlanta is given by the following formula:

$$h = 25d - \frac{d^2}{100}$$

Complete the square to find the maximum height the aircraft attains.

39. The amount of light you'll find as you enter the Springfield tunnel is a function of the general level of daylight that day, D_0, and the distance that you go into the tunnel, x meters. This function is

$$\text{light} = D_0 - \frac{D_0}{50}x + \frac{D_0 x^2}{5000}$$

Where in the tunnel is the light the dimmest? At that point, by what percent is the level of light decreased from the daylight level outside?

40. When a nerve cell is stimulated, it sends a pulse of electrical energy down its axon. For a particular nerve, the intensity of this pulse, I, can be

expressed as a function of time, t (in milliseconds), by the formula

$$I - 4 = 23t - t^2$$

How many milliseconds after the pulse begins does it reach its maximum?

41. The height in feet of an object thrown upward with an initial velocity of 80 feet per second is given by the expression $80t - 16t^2$, where t is the time the object is in the air. Complete the square to find the maximum height that the object will obtain.

42. The number of people in Sam's Country Store at any given time is given by the expression $-4t^2 + 12t + 7$, where t is the number of hours since the store opened. By completing the square, find the maximum number of people in Sam's store at the same time.

CHAPTER 9 REVIEW

Simplify:
1. $7b + \frac{7}{2}b$
2. $\frac{5}{2}d - 8d^4 + 2d^3 + d^2 + 3d^4 + cd + ad^3$
3. $(y^3 + 2y - y^2) - (xy^2 - y + y^3) + x^2y^2$
4. $2ac + \frac{2}{3}a^2c^2 + 4a^2c + 6ac^2 - 4ac$
5. $BCD^2 + \frac{1}{3}BC^2D + AC^2D$

Factor:
6. $6x^5 + 2x^2$
7. $5a^2b + 25ab^2$
8. $3M^2 - 2M^5$
9. $-7t^2 - 49t^3$
10. $-x^5 + x^3 - x^3a$

Multiply out:
11. $(x + 3)(3 + x^2 + 2x)$
12. $(2MN + 3)(4 - MN)$
13. $(M + N^2 + p)^2$
14. $(6B - 5 + B^2)(B + 2)$
15. $(2x + y + z)(2x - y + z)$
16. $(2G + 3H)^3$
17. $(\frac{3}{4}c + 2d)(\frac{3}{4}c - 2d)$
18. $(\frac{1}{3}x + \frac{1}{2}y)(\frac{1}{4}x + y)$
19. $(\sqrt{3}x - \sqrt{2})^2$
20. $(-3a^2c)(-4a^3b)$
21. $(5a^2cd^3)^4$
22. $\left(\dfrac{-4xy}{z^2}\right)^3$

Perform the indicated division:
23. $\dfrac{r^3s^2 + sr^2}{r^2s}$
24. $\dfrac{a^3 - 8}{a - 2}$

Write as a single polynomial:
25. $\dfrac{2 + 7x^2 - 9x}{x - 1}$
26. $\dfrac{2 - 3x - 5x^2}{1 + x}$

27. The dividend is $56a^2 - 3 - 17a$ and the quotient is $7a - 3$. What is the divisor?

28. Divide, and give the remainder if any: $\dfrac{x^2 + 4x - 16}{x - 3}$

29. Divide $(y^3 - 2y^2 + 3y - 6)$ by $(y - 2)$.

30. Divide $(3x - 1)$ into $(3x^3 + 6x - x^2 - 2)$.

Simplify:

31. $(4t)^2 - (-4t)^2$

32. $(\sqrt{2}j)^2 - 2\left(\dfrac{j}{\sqrt{2}}\right)^2$

33. $s - \{s - [s - (s - t)]\}$

34. $(a + c) - 4\{3 + 2\,[a - c\,(a - c)] + a\}$

35. $2x(x - 2) - x(2x - 5)$

Find the largest common factor of:

36. $3bc$ and $6c$

37. $-38A^3DG^2$ and $57A^2D^3G$

38. $x^3 - x$ and $x^3 + x$

39. $1 - 4a^2$ and $1 - 4a + 4a^2$

40. $7 - 7x$ and $ax - a$

Factor:

41. $x^2 - 25$

42. $a^2 - 7a + 12$

43. $y^2 - 99y - 100$

44. $2p^2 - 5p + 2$

45. $3x^2 + 5x - 2$

46. $1 - 3x + 2x^2$

47. $4p^2 - 12pq + 9q^2$

48. $25E^4 - 60E^2 + 36$

49. $49 - 64c^2$

50. $12x^2 + 34xz + 34z^2$

Complete the square:

51. $x^2 - 14x$

52. $a^2 + a$

53. $z + 2z^2$

54. $4y^2 + y$

55. $-2p^2 - 4p$

Simplify:

56. $K - \{LK - 2\,[K(L - K)^2]^2 + L^2\} + 1$

57. $x^2 + \frac{1}{2}y - \{y - x\,[y - (3y - 1) + 1] - 1\}$

58. $J^3 - K^3 - (J - K)^3$

59. $x - yz\,[x - z(y - x)] - z\{x + (-y)\,(x - y)\,(z)\,[x - z(y + 1)]\}$

60. $A(A + 3B) - (A + B)\,(2B + A)$

Factor:

61. $16A^2B^3C^2 + 8AB^2C^3 + 8A^3B^2C^2 + 4A^2BC^3$

62. $R^2 - 3A^2 - 2AR$

63. $KL^2 + K^2L + NL + LKN$

64. $17AB + 6A^2 - 3B^2$

65. $4x^2 - 8xy + 4y^2 - b^2$

66. $7A - \dfrac{AB^2D^2}{7}$

67. $C^2 + 2CD + D^2 + 2C + 2D + 1$

68. $(x - 2y)^2 - (z - 2)^2$

69. $\dfrac{G^3}{6} + G^2H + \dfrac{GH^2}{2}$

70. $q^2t + qr + qt^2 + tr$

71. $P^2 - 4PQ + 4Q^2 - R^2 + 4SR - 4S^2$

72. $A(X^2 - A^2) - X(X - A)$

73. $3000 + 15q^2 - 550q$

74. $C^{12} - D^{12}$

75. $-x^2 + x^3 + 24 - 14x$

76. $2x^2yz + xyz^2 + y^2z^2$

77. $X^2 + V^2 + W^2 + 2XV + 2XW + 2VW$

78. $f^3 + f^2 - 17f + 15$

79. $x^2 + 6x + 9 - y^2 - 4y - 4$

80. $5a^2 - 2a + 12a^3$

81. $12(C - R)^2 + 5R - 2 - 5C$

82. $10^{2B} - 7 \cdot 10^B + 12$

83. The factors of a polynomial are

$$(7R + 2S) \quad \text{and} \quad \left(\dfrac{3RS^2}{2} + 1\right). \quad \text{What is the polynomial?}$$

84. One factor of $-10 + x^4 + 2x - 5x^3$ is $(x - 5)$. What is the other?

85. $2m^3 - 3m^2 - 29m + 60$ has three linear factors. Two of them are $(m - 3)$ and $(m + 4)$. What is the third?

86. $x^5 + 2x^2 - 15x^3 - 2x^4 - 4x - 30$ has three linear factors. $(x - 5)$ is one. What are the other two?

Complete the square:

87. $D^2 - \frac{3}{4}D$

88. $-cy + gy^2$

Find the minimum value of:

89. $p^2 - 4p + 7$

90. $3s^2 - 5s + 2$

Find the maximum value of:

91. $-y^2 + 2y - 5$

92. $-3s^2 + 5s - 2$

By multiplying out each one separately, decide whether each pair of expressions are equal or unequal:

93. $(x^2 + y^2 + z^2)^2$ and $(x^2 + y^2)^2 + (z^2)^2$

94. $(q^2 + 2)(q + 3)$ and $(q^2 + 3)(q + 2)$

95. $(M^3 + N^3)$ and $(M + N)^3$

96. $(4a + b - c)(4a - b - c)$ and $(4a - c)^2 - b^2$

97. The height above the ground of a leaping frog is $-x^2 + 2x + 15$, where x is the horizontal distance he has moved. What is the maximum height he reaches?

98. The speed of a roller coaster car decreases as it climbs. If x is its height

above the ground in feet, then its velocity $= x^2 - 40x + 405$ feet per second. What is the minimum velocity, and at what height does it occur?

99. A PNU is a new atomic dishwasher. The manufacturer's revenue from PNU's depends on their price, p, and is given by:

$$\text{revenue} = p\left(20 - \frac{p}{20}\right)$$

What price gives the maximum possible revenue, and what is that maximum revenue?

100. The rate of growth of the population on a crowded island is

$$N(L - N)$$

where N is the number of people already on the island, and L is a constant. At what population is the growth rate a maximum?

10 ALGEBRAIC FRACTIONS

10.1   ALGEBRAIC FRACTIONS: WHAT THEY ARE

An algebraic fraction is a fraction in which the numerator and /or the denominator are algebraic expressions. For example,

$$\frac{x}{5} \qquad \frac{5y^2 + 2}{5ab + 3c} \qquad \frac{\dfrac{2}{x} + \dfrac{3}{x-1}}{\dfrac{4}{x+1} + \dfrac{x}{x-1}}$$

Until Section 10.6, we will deal only with fractions whose numerators and denominators are polynomials, like the first two shown above. Section 10.6 is about complex fractions, of which the third one is an example.

The thing to remember when working with algebraic fractions is that although they look much worse, they behave *exactly* the same as numerical fractions. In particular, they can be cancelled, added, subtracted, multiplied, and divided by precisely the same methods as used for numerical fractions.

10.2   MULTIPLYING FRACTIONS

Numerical fractions are multiplied by multiplying the numerators and the denominators separately. For example,

$$\frac{2}{3} \cdot \frac{5}{7} = \frac{2 \cdot 5}{3 \cdot 7} = \frac{10}{21}$$

or

$$4 \cdot \frac{1}{3} \cdot \frac{2}{9} = \frac{4}{1} \cdot \frac{1}{3} \cdot \frac{2}{9} = \frac{4 \cdot 1 \cdot 2}{1 \cdot 3 \cdot 9} = \frac{8}{27}$$

Algebraic fractions represent numerical fractions (for example, $\frac{x}{x+3}$ repre-

sents $\frac{2}{5}$ when x is 2, or $\frac{7}{10}$ when x is 7, etc.), and therefore they *must* be mul-

tiplied by the same rule as numerical fractions.

EXAMPLE: *Find* $\left(\dfrac{x}{x+2}\right) \cdot \left(\dfrac{x-3}{5}\right).$

$$\left(\frac{x}{x+2}\right) \cdot \left(\frac{x-3}{5}\right) = \frac{x(x-3)}{5(x+2)} \qquad \text{(multiplying tops and bottoms together)}$$

$$= \frac{x^2 - 3x}{5x + 10} \qquad \text{(multiplying out tops and bottoms separately)}$$

EXAMPLE: *Multiply* $4\left(\dfrac{2at}{t^2-5}\right)$ *by* $\left(\dfrac{t+a}{t+1}\right)$

$$4\left(\frac{2at}{t^2-5}\right) \cdot \left(\frac{t+a}{t+1}\right) = \frac{4}{1} \cdot \left(\frac{2at}{t^2-5}\right) \cdot \left(\frac{t+a}{t+1}\right)$$

$$= \frac{4 \cdot (2at) \cdot (t+a)}{(t^2-5)(t+1)}$$

$$= \frac{8at^2 + 8a^2t}{t^3 + t^2 - 5t - 5}$$

Multiplication is more usually combined with simplification, and so many more examples are given at the end of the next section.

PROBLEM SET   10.2

Multiply:

1. $\dfrac{4}{7} \cdot \dfrac{21}{10}$

2. $\dfrac{1}{3} \cdot \dfrac{6}{5} \cdot \dfrac{9}{8}$

3. $\dfrac{0}{12} \cdot \dfrac{24}{3}$

4. $\dfrac{\sqrt{3}}{2} \cdot \dfrac{\sqrt{12}}{3}$

5. $\dfrac{a}{(b+3)} \cdot \dfrac{b}{(a+b)}$

6. $\left(\dfrac{x+5}{x}\right)\left(\dfrac{x-2}{3}\right)$

7. $2\dfrac{(2a^2b^3)^5}{5} \cdot a^4$

8. $\left(\dfrac{2a+5}{4b-1}\right)\left(\dfrac{c-2}{d+1}\right)$

9. $\dfrac{(2x-1)}{2x} \cdot \dfrac{(5-x)}{3x}$

10. $-3 \cdot \dfrac{(x-1)}{(x+1)} \cdot \dfrac{(x+2)}{(x-2)}$

11. $-2\left(\dfrac{x-y}{x+y}\right)$

12. $\dfrac{5a^5c^2}{d} \cdot \dfrac{c}{d^2}$

13. $2\left(\dfrac{3b^7c^3}{d^2}\right)$

14. $\left(\dfrac{x+y}{z-x}\right)\left(\dfrac{y+x}{c+d}\right)$

15. $\left(\dfrac{2}{x-z}\right)\left(\dfrac{3z+x}{z-x}\right)$

16. $\dfrac{25p(rs)^2}{q^3} \cdot \dfrac{r^3}{5^3}$

17. $-1 \cdot \left[\dfrac{a-b}{(a+b)^2}\right]$

18. $\dfrac{p \cdot 7 \cdot x^4 z^2}{l} \cdot x$

19. $\left(\dfrac{x-y}{p^2r}\right)\left(\dfrac{y-x}{2p^3r}\right)$

20. $\left(\dfrac{2a^2}{b}\right)\left(\dfrac{ab}{Ac}\right)$

21. $\left(\dfrac{6a^5}{p^2}\right)\left(\dfrac{2a^3}{p^3+p^2}\right)$

22. $\left(\dfrac{x+1}{3}\right)^3$

23. $\dfrac{[5(a+x)^2+2]}{(c-d)} \cdot (a+x)$

24. $\dfrac{4a^2bx}{9cy} \cdot \dfrac{-12b^2x^2}{c^2y^2}$

25. $\dfrac{c-d}{c^2+d^2} \cdot \dfrac{d-c}{d}$

10.3   SIMPLIFYING FRACTIONS: CANCELLING

Numerical fractions are simplified by factoring the numerator and the denominator and then cancelling factors that occur in both places. This involves using the multiplication method backwards. For example,

$$\frac{42}{140} = \frac{7 \cdot 3 \cdot 2}{7 \cdot 5 \cdot 2 \cdot 2} = \left(\frac{3}{5 \cdot 2}\right)\left(\frac{7 \cdot 2}{7 \cdot 2}\right) = \frac{3}{5 \cdot 2} \cdot 1 = \frac{3}{5 \cdot 2} = \frac{3}{10}$$

$$\xleftarrow{\text{multiplying}}$$
$$\xrightarrow{\text{cancelling}}$$

In practice, the problem could be written as

$$\frac{42}{140} = \frac{\not7 \cdot 3 \cdot \not2}{\not7 \cdot 5 \cdot 2 \cdot \not2} = \frac{3}{10}$$

Algebraic fractions are simplified exactly the same way. For the moment we will assume that the numerator and the denominator are nothing worse than polynomials, which can probably be factored. Any factors occuring in both top and bottom are then cancelled.

EXAMPLE: *Simplify*

$$\frac{2x + 2y}{2a + 4b}$$

$$\frac{2x + 2y}{2a + 4b} = \frac{2(x + y)}{2(a + 2b)}$$ (factoring both the numerator and the denominator)

$$= \frac{2}{2}\left(\frac{x + y}{a + 2b}\right)$$ (writing the fraction as a product)

$$= 1 \cdot \left(\frac{x + y}{a + 2b}\right)$$

$$= \frac{x + y}{a + 2b}$$

or, as it is more usually (and briefly) written:

$$\frac{2x + 2y}{2a + 4b} = \frac{\cancel{2}(x + y)}{\cancel{2}(a + 2b)} = \frac{x + y}{a + 2b}$$

EXAMPLE: *Simplify*

$$\frac{3x + 9}{x^2 + 3x}$$

$$\frac{3x + 9}{x^2 + 3x} = \frac{3\cancel{(x + 3)}}{x\cancel{(x + 3)}}$$ [factoring top and bottom and cancelling common factor of $(x + 3)$]

$$= \frac{3}{x}$$

EXAMPLE: *Simplify*

$$\frac{p}{p^2 + pq}$$

$$\frac{p}{p^2 + pq} = \frac{\cancel{p}}{\cancel{p}(p + q)}$$

$$= \frac{1}{p + q}$$

The point of this example is to show that if everything in the numerator (or the denominator, for that matter) cancels, then you are left with a 1 in the numerator (or denominator).

EXAMPLE: *Simplify*

$$\frac{x^2 + 3x + 2}{x^2 + 5x + 4}.$$

Here the numerator and the denominator are both quadratics that can be factored by the usual methods:

$$\frac{x^2 + 3x + 2}{x^2 + 5x + 4} = \frac{\cancel{(x+1)}\,(x+2)}{\cancel{(x+1)}\,(x+4)} = \frac{x+2}{x+4}$$

EXAMPLE: *Simplify*

$$\frac{m^2 - n^2}{m^2 + 2mn + n^2}$$

The numerator of this is the difference of squares and the denominator is a perfect square:

$$\frac{m^2 - n^2}{m^2 + 2mn + n^2} = \frac{(m+n)\,(m-n)}{(m+n)^2} = \frac{\cancel{(m+n)}\,(m-n)}{\cancel{(m+n)}\,(m+n)} = \frac{m-n}{m+n}$$

EXAMPLE: *Simplify*

$$\frac{(3a + 3b)(6a + b)}{a^2 + 7ab + 6b^2}$$

$$\frac{(3a + 3b)\,(6a + b)}{a^2 + 7ab + 6b^2} = \frac{3\cancel{(a+b)}\,(6a + b)}{(a + 6b)\,\cancel{(a+b)}}$$

$$= \frac{3(6a + b)}{(a + 6b)}$$

The fact that the numerator is given to you as a product

$$(3a + 3b)\,(6a + b)$$

means that the factoring is partly done for you. You should definitely *not* multiply out the $(3a + 3b)\,(6a + b)$, because you would only have to refactor it. Also note that, as with numerical fractions, you must be sure to factor completely or you may not notice just how much you can cancel. For example, here you have to remember to factor $(3a + 3b)$ before you can use it to cancel the $(a + b)$ from the denominator.

Frequently helpful is the

> **Useful Trick:**
>
> $(a - b) = -(b - a)$
>
> or
>
> $(a - b) = (-1)(b - a)$

This comes from the distributive law, which says that

$$-(b - a) = -b - (-a) = -b + a = a - b$$

The trick is useful when you have a factor of $(a - b)$ in the numerator, and $(b - a)$ in the denominator. Then, by rewriting one as the negative of the other, you can cancel them, leaving a factor of -1.

EXAMPLE: *Simplify*

$$\frac{a - 3}{3 - a}$$

$$\frac{a - 3}{3 - a} = \frac{-(3 - a)}{(3 - a)} = \frac{(-1)\,\cancel{(3 - a)}}{\cancel{(3 - a)}} = -1$$

EXAMPLE: *Simplify*

$$\frac{x - 5}{25 - x^2}$$

Factoring gives

$$\frac{x - 5}{25 - x^2} = \frac{x - 5}{(5 + x)(5 - x)}$$

Now, since we have an $(x - 5)$ in the numerator and a $(5 - x)$ in the denominator,

we will use $(x - 5) = (-1)(5 - x)$:

$$\frac{x - 5}{25 - x^2} = \frac{x - 5}{(5 + x)(5 - x)} = \frac{(-1)\,\cancel{(5 - x)}}{(5 + x)\,\cancel{(5 - x)}} = \frac{-1}{(5 + x)}$$

But what if, instead of $(x - 5) = (-1)(5 - x)$, we had used $(5 - x) = (-1)(x - 5)$? Then we would have had:

$$\frac{x - 5}{(5 + x)(5 - x)} = \frac{\cancel{(x - 5)}}{(5 + x)(-1)\,\cancel{(x - 5)}} = \frac{1}{(-1)(5 + x)}$$

Now remember the "golden rule of fractions" (Section 4.7), which says that you can multiply a fraction top and bottom by any number. If we multiply top and bottom by (-1) we get:

$$\frac{1}{(-1)(5+x)} = \frac{(-1)}{(-1)(-1)(5+x)} = \frac{-1}{(5+x)}$$

which is the same answer as before. So it doesn't matter which way you do the problem.

EXAMPLE: *Simplify*

$$\frac{(b^2 - 4a^2)(3b - 9a)}{(6a^2 - 5ab + b^2)}$$

Factoring

$$\frac{(b^2 - 4a^2)(3b - 9a)}{6a^2 - 5ab + b^2} = \frac{(b + 2a)(b - 2a)\,3(b - 3a)}{(2a - b)(3a - b)}$$

We now use $(b - 2a) = (-1)(2a - b)$ and $(b - 3a) = (-1)(3a - b)$:

$$\frac{(b + 2a)(b - 2a)\,3(b - 3a)}{(2a - b)(3a - b)} = \frac{(b + 2a)(-1)\,\cancel{(2a - b)}\,3(-1)\,\cancel{(3a - b)}}{\cancel{(2a - b)}\,\cancel{(3a - b)}}$$

$$= \frac{(-1)(-1)\,3(b + 2a)}{1}$$

$$= 3(b + 2a)$$

Multiplication and Simplification: How to Save Time *and* Get the Right Answer

When you multiply $\frac{2}{3}$ by $\frac{15}{8}$, you can do all the multiplication first:

$$\frac{2}{3} \cdot \frac{15}{8} = \frac{30}{24}$$

and then the cancelling:

$$\frac{2}{3} \cdot \frac{15}{8} = \frac{30}{24} = \frac{\cancel{3} \cdot \cancel{2} \cdot 5}{\cancel{3} \cdot \cancel{2} \cdot 2 \cdot 2} = \frac{5}{4}$$

However, it is quicker (and easier) to do all the cancelling first:

$$\frac{2}{3} \cdot \frac{15}{8} = \frac{\cancel{2}}{\cancel{3}} \cdot \frac{\cancel{3} \cdot 5}{\cancel{2} \cdot 2 \cdot 2}$$

and then multiply out what is left:

$$\frac{2}{3} \cdot \frac{15}{8} = \frac{\cancel{2}}{\cancel{3}} \cdot \frac{\cancel{3} \cdot 5}{\cancel{2} \cdot 2 \cdot 2} = \frac{5}{2 \cdot 2} = \frac{5}{4}$$

The first way you do a great deal of unnecessary multiplying because as soon as you have multiplied everything together you have to factor again so that you can cancel. The second way, you have to do much less multiplying because, with any luck, you have got rid of a good deal of the fraction by cancelling before you get to the multiplying stage.

The same thing applies to algebraic fractions. Suppose you have to multiply

$$\frac{x}{x+1} \quad \text{by} \quad \frac{x^2 + 3x + 2}{x^2 + 3x}$$

If you multiply first and then cancel, the calculation looks like this:

$$\left(\frac{x}{x+1}\right)\left(\frac{x^2 + 3x + 2}{x^2 + 3x}\right) = \frac{x(x^2 + 3x + 2)}{(x+1)(x^2 + 3x)}$$

$$= \frac{x^3 + 3x^2 + 2x}{x^3 + 4x^2 + 3x} \quad \text{a tiresome thing to multiply out}$$

$$= \frac{\cancel{x(x+1)}(x+2)}{\cancel{x(x+1)}(x+3)} \quad \text{a \textit{great} deal of factoring}$$

$$= \frac{x+2}{x+3}$$

If you factor and cancel first and then multiply, the calculation looks like this:

$$\left(\frac{x}{x+1}\right)\left(\frac{x^2 + 3x + 2}{x^2 + 3x}\right) = \frac{\cancel{x} \cdot \cancel{(x+1)}(x+2)}{\cancel{(x+1)} \cdot \cancel{x}(x+3)}$$

$$= \frac{x+2}{x+3}$$

which is *much* easier. So the moral for multiplying fractions is this:

Factor and cancel everything you can *before* doing any multiplication.

EXAMPLE: *Multiply*

$$\frac{\pi a + 4\pi}{2a^2 - 14a} \quad \text{by} \quad \frac{2a\pi^2}{a^2 - 3a - 28}$$

Factor and cancel *first*:

$$\left(\frac{\pi a + 4\pi}{2a^2 - 14a}\right)\left(\frac{2a\pi^2}{a^2 - 3a - 28}\right) = \frac{\pi\,\cancel{(a+4)}\,\cancel{2a}\,\pi^2}{\cancel{2a}(a-7)\cdot\cancel{(a+4)}\,(a-7)}$$

$$= \frac{\pi\cdot\pi^2}{(a-7)\,(a-7)}$$

$$= \frac{\pi^3}{(a-7)^2} \qquad \text{(you can leave the answer this way)}$$

$$= \frac{\pi^3}{a^2 - 14a + 49} \qquad \text{(or you can multiply it out, like this)}$$

EXAMPLE: *Multiply*

$$\left(\frac{16p^2 - q^2}{2p + q}\right)\cdot\left(\frac{8p^2 + 2pq - q^2}{16p^2 - 8pq + q^2}\right)$$

Factor and cancel first:

$$\left(\frac{16p^2 - q^2}{2p + q}\right)\cdot\left(\frac{8p^2 + 2pq - q^2}{16p^2 - 8pq + q^2}\right)$$

$$= \frac{(4p+q)\,\cancel{(4p-q)}\cdot\cancel{(4p-q)}\,\cancel{(2p+q)}}{\cancel{(2p+q)}\,\cancel{(4p-q)}\,\cancel{(4p-q)}}$$

$$= \frac{(4p+q)}{1}$$

$$= 4p + q$$

PROBLEM SET 10.3

Simplify:

1. $\dfrac{12}{70}$

2. $\dfrac{0.7 + 0.2 + 0.6}{1 - 0.25}$

3. $\dfrac{115}{65}$

4. $\dfrac{2a - 6}{3 - a}$

5. $\dfrac{b^2 - 4}{2 + b}$

6. $\dfrac{x^2 - y^2}{y^2 - x^2}$

7. $\dfrac{a^2 + 8a + 15}{7a + 12 + a^2}$

8. $\dfrac{10rs}{5s^2 + 20s}$

9. $\dfrac{8g + 4h}{12g}$

10. $\dfrac{2p^2 + p - 6}{p + 2}$

11. $\dfrac{km + kn}{n^2 + nm}$

12. $\dfrac{4 - z^2}{(2 - z)^2}$

13. $\dfrac{3x^2 - x}{3x - 1}$

14. $\dfrac{b^3 - z^3}{z^3 - b^3}$

15. $\dfrac{4(ab)^2 + 4ab + 1}{4(ab)^2 - 1}$ 16. $\dfrac{(x+2)^2 + 3(x+2) + 2}{x+3}$

Multiply and simplify:

17. $\dfrac{(x-1)}{(x^2 - 3x + 2)} \cdot \dfrac{(6 - 5x + x^2)}{(4-x)}$

18. $\dfrac{c-d}{c+d} \cdot (d^2 - c^2)$

19. $6s\left(\dfrac{k}{3} + 1 - \dfrac{3k}{2s}\right)$

20. $\left(\dfrac{5x}{15a - 20b}\right) \cdot \left(\dfrac{3a - 4b}{2x^3}\right)$

21. $\left(\dfrac{a-b}{d+c}\right) \cdot \left(\dfrac{a+c}{b-a}\right)$

22. $-\dfrac{1-x}{x-2} \cdot \dfrac{x^2 - 4}{x^2 - 1}$

23. $\dfrac{pqr}{abc}\left[a\left(\dfrac{1}{pqr}\right)^2\right]$

24. $\dfrac{a^2 - 6a + 9}{a^2 - 9} \cdot \dfrac{3a + 9}{a - 3}$

25. $\dfrac{6y^2 - 5y - 6}{-11y - 10 + 6y^2} \cdot \dfrac{15 - 6y}{9 + 4y^2 - 12y}$

26. $(x+y) \cdot \left(\dfrac{2x - y}{x^2 - y^2}\right)$

27. $\dfrac{4r^3 s^2}{7ts} \cdot \dfrac{21t^3 r^2 s}{-6rs^3}$

28. $\dfrac{4k - 1}{8} \cdot \dfrac{2k + 12}{3k - 1 + 4k^2}$

Simplify:

29. $\dfrac{(x-1)a - (1-x)c}{(x-1)ac}$

30. $\dfrac{d^2 + dg - 2g^2}{d^3 - g^3}$

31. $\dfrac{ax + bx - cx}{ay + by - cy}$

32. $\dfrac{2c(c-3) - 8}{2c(c-3) + 8c}$

33. $\dfrac{6(ab)^2 - 7ab - 3}{2(ab)^2 - 9ab + 9}$

34. $\dfrac{x^6 - 5x^4 - x^3}{x^2}$

35. $\dfrac{ars - art + as - at}{a^2 r + a^2}$

36. $\dfrac{x^8 - z^8}{(x^4 + z^4)(x - z)^3}$

37. $\dfrac{-18x^3 y^2 - 6x^2 y^3 + 24x^2 y^4}{-6x^2 y^2}$

38. $\dfrac{6x^{2n+1} - 13x^{2n} + 6x^{2n-1}}{3x^{n+1} - 2x^n}$

Multiply and simplify:

39. $\dfrac{4m^2 - n^4}{2m^2 + 9mn^2 - 5n^4} \cdot \dfrac{m^2 + 5mn^2}{2m + n^2}$

40. $\dfrac{3 - h}{(h-3)^2} \cdot \dfrac{h^2 - 5h + 6}{(-2)(h-3)}$

41. $\dfrac{p^2 - 4}{p^2 + 1} \cdot \dfrac{p^4 - 3p^2 - 4}{p^2 - 6p + 8}$

42. $\dfrac{2}{x-y} \cdot \dfrac{x^3 - y^3}{xy - y^2} \cdot \dfrac{y}{2x^2 + 2xy + 2y^2}$

43. $\dfrac{4x^2 - 1}{a^2 - 9} \cdot \dfrac{3a - 9}{4x^2 + 4x + 1}$

44. $\dfrac{5w^3 + 40}{2abc^3} \cdot \dfrac{ab^2 - a^2b}{(b^2 - a^2)(w + 2)^2}$

45. $\left[\dfrac{5m^2n^2}{8pq(r + 3)}\right]\left[\dfrac{2p^2(q + 3)}{mn^5(r + 3)^2}\right]$

10.4   ADDING AND SUBTRACTING ALGEBRAIC FRACTIONS

Remember from Section 4.5 that long process of finding and using a least common denominator [L.C.D.]; the same method is used here.

To Add or Subtract Algebraic Fractions:

1. Factor each denominator into prime factors.
2. List all the different prime factors occurring in any of the denominators. For each of these factors, pick the highest power to which it occurs in any one denominator, and write the factor to that power. Multiply together the factors to these powers. This is the L.C.D.
3. Convert each fraction into one with the L.C.D. for a denominator. For each fraction pick out those factors in the L.C.D. but not in the denominator and multiply top and bottom by these factors.
4. Replace old fractions by new ones with the same value, but over the L.C.D. Since the denominators are now all the same, we can add or subtract the numerators.

EXAMPLE: *Find*

$$\frac{5}{54} - \frac{1}{180}$$

STEP 1.   Factoring the denominators:

$$54 = 2 \cdot 27 = 2 \cdot 3^3$$

$$180 = 18 \cdot 10 = 9 \cdot 2 \cdot 2 \cdot 5 = 2^2 \cdot 3^2 \cdot 5$$

STEP 2.   Different prime factors are 2, 3, 5.

Highest power of 2 occurring is 2 (in $180 = 2^{②} \cdot 3^2 \cdot 5$)

Highest power of 3 occurring is 3 (in $54 = 2 \cdot 3^{③}$)

Highest power of 5 occurring is 1 (in $180 = 2^2 \cdot 3^2 \cdot 5^{\text{①}}$)

So write down 2^2, 3^3, 5^1 and multiply them to give L.C.D. $= 2^2 \cdot 3^3 \cdot 5$ ($= 540$ if you want to multiply it out, but it's better not to).

STEP 3.   Since $54 = 2 \cdot 3^3$ and L.C.D. $= 2^2 \cdot 3^3 \cdot 5$, 54 has all the factors in the L.C.D. except a 2 and a 5. Therefore you must multiply $\frac{5}{54}$ top and bottom by $2 \cdot 5 = 10$ so that the denominator becomes the L.C.D.

$$\frac{5}{54} = \frac{5}{54} \cdot \frac{10}{10} = \frac{50}{540}$$

Since $180 = 2^2 \cdot 3^2 \cdot 5$ and L.C.D. $= 2^2 \cdot 3^3 \cdot 5$, 180 has all the factors in the L.C.D. except a 3. Therefore you must multiply $\frac{1}{180}$ top and bottom by 3 so that the denominator becomes the L.C.D.

$$\frac{1}{180} = \frac{1}{180} \cdot \frac{3}{3} = \frac{3}{540}$$

STEP 4. $$\frac{5}{54} - \frac{1}{180} = \frac{50}{540} - \frac{3}{540} = \frac{47}{540}$$

Algebraic fractions, needless to say, work exactly the same way. First, using the methods of Chapter 9, you must factor all the denominators into prime factors (expressions that can be factored no further). Then you create and use the L.C.D. exactly as above.

EXAMPLE: *Add*

$$\frac{1}{2x^2} + \frac{3}{4x}$$

STEP 1.   Factor the denominators:

$$2x^2 = 2 \cdot x^2$$

$$4x = 2^2 \cdot x$$

STEP 2.   Different prime factors are 2 and x.
Highest power of 2 occurring is 2 (in $4x = 2^{\text{②}} \cdot x$)
Highest power of x occurring is 2 (in $2x^{\text{②}}$)
So write down 2^2 and x^2 and multiply them to give
L.C.D. $= 2^2 x^2 = 4x^2$

STEP 3.   Since L.C.D. $= 4x^2 = 2^2 \cdot x^2$, the only factor in the L.C.D. but not in $2x^2$ is a 2. Therefore we must multiply $\dfrac{1}{2x^2}$ top and bottom by

2 to get a denominator of $4x^2$:

$$\frac{1}{2x^2} = \frac{1}{2x^2} \cdot \frac{2}{2} = \frac{2}{4x^2}$$

Since L.C.D. $= 4x^2 = 2^2 \cdot x^2$ and $4x = 2^2 \cdot x$, the only factor in the L.C.D. but not in $4x$ is an x. Therefore we must multiply $\frac{3}{4x}$ top and bottom by x to get a denominator of $4x^2$:

$$\frac{3}{4x} = \frac{3}{4x} \cdot \frac{x}{x} = \frac{3x}{4x^2}$$

STEP 4.

$$\frac{1}{2x^2} + \frac{3}{4x} = \frac{2}{4x^2} + \frac{3x}{4x^2} = \frac{2 + 3x}{4x^2}$$

EXAMPLE: *Combine*

$$\frac{a}{3b^2c} + \frac{c^2}{6ab}$$

Combine means combine into one fraction, or, in other words, add $\frac{a}{3b^2c}$ and $\frac{c^2}{6ab}$.

STEP 1.

Factor:

$$3b^2c = 3 \cdot b^2 \cdot c$$

$$6ab = 2 \cdot 3 \cdot a \cdot b$$

STEP 2.

Different prime factors are 2, 3, a, b, c.
Highest power of 2 occurring is 1 (in $6ab = 2^① \cdot 3 \cdot a \cdot b$)
Highest power of 3 occurring is 1 (occurs in both denominators)
Highest power of a occurring is 1 (in $6ab = 2 \cdot 3 \cdot a^① \cdot b$)
Highest power of b occurring is 2 (in $3b^2c = 3 \cdot b^② \cdot c$)
Highest power of c occurring is 1 (in $3b^2c = 3 \cdot b^2 \cdot c^①$)
Write down 2^1, 3^1, a^1, b^2, c^1 and multiply them to give
L.C.D. $= 2 \cdot 3 \cdot a \cdot b^2 \cdot c = 6ab^2c$

STEP 3.

The only factors that are in the L.C.D. but not in $3b^2c$ are a 2 and an a. Therefore we multiply $\frac{a}{3b^2c}$ top and bottom by $2a$:

$$\frac{a}{3b^2c} = \frac{a}{3b^2c} \cdot \frac{2a}{2a} = \frac{2a^2}{6ab^2c}$$

The only factors in the L.C.D. but not in $6ab$ are a b and a c.

Therefore we multiply $\dfrac{c^2}{6ab}$ top and bottom by bc:

$$\frac{c^2}{6ab} = \frac{c^2}{6ab} \cdot \frac{bc}{bc} = \frac{bc^3}{6ab^2c}$$

STEP 4.

$$\frac{a}{3b^2c} + \frac{c^2}{6ab} = \frac{2a^2}{6ab^2c} + \frac{bc^3}{6ab^2c} = \frac{2a^2 + bc^3}{6ab^2c}$$

EXAMPLE: *Combine*

$$\frac{5}{x^2y^2} - \frac{10}{x^2} - \frac{3}{y^2}$$

The fact that there are three terms here, and subtractions instead of additions, makes no difference to the method.

STEP 1.   The denominators, x^2y^2, x^2, and y^2 are already factored.

STEP 2.   Highest power of x occurring is the second; highest power of y occurring is the second; therefore the L.C.D. is x^2y^2.

STEP 3. $\dfrac{5}{x^2y^2}$ already has the right denominator so nothing needs to be done to it

$\dfrac{10}{x^2}$ needs to be multiplied top and bottom by y^2 to make the denominator x^2y^2

$$\frac{10}{x^2} = \frac{10y^2}{x^2y^2}$$

$\dfrac{3}{y^2}$ needs to be multiplied top and bottom by x^2 to make the denominator x^2y^2

$$\frac{3}{y^2} = \frac{3x^2}{x^2y^2}$$

STEP 4.

$$\frac{5}{x^2y^2} - \frac{10}{x^2} - \frac{3}{y^2} = \frac{5}{x^2y^2} - \frac{10y^2}{x^2y^2} - \frac{3x^2}{x^2y^2}$$

$$= \frac{5 - 10y^2 - 3x^2}{x^2y^2}$$

EXAMPLE: *Add*

$$\frac{1}{x+1} + \frac{1}{x}$$

STEP 1.   Denominators $(x + 1)$, x are already factored.

STEP 2.   Highest power of $(x + 1)$ occurring is the first; highest power of x occurring is the first; therefore
L.C.D. $= x(x + 1)$

STEP 3. $\dfrac{1}{x + 1}$ must be multiplied top and bottom by x to make the denominator $x(x + 1)$:

$$\frac{1}{x + 1} = \frac{1}{x + 1} \cdot \frac{x}{x} = \frac{x}{x(x + 1)}$$

Similarly, $\dfrac{1}{x}$ must be multiplied top and bottom by $(x + 1)$ to make the denominator $x(x + 1)$:

$$\frac{1}{x} = \frac{1}{x} \cdot \frac{(x + 1)}{(x + 1)} = \frac{x + 1}{x(x + 1)}$$

STEP 4. $$\frac{1}{x + 1} + \frac{1}{x} = \frac{x}{x(x + 1)} + \frac{(x + 1)}{x(x + 1)}$$

$$= \frac{x + (x + 1)}{x(x + 1)}$$

$$= \frac{2x + 1}{x(x + 1)}$$

Notice: The parentheses around the $(x + 1)$ in the numerator aren't really necessary, and so the problem is often written without them:

$$\frac{1}{x + 1} + \frac{1}{x} = \frac{x}{x(x + 1)} + \frac{x + 1}{x(x + 1)}$$

$$= \frac{x + x + 1}{x(x + 1)}$$

$$= \frac{2x + 1}{x(x + 1)}$$

EXAMPLE: *Subtract*

$$\frac{1}{x + 1} - \frac{1}{x}$$

Most of the steps in this problem are exactly the same as those in the problem above, but it also illustrates a very important point about watching parentheses and signs in a subtraction problem. Up to the fourth step, the calculations are exactly the same as in the example above, so we will start from there:

STEP 4.

$$\frac{1}{x+1} - \frac{1}{x} = \frac{x}{x(x+1)} - \frac{(x+1)}{x(x+1)}$$

$$= \frac{x - (x+1)}{x(x+1)}$$

$$= \frac{x - x - 1}{x(x+1)} \quad \text{(removing parentheses by the distributive law)}$$

$$= \frac{-1}{x(x+1)}$$

Notice: Here we *cannot* leave out the parentheses around the $(x+1)$ in the numerator as we could in the last example. Without the parentheses we would have had:

$$\frac{1}{x+1} - \frac{1}{x} = \frac{x}{x(x+1)} - \frac{x+1}{x(x+1)} \quad \text{(O.K. so far)}$$

$$= \frac{x - x + 1}{x(x+1)} \quad \text{(wrong)}$$

This last expression is wrong because we are meant to be subtracting the *whole* of the second fraction, and therefore the *whole* of its numerator, that is, both the x *and* the 1. As it stands we have subtracted only the x. Therefore the parentheses in the numerator are essential because they make us treat $x+1$ as one quantity. Without them, you end up with

$$\frac{x - x + 1}{x(x+1)} \quad \text{i.e.} \quad \frac{+1}{x(x+1)}$$

instead of the correct answer,

$$\frac{-1}{x(x+1)}$$

EXAMPLE: *Add*

$$\frac{2}{x^2 + x} + \frac{3}{x^3 - x^2}$$

STEP 1.   Here the denominators are not factored, so we must factor them:

$$x^2 + x = x(x+1)$$

$$x^3 - x^2 = x^2(x-1)$$

STEP2.

Highest power of x occurring is the second;
highest power of $(x + 1)$ occurring is the first;
highest power of $(x - 1)$ occurring is the first;
therefore
L.C.D. $= x^2(x + 1)\,(x - 1)$

Always leave the L.C.D. in factored form. Multiplying out takes time,
and anyway you can't simplify the fraction unless everything is factored.

STEP3.

$x^2 + x = x(x + 1)$ needs multiplying by $x(x - 1)$ to make the
L.C.D. Therefore

$$\frac{2}{x^2 + x} = \frac{2}{x(x + 1)} \cdot \frac{x(x - 1)}{x(x - 1)} = \frac{2x(x - 1)}{x^2(x + 1)\,(x - 1)}$$

$x^3 - x^2 = x^2(x - 1)$ needs multiplying by $(x + 1)$ to make the
L.C.D. Therefore

$$\frac{3}{x^3 - x^2} = \frac{3}{x^2(x - 1)} \cdot \frac{(x + 1)}{(x + 1)} = \frac{3(x + 1)}{x^2(x + 1)\,(x - 1)}$$

STEP4.

$$\frac{2}{x^2 + x} + \frac{3}{x^3 - x^2} = \frac{2x(x - 1)}{x^2(x + 1)\,(x - 1)} + \frac{3(x + 1)}{x^2(x + 1)\,(x - 1)}$$

$$= \frac{2x(x - 1) + 3(x + 1)}{x^2(x + 1)\,(x - 1)}$$

In order to simplify the numerator of this fraction, we have to
multiply out each term, then combine like terms if possible (see
Section 9.1).

$$\frac{2}{x^2 + x} + \frac{3}{x^3 - x^2} = \frac{2x(x - 1) + 3(x + 1)}{x^2(x + 1)\,(x - 1)}$$

$$= \frac{2x^2 - 2x + 3x + 3}{x^2(x + 1)\,(x - 1)} \qquad \text{(by the distributive law)}$$

$$= \frac{2x^2 + x + 3}{x^2(x + 1)\,(x - 1)} \qquad \text{(collecting like terms)}$$

EXAMPLE: *Add*

$$k + \frac{1}{k + 3}$$

To do this problem we have to look at k as the fraction $\dfrac{k}{1}$.

STEP 1. The denominators 1, $(k + 3)$ are already factored.

STEP 2. L.C.D. $= 1 \cdot (k + 3) = (k + 3)$

STEP 3. Now $\dfrac{k}{1}$ must be multiplied top and bottom by $(k + 3)$:

$$\frac{k}{1} = \frac{k}{1} \cdot \frac{(k + 3)}{(k + 3)} = \frac{k(k + 3)}{(k + 3)}$$

whereas $\dfrac{1}{k + 3}$ already has the right denominator.

STEP 4.
$$k + \frac{1}{k + 3} = \frac{k}{1} + \frac{1}{(k + 3)}$$

$$= \frac{k(k + 3)}{(k + 3)} + \frac{1}{(k + 3)}$$

$$= \frac{k(k + 3) + 1}{(k + 3)}$$

$$= \frac{k^2 + 3k + 1}{k + 3} \qquad \text{(using the distributive law)}$$

EXAMPLE: *Combine*

$$\frac{1}{p + q} - \frac{1}{p - q} - \frac{2p}{p^2 - q^2}$$

STEP 1. Denominators are

$(p + q)$

$(p - q)$

$p^2 - q^2 = (p + q)(p - q)$

STEP 2. L.C.D. $= (p + q)(p - q)$ (since highest powers of $(p + q)$ and $(p - q)$ occurring are 1)

STEP 3. Now $(p + q)$ must be multiplied by $(p - q)$ to make the L.C.D., so

$$\frac{1}{p + q} = \frac{1}{p + q} \cdot \frac{p - q}{p - q} = \frac{p - q}{(p + q)(p - q)}$$

And $(p - q)$ must be multiplied by $(p + q)$ to make the L.C.D., so

$$\frac{1}{p-q} = \frac{1}{p-q} \cdot \frac{p+q}{p+q} = \frac{p+q}{(p+q)(p-q)}$$

Now

$$\frac{2}{p^2 - q^2} = \frac{2}{(p+q)(p-q)}$$

already has the right denominator, and so nothing needs to be done to it.

STEP 4.

$$\frac{1}{p+q} - \frac{1}{p-q} - \frac{2p}{p^2 - q^2}$$

$$= \frac{p-q}{(p+q)(p-q)} - \frac{(p+q)}{(p+q)(p-q)} - \frac{2p}{(p+q)(p-q)}$$

$$= \frac{p - q - (p+q) - 2p}{(p+q)(p-q)} \qquad \text{(remember parentheses in numerator, please!)}$$

$$= \frac{p - q - p - q - 2p}{(p+q)(p-q)} \qquad \text{(removing parentheses using distributive law)}$$

$$= \frac{-2q - 2p}{(p+q)(p-q)}$$

But we aren't finished yet, because this time our fraction can be simplified by cancelling:

$$\frac{-2q - 2p}{(p+q)(p-q)} = \frac{-2\cancel{(q+p)}}{\cancel{(p+q)}(p-q)} = \frac{-2}{(p-q)} \qquad \begin{array}{l}[(q+p) \text{ is the} \\ \text{same as } (p+q)]\end{array}$$

EXAMPLE: *Combine*

$$\frac{2}{1 - 4a^2} + \frac{1}{2a - 1} - \frac{1}{2a + 1}$$

STEP 1.   Factor denominators:

$$1 - 4a^2 = (1 + 2a)(1 - 2a)$$

$(2a - 1)$ and $(2a + 1)$ are already factored.

STEP 2.   You will notice that this problem seems to be very tiresome because $1 - 4a^2$ has factors of $(1 + 2a)$ and $(1 - 2a)$, whereas the other fractions have denominators of $(2a - 1)$ and $(2a + 1)$. Now $(1 + 2a)$ and $(2a + 1)$ are in fact the same, but $(1 - 2a)$ and $(2a - 1)$, although they look similar, are not the same. Regarding $(1 - 2a)$ and $(2a - 1)$ as different factors gives us

$$(1 + 2a)(1 - 2a)(2a - 1)$$

as the common denominator—which is rather large. The problem can be made much easier by the Useful Trick of Section 10.3:

$$(2a - 1) = (-1)(1 - 2a)$$

Using this and the fact that $(2a + 1) = (1 + 2a)$ to rewrite the problem, we get:

$$\frac{2}{1 - 4a^2} + \frac{1}{2a - 1} - \frac{1}{2a + 1}$$

$$= \frac{2}{(1 + 2a)(1 - 2a)} + \frac{1}{(-1)(1 - 2a)} - \frac{1}{(1 + 2a)}$$

$$= \frac{2}{(1 + 2a)(1 - 2a)} - \frac{1}{(1 - 2a)} - \frac{1}{(1 + 2a)}$$

Now the factor $(1 - 2a)$ occurs in the first two denominators, and the factor $(2a - 1)$ occurs nowhere. The only other thing that has changed is that the addition has become a subtraction, which is hardly a problem.

Now the L.C.D. is $(1 + 2a)(1 - 2a)$ [because each factor occurs only to the first power]. Since the L.C.D is smaller than the other common denominator, we have to multiply the original fractions by less to get them over the new denominator. This means that the new numerators are smaller—which is always a good thing.

STEP 3   Convert each fraction to one with $(1 + 2a)(1 - 2a)$ in the denominator:

$$\frac{2}{1 - 4a^2} = \frac{2}{(1 + 2a)(1 - 2a)}, \text{ so nothing needs doing here.}$$

$$\frac{1}{1 - 2a} = \frac{1}{(1 - 2a)} \cdot \frac{(1 + 2a)}{(1 + 2a)} = \frac{1 + 2a}{(1 + 2a)(1 - 2a)}$$

$$\frac{1}{1 + 2a} = \frac{1}{(1 + 2a)} \cdot \frac{(1 - 2a)}{(1 - 2a)} = \frac{1 - 2a}{(1 + 2a)(1 - 2a)}$$

STEP 4 $$\frac{2}{1 - 4a^2} + \frac{1}{2a - 1} - \frac{1}{2a + 1}$$

$$= \frac{2}{(1 + 2a)(1 - 2a)} - \frac{1}{(1 - 2a)} - \frac{1}{(1 + 2a)} \qquad \text{(rewriting using the Useful Trick)}$$

$$= \frac{2}{(1 + 2a)(1 - 2a)} - \frac{(1 + 2a)}{(1 + 2a)(1 - 2a)} - \frac{(1 - 2a)}{(1 + 2a)(1 - 2a)}$$

$$= \frac{2 - (1 + 2a) - (1 - 2a)}{(1 + 2a)(1 - 2a)}$$

$$= \frac{2 - 1 - 2a - 1 + 2a}{(1 + 2a)(1 - 2a)} \qquad \text{(using the distributive law on the numerator)}$$

$$= \frac{0}{(1 + 2a)(1 - 2a)}$$

$$= 0 \qquad \text{(because 0 over anything except zero is 0)}$$

Therefore

$$\frac{2}{1 - 4a^2} + \frac{1}{2a - 1} - \frac{1}{2a + 1} = 0$$

EXAMPLE: *Combine into one fraction and simplify*

$$\sqrt{x^2 + 1} - \frac{1}{\sqrt{x^2 + 1}}$$

This looks different from the other examples because $\sqrt{x^2 + 1}$ is not a polynomial; however, it can be done by exactly the same methods. If you think of the first term as $\dfrac{\sqrt{x^2 + 1}}{1}$ then both terms look like fractions and can be put over a common denominator, in this case $\sqrt{x^2 + 1}$.

To convert $\dfrac{\sqrt{x^2 + 1}}{1}$ to a fraction with $\sqrt{x^2 + 1}$ as denominator, multiply top and bottom by $\sqrt{x^2 + 1}$:

$$\frac{\sqrt{x^2 + 1}}{1} = \frac{\sqrt{x^2 + 1}}{1} \cdot \frac{\sqrt{x^2 + 1}}{\sqrt{x^2 + 1}} = \frac{\sqrt{x^2 + 1} \cdot \sqrt{x^2 + 1}}{\sqrt{x^2 + 1}} = \frac{x^2 + 1}{\sqrt{x^2 + 1}}$$

(since by the definition of a square root,
$\sqrt{x^2 + 1} \cdot \sqrt{x^2 + 1} = x^2 + 1$)

Therefore,

$$\sqrt{x^2 + 1} - \frac{1}{\sqrt{x^2 + 1}} = \frac{x^2 + 1}{\sqrt{x^2 + 1}} - \frac{1}{\sqrt{x^2 + 1}} = \frac{x^2 + 1 - 1}{\sqrt{x^2 + 1}}$$

$$= \frac{x^2}{\sqrt{x^2 + 1}}$$

PROBLEM SET   10.4

Perform the indicated operation and simplify:

1. $\dfrac{2}{m} + \dfrac{3}{2m}$

2. $\dfrac{1}{x} + \dfrac{1}{y}$

3. $1 + \dfrac{1}{p+1}$

4. $\dfrac{3}{y^2} + \dfrac{1}{y}$

5. $\dfrac{2}{xy^2} - \dfrac{1}{x^2y}$

6. $\dfrac{3x}{z^2} - \dfrac{2z}{x^2}$

7. $\dfrac{1}{p-1} - \dfrac{1}{p+1}$

8. $\dfrac{2}{w^2+w} + \dfrac{3}{4w+4}$

9. $\dfrac{1}{x+2} - \dfrac{2}{x+3}$

10. $\dfrac{2}{3x} - \dfrac{1}{6}$

11. $\dfrac{x}{24} - \dfrac{y}{30}$

12. $\dfrac{1}{ax^2} - \dfrac{2}{bx}$

13. $\dfrac{a}{b-c} + \dfrac{a}{c-b}$

14. $\dfrac{2}{xT^2} - \dfrac{1}{T}$

15. $\dfrac{1}{r} - \dfrac{1}{s} + \dfrac{1}{t}$

16. $\dfrac{x}{x-1} + \dfrac{2x}{1-x}$

17. $\dfrac{b}{l^2} - \dfrac{c}{l} + 2l$

18. $\dfrac{x-1}{x+1} - \dfrac{x+1}{x-1}$

19. $\dfrac{a-b+c}{x-y} + \dfrac{a-b-c}{y-x}$

20. $\dfrac{x}{ay} - \dfrac{2x}{3y} - \dfrac{7x}{y}$

21. $\dfrac{1}{x^5y^2} - \dfrac{1}{xy^4z}$

22. $\dfrac{1}{a^2(a+1)} - \dfrac{1}{a^5(a-1)}$

23. $\dfrac{x}{x^2-4} + \dfrac{2}{2+x}$

24. $\dfrac{1}{m^2-1} - \dfrac{1}{(m-1)^2}$

25. $\dfrac{a-b}{c-d} + \dfrac{a-b}{d-c}$

26. $\dfrac{3}{y^2} + \dfrac{1}{y-b}$

27. $\dfrac{x+a}{3x+6a} + \dfrac{x}{4a-2x}$

28. $\dfrac{r}{t+t^2} - \dfrac{1}{t(t+t^2)}$

29. $\dfrac{1}{4a^2+4a+1} + \dfrac{1-a}{1-2a}$

30. $\dfrac{3(z+1)}{z^2+5z-14} - \dfrac{1}{z-2}$

31. $\dfrac{x^2}{a^2b^2} + \dfrac{y^2}{a^2c^2} - \dfrac{z^2}{b^2c^2}$

32. $\dfrac{f-1}{f+2} + \dfrac{f-3}{f-2} - \dfrac{f^2}{f^2-4}$

33. $\dfrac{k}{k^2-5k+6} + \dfrac{2}{k-2} - \dfrac{3}{k-3}$

34. $\dfrac{r^2+s^2}{r^2-s^2} - \dfrac{r}{r+s} + \dfrac{s}{s-r}$

35. $\dfrac{1}{a+b} - \left(\dfrac{1}{a-b} - \dfrac{1}{a^2-b^2} \right)$

36. $\dfrac{2}{t^2-11t+30} - \dfrac{1}{t^2-25} - \dfrac{1}{t^2-36}$

37. $2 - \left(\dfrac{3}{x} + \dfrac{4}{x-2} \right)$

38. $a\left(\dfrac{1}{x} + \dfrac{1}{x-1}\right) - \left(\dfrac{1}{x-1} - \dfrac{1}{x-2}\right)$

39. $5y - \dfrac{4y^2 + 7y - 1}{2 + y} - 6$

40. $w^2 - \dfrac{1}{w} + \dfrac{1}{w^2}$

41. $\left(\dfrac{1}{x^2 - 3x + 2} - \dfrac{2}{x^2 - 1}\right)\left(x + \dfrac{2(2x - 1)}{x - 5}\right)$

42. $\left(x + \dfrac{1}{x} + 2\right)\left(1 - \dfrac{1}{x+1}\right)$

43. $\left(\dfrac{x}{y} + \dfrac{y}{z} + \dfrac{z}{x}\right)\left(\dfrac{xyz}{2}\right)$

44. $\left[1 - \left(\dfrac{a}{a-b}\right)^2\right]\left[1 - \dfrac{a}{b}\right]$

45. $\left(\dfrac{s}{s-r} - \dfrac{2sr}{s^2 - r^2}\right)\left(\dfrac{1}{s} - \dfrac{2}{s-r}\right)$

10.5   DIVIDING FRACTIONS

Numerical fractions are divided by using the "golden rule of fractions," which says that you can multiply a fraction by anything (except zero) *provided that you do it to both the top and the bottom* (see Section 4.7). To divide one fraction into another, multiply top and bottom by the *reciprocal* of the bottom fraction. The point of multiplying by the reciprocal of the bottom fraction is that it makes the bottom of the expression 1, and so the division problem becomes a multiplication problem. For example,

$$\dfrac{\dfrac{6}{7}}{\dfrac{5}{4}} = \dfrac{\dfrac{6}{7} \cdot \dfrac{4}{5}}{\dfrac{5}{4} \cdot \dfrac{4}{5}} = \dfrac{\dfrac{6}{7} \cdot \dfrac{4}{5}}{\dfrac{5}{4} \cdot \dfrac{4}{5}} = \dfrac{\dfrac{6 \cdot 4}{7 \cdot 5}}{1} = \dfrac{24}{35}$$

$$\left(\dfrac{4}{5} \text{ is the } reciprocal \text{ of } \dfrac{5}{4}\right)$$

The reciprocal of an algebraic fraction $\dfrac{a}{b}$, *where a and b are expressions, is* defined the same way as the reciprocal of a numerical fraction, that is, as the fraction "turned upside down," or $\dfrac{b}{a}$.

The reciprocal of an algebraic fraction has the property that when it is multiplied by the original fraction, you get 1. The "golden rule" applies equally well to algebraic fractions, and so to divide two fractions you again multiply top and bottom by the reciprocal of the bottom fraction. For example, suppose we want to divide $\dfrac{x}{x+3}$ into $\dfrac{x^2+5}{x-2}$, which means:

$$\dfrac{\dfrac{x^2+5}{x-2}}{\dfrac{x}{x+3}}$$

The reciprocal of the bottom fraction, $\dfrac{x}{x+3}$, is $\dfrac{x+3}{x}$. Therefore multiplying top and bottom by this gives:

$$\dfrac{\dfrac{x^2+5}{x-2}}{\dfrac{x}{x+3}} \cdot \dfrac{\dfrac{x+3}{x}}{\dfrac{x+3}{x}} = \dfrac{\dfrac{x^2+5}{x-2} \cdot \dfrac{x+3}{x}}{\dfrac{x}{x+3} \cdot \dfrac{x+3}{x}} = \dfrac{\dfrac{x^2+5}{x-2} \cdot \dfrac{x+3}{x}}{①}$$

Because
$\dfrac{x}{x+3} \cdot \dfrac{x+3}{x} = 1$

Notice that nothing can be cancelled here to make the multiplication simpler.

$$= \dfrac{x^2+5}{x-2} \cdot \dfrac{x+3}{x}$$

$$= \dfrac{(x^2+5)\,(x+3)}{(x-2)x}$$

EXAMPLE: *Simplify*

$$\dfrac{\dfrac{2x^2+x}{3x-2}}{\dfrac{2x+1}{6x-4}}$$

Multiply top and bottom by $\dfrac{6x-4}{2x+1}$:

$$\dfrac{\dfrac{2x^2+x}{3x-2}}{\dfrac{2x+1}{6x-4}} \cdot \dfrac{\dfrac{6x-4}{2x+1}}{\dfrac{6x-4}{2x+1}} = \dfrac{\dfrac{2x^2+x}{3x-2} \cdot \dfrac{6x-4}{2x+1}}{\dfrac{2x+1}{6x-4} \cdot \dfrac{6x-4}{2x+1}}$$

$$= \dfrac{\dfrac{2x^2+x}{3x-2} \cdot \dfrac{6x-4}{2x+1}}{1} \qquad \left(\text{since } \dfrac{2x+1}{6x-4} \cdot \dfrac{6x-4}{2x+1} = 1\right)$$

$$= \dfrac{2x^2+x}{3x-2} \cdot \dfrac{6x-4}{2x+1}$$

Now, keeping in mind Section 10.3· on multiplying and cancelling, which I hope convinced you that it is much more efficient to cancel first and then multiply, we continue as follows:

$$\frac{2x^2 + x}{3x - 2} \cdot \frac{6x - 4}{2x + 1} = \frac{x(2x + 1) \cdot 2(3x - 2)}{(3x - 2)(2x + 1)}$$

$$= 2x$$

EXAMPLE: *Simplify*

$$\frac{\dfrac{1}{x + 2}}{x + 3}$$

Think of the denominator as $\dfrac{x + 3}{1}$. Then you can apply the

usual method by multiplying top and bottom by $\dfrac{1}{x + 3}$:

$$\frac{\dfrac{1}{x + 2}}{x + 3} = \frac{\dfrac{1}{x + 2}}{\dfrac{x + 3}{1}} \cdot \frac{\dfrac{1}{x + 3}}{\dfrac{1}{x + 3}} = \frac{\dfrac{1}{x + 2} \cdot \dfrac{1}{x + 3}}{\dfrac{x + 3}{1} \cdot \dfrac{1}{x + 3}} = \frac{\dfrac{1}{x + 2} \cdot \dfrac{1}{x + 3}}{1}$$

$$= \frac{1}{(x + 2)} \cdot \frac{1}{(x + 3)} = \frac{1}{(x + 2)(x + 3)}$$

EXAMPLE: *Simplify*

$$\frac{\dfrac{x^2 + x}{x}}{x + 1}$$

Multiply top and bottom by $\dfrac{x + 1}{x}$, and think of the numerator

as $\dfrac{x^2 + x}{1}$.

$$\frac{\dfrac{x^2 + x}{x}}{x + 1} = \frac{\dfrac{\dfrac{x^2 + x}{1}}{\dfrac{x}{x + 1}} \cdot \dfrac{\dfrac{x + 1}{x}}{\dfrac{x + 1}{x}}} = \frac{\dfrac{x^2 + x}{1} \cdot \dfrac{x + 1}{x}}{1}$$

$$= \frac{x^2 + x}{1} \cdot \frac{x + 1}{x}$$

$$= \frac{x(x + 1)(x + 1)}{x} \qquad \text{(remember to factor and cancel before multiplying out)}$$

$$= (x + 1)^2$$

EXAMPLE: *Simplify*

$$\frac{\dfrac{1}{\sqrt{x^2+1}}}{\sqrt{x^2+1}}$$

Think of the denominator, $\sqrt{x^2+1}$, as $\dfrac{\sqrt{x^2+1}}{1}$; then multiply

by its reciprocal, $\dfrac{1}{\sqrt{x^2+1}}$.

$$\frac{\dfrac{1}{\sqrt{x^2+1}}}{\sqrt{x^2+1}} = \frac{\dfrac{1}{\sqrt{x^2+1}}}{\dfrac{\sqrt{x^2+1}}{1}} \cdot \frac{\dfrac{1}{\sqrt{x^2+1}}}{\dfrac{1}{\sqrt{x^2+1}}} = \frac{\dfrac{1}{\sqrt{x^2+1}} \cdot \dfrac{1}{\sqrt{x^2+1}}}{1}$$

$$= \frac{1}{\sqrt{x^2+1} \cdot \sqrt{x^2+1}} = \frac{1}{x^2+1}$$

(since, by definition of a square root,

$$\sqrt{x^2+1} \cdot \sqrt{x^2+1} = x^2+1).$$

It is of course helpful to be able to divide fractions more quickly than has been done in these examples. In practice, therefore, one does not bother to write down the multiplication by the reciprocal top and bottom. Since the net effect of division by a certain fraction is multiplication by its reciprocal, division then looks as follows:

EXAMPLE: *Simplify*

$$\frac{\dfrac{x+2}{x^2+x}}{\dfrac{x^2+4x+4}{x^2-3x}}$$

$$= \frac{x+2}{x^2+x} \cdot \frac{x^2-3x}{x^2+4x+4} \qquad \text{(division by fraction converted to multiplication by reciprocal)}$$

$$= \frac{\cancel{(x+2)}}{\cancel{x}(x+1)} \cdot \frac{\cancel{x}(x-3)}{(x+2)^{\cancel{2}}} \qquad \text{(because it pays to factor and cancel before multiplying)}$$

$$= \frac{(x-3)}{(x+1)(x+2)}$$

EXAMPLE: *Simplify*

$$\frac{\dfrac{4m^3 - m}{\pi m^2 + \pi}}{\dfrac{4m^2 - 4m + 1}{m^3 + m}} = \frac{4m^3 - m}{\pi m^2 + \pi} \cdot \frac{m^3 + m}{4m^2 - 4m + 1}$$

$$= \frac{m(4m^2 - 1)}{\pi(m^2 + 1)} \cdot \frac{m(m^2 + 1)}{(2m - 1)^2}$$

$$= \frac{m(2m + 1)\,(2m - 1) \cdot m(m^2 + 1)}{\pi(m^2 + 1)\,(2m - 1)^2}$$

$$= \frac{m^2[2m + 1]}{\pi[(2m - 1)}$$

PROBLEM SET   10.5

Simplify:

1. $\dfrac{\dfrac{-20}{3}}{\dfrac{5}{12}}$

2. $\dfrac{\dfrac{-12}{49}}{\dfrac{24}{35}}$

3. $\dfrac{\dfrac{\frac{1}{6}}{7}}{\dfrac{1}{\frac{5}{12}}}$

4. $\dfrac{\dfrac{4x^2}{y}}{\dfrac{5}{z}}$

5. $\dfrac{\dfrac{2a}{3b}}{\dfrac{14a^3}{9b^2}}$

6. $\dfrac{\dfrac{6x}{x - 3}}{\dfrac{4x}{x^2 - 9}}$

7. $\dfrac{\dfrac{r^2 s}{3t}}{\dfrac{rt}{30s}}$

8. $\dfrac{\dfrac{a^2}{b}}{\dfrac{1}{abc^3}}$

9. $\dfrac{\dfrac{3x + z}{1 - y}}{\dfrac{3x + z}{y - 1}}$

10. $\dfrac{\dfrac{a^2 bc}{c^2 db}}{\dfrac{ab^2}{cd^2}}$

11. $\dfrac{\dfrac{1}{x}}{\dfrac{2x}{x^2 + x}}$

12. $\dfrac{\dfrac{x - 1}{x + 2}}{x - 1}$

13. $\dfrac{\dfrac{-34r^3 st^2}{8x^2 y^2}}{\dfrac{-17r^2 t^2 s^3}{24xy^3}}$

14. $\dfrac{\dfrac{x}{y}}{\dfrac{x^2 + x}{y^2 - y}}$

15. $\dfrac{\dfrac{3r + 6}{4}}{\dfrac{4 + r^2 + 4r}{4 - r^2}}$

16. $\dfrac{\dfrac{2}{3x + 3y}}{\dfrac{3}{5x + 5y}}$

25. $\dfrac{\dfrac{x}{x^2 - 2x - 15}}{\dfrac{5x^2}{x + 3}}$

34. $\dfrac{\dfrac{t^3 + t + t^2 + 1}{t^3 - 1}}{\dfrac{t(t + 1)}{t(t + 1) + 1}}$

17. $\dfrac{\dfrac{6x}{x - 3}}{\dfrac{4x}{x^2 - 9}}$

26. $\dfrac{\dfrac{c - d}{c^2 + d^2}}{\dfrac{d^2 - c^2}{d^2 + cd}}$

35. $\dfrac{\dfrac{r^3 - r^2 a}{rb - r}}{\left(\dfrac{r - a}{b - 1}\right)^2}$

18. $\dfrac{\dfrac{(w + 1)^2}{w - 3}}{\dfrac{w - 3}{w + 1}}$

27. $\dfrac{\dfrac{a^2 b}{a + b}}{\dfrac{ab^2}{a^2 - b^2}}$

36. $\dfrac{\dfrac{x^3 - y^3}{x^4 - y^4}}{\dfrac{1}{y + x}}$

19. $\dfrac{\dfrac{4x}{x^2 + 2x - 8}}{\dfrac{x}{x - 2}}$

28. $\dfrac{\dfrac{1 - x^2}{x^2 - 3x - 10}}{\dfrac{x^2 - 12x + 35}{x^2 + 3x + 2}}$

37. $\dfrac{\dfrac{3a^4 - 27b^4}{6 + 6a^2}}{\dfrac{b^3 + 3b^2}{3a^2 + 3}}$

20. $\dfrac{\dfrac{(mx - nx)}{m^2 - n^2}}{4x}$

29. $\dfrac{\dfrac{3a + 3b}{a^2 - b^2}}{\dfrac{3a - 3b}{(b - a)^2}}$

38. $\dfrac{\dfrac{2x - y}{x - y} \cdot \dfrac{3x - 2y}{x^2 - 4xy}}{\dfrac{2y - 3x}{x^2 - 5xy + 4y^2}}$

21. $\dfrac{\dfrac{(4\pi^2 m^2 - n^2)}{2\pi^2 m + \pi n}}{2\pi m - n}$

30. $\dfrac{\dfrac{2by - 2b}{y^2 + 1}}{\dfrac{4b - 4y}{(y^2 + 1)^2}}$

39. $\dfrac{\dfrac{27 - r^3}{(r - 1)^2 + 3}}{\dfrac{3r^2 - 9r}{3r^2 - 6r}}$

22. $\dfrac{\dfrac{2xy^3 z}{z^2 - zx}}{4y^2 z}$

31. $\dfrac{\dfrac{4 - x^4}{3xy}}{(2 + x^2)}$

40. $\dfrac{\dfrac{4m^2 - n^4}{2m^2 + 9mn^2 - 5n^4}}{\dfrac{2m + n^2}{m^2 + 5mn^2}}$

23. $\dfrac{\dfrac{2z^2 - 10z - 28}{49 - z^2}}{\dfrac{z^2 - 6 - z}{3 - 4z + z^2}}$

32. $\dfrac{\dfrac{(a + b)^5}{4x + 2x^2}}{\dfrac{(a + b)(b + a)^5}{2x}}$

24. $\dfrac{\dfrac{6r^2 - 5rs + s^2}{4r^2 + s^2 - 4rs}}{\dfrac{9r^2 - s^2}{2rs - s^2}}$

33. $\dfrac{\dfrac{(a - b)(a + b)}{c}}{\dfrac{b - a}{c}}$

10.6   COMPLEX FRACTIONS

Complex fractions are those in which either the numerator or the denominator (or both) contains a fraction. For example,

$$\frac{\dfrac{1}{a}}{\dfrac{2}{a+1}+\dfrac{1}{a}} \quad \text{and} \quad \frac{\dfrac{x-1}{x+1}-\dfrac{x+1}{x-1}}{\dfrac{2}{x-1}-\dfrac{1}{x+1}} \quad \text{are complex fractions.}$$

Complex fractions do, indeed, look complicated, although actually they are not much worse than the usual sort. It is always possible to simplify them, and to convert them to ordinary (and uncomplex) fractions.

EXAMPLE: *Simplify*

$$\frac{\dfrac{1}{a}}{\dfrac{2}{a+1}+\dfrac{1}{a}}$$

The best method for doing this is to convert the top and bottom each into a single fraction, and then divide these two fractions. Here goes:

$$\text{Top} = \frac{1}{a}$$

this is already a single fraction.

$$\text{Bottom} = \frac{2}{a+1} + \frac{1}{a}$$

This can be made into one fraction by using the L. C. D., which is $a(a+1)$:

$$\text{Bottom} = \frac{2}{a+1} + \frac{1}{a} = \frac{2}{a+1} \cdot \frac{a}{a} + \frac{1}{a} \cdot \frac{a+1}{a+1}$$

$$= \frac{2a}{a(a+1)} + \frac{a+1}{a(a+1)}$$

$$= \frac{2a+(a+1)}{a(a+1)}$$

$$= \frac{3a+1}{a(a+1)}$$

Therefore,

$$\text{Whole fraction} = \frac{\dfrac{1}{a}}{\dfrac{2}{a+1}+\dfrac{1}{a}} = \frac{\text{top}}{\text{bottom}} = \frac{\dfrac{1}{a}}{\dfrac{3a+1}{a(a+1)}}$$

Now convert the division to multiplication by the reciprocal of the bottom fraction, and then cancel before doing the final multiplying.

$$\text{Whole fraction} = \frac{\dfrac{1}{a}}{\dfrac{3a+1}{a(a+1)}}$$

$$= \frac{1}{a}\cdot\frac{a(a+1)}{3a+1} \quad \left[\text{reciprocal of bottom is } \tfrac{a(a+1)}{3a+1}\right]$$

$$= \frac{\cancel{a}(a+1)}{\cancel{a}(3a+1)}$$

$$= \frac{a+1}{3a+1}$$

Method for Simplifying Complex Fractions

1. Add or subtract the top fractions.
2. Add or subtract the bottom fractions.
3. Put the top and bottom back together.
4. Invert the bottom fraction and multiply (do not multiply out).
5. Cancel everything possible.

EXAMPLE: *Simplify*

$$\frac{\dfrac{x-1}{x+1}-\dfrac{x+1}{x-1}}{\dfrac{2}{x-1}-\dfrac{1}{x+1}}$$

STEP 1. $\text{Top} = \dfrac{x-1}{x+1}-\dfrac{x+1}{x-1} \qquad$ has L.C.D. of $(x+1)(x-1)$

Therefore,

$$\text{Top} = \frac{x-1}{x+1}\cdot\frac{x-1}{x-1}-\frac{x+1}{x-1}\cdot\frac{x+1}{x+1}$$

$$= \frac{(x-1)^2}{(x+1)(x-1)}-\frac{(x+1)^2}{(x+1)(x-1)}$$

$$= \frac{(x-1)^2 - (x+1)^2}{(x+1)(x-1)}$$

$$= \frac{(x^2 - 2x + 1) - (x^2 + 2x + 1)}{(x+1)(x-1)}$$

$$= \frac{x^2 - 2x + 1 - x^2 - 2x - 1}{(x+1)(x-1)}$$

$$= \frac{-4x}{(x+1)(x-1)}$$

STEP 2. $\text{Bottom} = \dfrac{2}{x-1} - \dfrac{1}{x+1}$ has L.C.D. of $(x+1)(x-1)$

Therefore,

$$\text{Bottom} = \frac{2}{(x-1)} \cdot \frac{(x+1)}{(x+1)} - \frac{1}{(x+1)} \cdot \frac{(x-1)}{(x-1)}$$

$$= \frac{2(x+1) - (x-1)}{(x+1)(x-1)}$$

$$= \frac{2x + 2 - x + 1}{(x+1)(x-1)}$$

$$= \frac{x+3}{(x+1)(x-1)}$$

STEP 3. $\text{Whole fraction} = \dfrac{\text{top}}{\text{bottom}}$

$$= \frac{\dfrac{-4x}{(x+1)(x-1)}}{\dfrac{(x+3)}{(x+1)(x-1)}}$$

STEP 4. $$= \frac{-4x}{(x+1)(x-1)} \cdot \frac{(x+1)(x-1)}{(x+3)}$$ [inverting $\dfrac{(x+3)}{(x+1)(x-1)}$ and multiplying]

$$= \frac{-4x}{x+3}$$

EXAMPLE: *Simplify*

$$\frac{\dfrac{1}{s^2} - \dfrac{1}{s^2 - 1}}{2 + \dfrac{1}{s-1} - \dfrac{1}{s+1}}$$

STEP 1. $\text{Top} = \dfrac{1}{s^2} - \dfrac{1}{s^2 - 1} = \dfrac{1}{s^2} - \dfrac{1}{(s-1)(s+1)}$

$= \dfrac{(s-1)(s+1)}{s^2(s-1)(s+1)} - \dfrac{s^2}{s^2(s-1)(s+1)}$ [L.C.D. $= s^2(s-1)(s+1)$]

$= \dfrac{(s-1)(s+1) - s^2}{s^2(s-1)(s+1)}$

$= \dfrac{s^2 - 1 - s^2}{s^2(s-1)(s+1)}$

$= \dfrac{-1}{s^2(s-1)(s+1)}$

STEP 2. $\text{Bottom} = 2 + \dfrac{1}{s-1} - \dfrac{1}{s+1}$

$= \dfrac{2(s-1)(s+1)}{(s-1)(s+1)} + \dfrac{s+1}{(s-1)(s+1)} - \dfrac{s-1}{(s-1)(s+1)}$

[L.C.D. $= (s-1)(s+1)$]

$= \dfrac{2(s-1)(s+1) + (s+1) - (s-1)}{(s-1)(s+1)}$

$= \dfrac{2s^2 - 2 + s + 1 - s + 1}{(s-1)(s+1)}$

$= \dfrac{2s^2}{(s-1)(s+1)}$

STEP 3. $\text{Whole fraction} = \dfrac{\text{top}}{\text{bottom}}$

$= \dfrac{\dfrac{-1}{s^2(s-1)(s+1)}}{\dfrac{2s^2}{(s-1)(s+1)}}$

STEP 4. $= \dfrac{-1}{s^2(s-1)(s+1)} \cdot \dfrac{(s-1)(s+1)}{2s^2}$ [inverting bottom fraction and multiplying]

STEP 5. $= \dfrac{-1}{s^2 \,(\cancel{s-1})(\cancel{s+1})} \cdot \dfrac{(\cancel{s-1})(\cancel{s+1})}{2s^2}$

$= -\dfrac{1}{2s^4}$

EXAMPLE: *Simplify*

$$\dfrac{\dfrac{1}{a^2 + ab} - \dfrac{1}{ab + b^2}}{\dfrac{1}{a} - \dfrac{1}{b}}$$

$$= \frac{\dfrac{b}{ab\,(a+b)} - \dfrac{a}{ab\,(a+b)}}{\dfrac{b}{ab} - \dfrac{a}{ab}} \cdot$$

[L. C. D. top $= ab\,(a+b)$]

[L.C.D. bottom $= ab$]

$$= \frac{\dfrac{b-a}{ab(a+b)}}{\dfrac{b-a}{ab}}$$

$$= \frac{(b-a)}{ab(a+b)} \cdot \frac{ab}{(b-a)}$$

[inverting bottom fraction and multiplying]

$$= \frac{1}{a+b}$$

PROBLEM SET   10.6

Simplify each complex fraction:

1. $\dfrac{\dfrac{1}{3} + \dfrac{1}{4}}{1 - \dfrac{1}{6}}$

7. $\dfrac{\dfrac{1}{x} + 1}{1 + \dfrac{1}{x^2}}$

12. $\dfrac{t}{1 - \dfrac{1}{1-t}}$

2. $\dfrac{\dfrac{3}{2} + \dfrac{1}{5}}{1 - \left(\dfrac{3}{2}\right)\left(\dfrac{1}{5}\right)}$

8. $\dfrac{1}{1 - \dfrac{3x}{2x+2}}$

13. $\dfrac{\dfrac{2a}{a+b} - 2}{\dfrac{a}{a+b} - 1}$

3. $\dfrac{1}{\dfrac{1}{p} + \dfrac{2}{p}}$

9. $\dfrac{\dfrac{a}{b} - \dfrac{b}{c}}{\dfrac{b}{a} - \dfrac{c}{b}}$

14. $\dfrac{\dfrac{xy}{z} - 2d}{x - \dfrac{2zd}{y}}$

4. $\dfrac{3}{2 - \dfrac{4}{x}}$

10. $\dfrac{\dfrac{3y+2}{8+x}}{\dfrac{y}{x}}$ $\dfrac{}{x+8}$

15. $\dfrac{\dfrac{c}{c+d} + \dfrac{d}{c-d}}{\dfrac{c}{c-d} - \dfrac{d}{c+d}}$

5. $\dfrac{\dfrac{4}{x} - x}{3}$

11. $\dfrac{1}{\dfrac{1}{r} + \dfrac{1}{s}}$

16. $\dfrac{\dfrac{a}{x-a} + \dfrac{x}{x+a}}{\dfrac{x+a}{x} - \dfrac{2a}{x+a}}$

6. $\dfrac{1 + \dfrac{1}{a}}{1 - \dfrac{1}{a^2}}$

17. $\dfrac{\dfrac{p^2q}{r} - 2t}{p^2 - \dfrac{2rt}{q}}$

21. $\dfrac{\dfrac{c - c^2}{c^2 - 1}}{\dfrac{c}{c + 1} - c}$

25. $\dfrac{\dfrac{1}{2f} - \dfrac{4}{j}}{\dfrac{1}{f} + \dfrac{2}{3j}}$

18. $\dfrac{2 + \dfrac{a - 2}{1 - a^2}}{2 - \dfrac{3}{1 + a}}$

22. $\dfrac{\dfrac{2}{x} - \dfrac{3}{y}}{\dfrac{4}{x^2} - \dfrac{9}{y^2}}$

26. $\dfrac{1 + b - \dfrac{2}{b}}{\dfrac{6}{b^2} + \dfrac{1}{b} - 1}$

19. $\dfrac{\dfrac{x}{y} - \dfrac{x - y}{x + y}}{\dfrac{y}{x} + \dfrac{x - y}{x + y}}$

23. $\dfrac{1}{\dfrac{3x + 1}{2x - 2} - 5 + \dfrac{3}{x + 1}}$

27. $\dfrac{x + y + \dfrac{x - 2y}{2}}{x + y + \dfrac{x - 3y}{3}}$

20. $\dfrac{\dfrac{b}{3} - \dfrac{2w^2}{3b - 3w}}{b - \dfrac{w(b - 4w)}{w - b}}$

24. $\dfrac{1 - \dfrac{1}{a}}{1 - \dfrac{1}{a^2}} - \dfrac{2}{a^2 - a}$

28. $\dfrac{1 + \dfrac{2x}{x^2 + 1}}{1 - \dfrac{2(x + 2)}{x^2 + 1}}$

29. $\dfrac{\dfrac{4a^5 - 3a^3 - 20a^2b + 15b}{(a^3 - 5b)^2}}{\dfrac{3b}{a^3 - 5b} - \dfrac{5}{a}}$

34. $\dfrac{2}{1 - \dfrac{1}{a^2}} - \dfrac{3}{1 - \dfrac{1}{a}}$

30. $\dfrac{6\left(\dfrac{3x - 1}{2} - \dfrac{4x - 3}{3}\right)}{1 + \dfrac{3}{x}}$

35. $\dfrac{\dfrac{1}{x}}{2x - 1} + \dfrac{\dfrac{1}{x}}{2x + 1}$

36. $\dfrac{\dfrac{-2 - y}{2 + y} + \dfrac{2 + y}{2 - y}}{\dfrac{1}{2 + y} + \dfrac{1}{2 - y}}$

31. $\dfrac{\dfrac{2x}{y} + 1 - \dfrac{y}{x}}{\dfrac{2x}{y} + \dfrac{y}{x} - 3}$

37. $\dfrac{\dfrac{1 + m^2}{1 - m^2} - \dfrac{2 + m}{1 - m}}{2 - \dfrac{1 - m}{1 + m}}$

32. $\dfrac{\left[\dfrac{(x + y)^3}{2x(x - y)^2}\right]\left[x - \dfrac{y^2}{x}\right]}{1 + \dfrac{y}{x}}$

38. $\dfrac{1}{a - \dfrac{1}{a + \dfrac{1}{a}}}$

33. $\dfrac{1 + \dfrac{x}{y}}{\dfrac{x^2 + xy}{y} + \dfrac{x^2 + xy}{x}}$

39. $$\dfrac{\dfrac{1}{x^2 - 3x + 2} - \dfrac{2}{x^2 - 1}}{x + \dfrac{2(2x - 1)}{x - 5}}$$

43. $$\dfrac{1}{1 - \dfrac{1}{1 - \dfrac{1}{1 - x}}}$$

40. $$\dfrac{\dfrac{1}{a(a + 1)} - \dfrac{1}{a(1 - a)}}{\dfrac{1}{a^2 - 1} - \dfrac{1}{1 + a^2}}$$

44. $$\dfrac{2a^2}{5a - \dfrac{4a - 1}{1 + \dfrac{2a + 5}{3a - 2}}}$$

41. $$\dfrac{t - \dfrac{t^2 - 1}{t}}{1 - \dfrac{t - 1}{t}}$$

45. $$\dfrac{1 + \dfrac{1}{x}}{x - \dfrac{1}{2x + \dfrac{x + 1}{x}}}$$

42. $$\dfrac{2 - \dfrac{7z + 2}{z^2 - 1}}{z - 3 - \dfrac{5}{z + 1}}$$

CHAPTER 10 REVIEW

Simplify to a fraction whose numerator and denominator contain no fractions and have no common factor:

1. $\dfrac{3x}{7} \cdot \dfrac{35}{9y}$

2. $\dfrac{-3A^4}{-A}$

3. $\dfrac{\dfrac{a}{\sqrt{3}}}{\dfrac{6}{\sqrt{2}a}}$

4. $\dfrac{(x - 4)(2x + 9)}{5} \cdot \dfrac{x + 4}{-(2x - 8)}$

5. $6 \cdot \dfrac{(b - 7)}{(b + 1)} \cdot \dfrac{1 - b^2}{b^2 - 14b + 49}$

6. $\dfrac{ABD + ABC}{ACD + AD^2}$

7. $\dfrac{14E - 7}{4E^2 - 1}$

8. $\dfrac{60 - 3x - 3x^2}{3x^2 - 48}$

9. $\dfrac{5x + 2}{10x^2 - x - 2}$

10. $\dfrac{6R^2 + 5s(R - 4s)}{2R - 5s}$

11. $\dfrac{x - y}{ax - ay - x + y}$

12. $\dfrac{r^2 - 8r + 12}{4 + r^2 - 4r}$

13. $\dfrac{a^2 + 2ab + b^2}{a - b} \cdot \dfrac{a^2 - 2ab + b^2}{a + b}$

14. $\dfrac{2x^2 - 6x + 4}{x^2 - 1} \cdot \dfrac{1 + x}{2 - x}$

15. $\dfrac{(a + b)^2 - c^2}{a + b - c}$

16. $\dfrac{x^4 + x^3}{y^2x + y^2}$

Perform the indicated operation:

17. $\dfrac{1}{x+4} - \dfrac{3}{x+2}$

18. $\dfrac{8x^2 - x}{6x^2 - x - 15} + \dfrac{10 + x}{5 - 3x} + \dfrac{3 - 2x}{3 + 2x}$

19. $\dfrac{9}{x+3} - \dfrac{8}{x-2} + 1$

20. $\dfrac{2}{z - z^2} - \dfrac{1}{z - z^3}$

21. $\dfrac{2}{a-1} + \dfrac{3}{a+1} - \dfrac{7}{4}$

22. $1 + \dfrac{2x}{2x-1} - \dfrac{8x^2}{4x^2 - 1}$

23. $\dfrac{x+1}{x^2 - 5x + 6} - \dfrac{3}{x-2}$

24. $3p - \dfrac{p^2}{2p+3} - 2$

25. $\dfrac{y}{x+y} - \dfrac{xy}{(x+y)^2} + \dfrac{xy^2}{(x+y)^3}$

26. $\dfrac{r}{r^2 - 7r + 12} - \dfrac{3}{4-r} + \dfrac{4}{r-3}$

Simplify:

27. $-\dfrac{(2-x)(x+2)}{(2+x)(x-3)}$

28. $\dfrac{\dfrac{a^2}{5} - 5}{1 + \dfrac{a}{5}}$

29. $\dfrac{\dfrac{x-y}{a+b}}{\dfrac{y-x}{a}}$

30. $\dfrac{\dfrac{f^2 - 9}{f^2 - 6f + 9}}{f - 3}$

31. $\dfrac{1 + A + CD + FA}{FA + A + CD + 1}$

32. $\dfrac{\dfrac{2b^2 - 7b}{15b - 6}}{\dfrac{4b^2 - 28b + 49}{5b^2 + 3b - 2}}$

33. $\left(\dfrac{x+y}{x} - 1\right)\left(\dfrac{x}{y} + 1\right)$

34. $\dfrac{\dfrac{2}{x} - \dfrac{x}{2}}{\dfrac{x}{2} - 4 + \dfrac{6}{x}}$

35. $\dfrac{\dfrac{1}{a} + b}{1 - ab} - \dfrac{\dfrac{1}{a} - b}{1 + ab}$

36. $\dfrac{\dfrac{C+D}{CD-1} - \dfrac{1}{D}}{1 + \dfrac{C+D}{D(CD-1)}}$

37. $\dfrac{\left(s + \dfrac{1}{s+1} - 1\right)\left(s + \dfrac{1}{s} + 2\right)}{s - \dfrac{1}{s}}$

38. $\dfrac{\dfrac{2}{q} - \dfrac{1}{q+1} + \dfrac{1}{1-q}}{\dfrac{1}{(q-1)} - \dfrac{1}{(q+1)}}$

39. $\dfrac{\dfrac{A}{3} - \dfrac{2x^2}{3A - 3x}}{A - \dfrac{x(A - 4x)}{x - A}}$

40. $\dfrac{\dfrac{B}{3} - \dfrac{2}{B+1}}{2 + \dfrac{3(B-1)}{B(B+1)}}$

41. $\dfrac{\dfrac{1}{x^2} + \dfrac{1}{x}}{1 + \dfrac{1}{x^3}}$

42. $-\dfrac{-\dfrac{A^2}{x^2}}{1 - \dfrac{A^2}{x^2}}$

43. $\dfrac{3x^2 - 3}{2x^2 + 4x - 6}$

44. $\dfrac{\dfrac{2x}{y + 5} - 1}{1 - \dfrac{2x}{y + 5}}$

45. $\dfrac{-[-(a - 5)]}{25 - a^2}$

46. $\dfrac{\dfrac{1 + \dfrac{4}{ac}}{a^2 c} + c}{a - \dfrac{1 - \dfrac{4}{c}}{ac}}$

47. $\dfrac{1 - \dfrac{1}{4A}}{1 + 4A} + \dfrac{1 + \dfrac{1}{4A}}{1 - 4A}$

48. $\dfrac{n + 2 + \dfrac{3n + 6}{n - 1}}{1 - \left(\dfrac{3}{n - 1}\right)^2}$

49. $1 - \dfrac{1}{R + \dfrac{1}{R - \dfrac{1}{R}}}$

50. $\dfrac{\dfrac{x - 2y}{4} + 2(x + y) - 1}{\dfrac{x - 2y}{4} + 2(x + y) + 1}$

51. $\dfrac{AX + BY + BX + AY}{AX + CZ + CX + AZ} \cdot \dfrac{A^2 + BC + AB + AC}{X^2 + YZ + XY + XZ}$

Show by simplifying each pair of expressions whether or not they are equal to one another:

52. $\dfrac{4 - \dfrac{2}{B + 1} - 1}{8 - \dfrac{4}{B + 1} + 1}$ and $\dfrac{1}{2} - \dfrac{2}{B + 1}$

53. $\dfrac{1}{\left(U - \dfrac{1}{k}\right)\left(V - \dfrac{1}{k}\right)}$ and k^2, where $k = \dfrac{1}{U} + \dfrac{1}{V}$

54. $\left(\dfrac{Y}{X} - \dfrac{X}{Y}\right)\left(\dfrac{AX + BY}{X + Y} - \dfrac{AX - BY}{X - Y}\right)$ and $2(A - B)$

Simplify:

55. $\dfrac{x-1}{x+4} - \dfrac{x^3 - 5x^2 + 20x - 16}{x^3 + 64}$

56. $\dfrac{x + 2y}{x^4 - y^2} \cdot \dfrac{x^2 - y}{x^3 + 2x^2 y - xy - 2y^2}$

Divide:

57. $(y^2 - 3.3y + 2)$ by $(0.5y - 0.4)$

58. $A^{2n} + Y^n$ into $A^{4n} - Y^{2n}$

59. If

$$A = \frac{2t}{1 + t^2} \quad \text{and} \quad B = \frac{1 - t^2}{1 + t^2}$$

show by simplifying that the value of $(A^2 + B^2)$ does not contain t.

60. Show that if

$$q = 1 - \frac{1}{r} \quad \text{and} \quad r = 1 + \frac{1}{s - 1}$$

then q is the reciprocal of s.

61. Simplify:

$$\left(1 + \frac{1}{n}\right)\left(1 + \frac{1}{n + 1}\right)\left(1 + \frac{1}{n + 2}\right)\left(1 + \frac{1}{n + 3}\right)$$

If the product were continued in the same pattern, what would be the kth factor? The product of the k factors?

62. Simplify:

(a) $\dfrac{a(a-1)\,(a-2)\,\cdots\,(a-n)}{2(2a-2)\,(2a-4)\,\cdots\,(2a-2n)}$

(b) $\dfrac{(a-1)\,(a-2)\,(a-3)\,\cdots\,(a-n)}{(a-1)\,(2a-4)\,(3a-9)\,\cdots\,(na-n^2)}$

11 EXPRESSIONS VS. EQUATIONS VS. IDENTITIES

11.1 THE DIFFERENCE BETWEEN EXPRESSIONS, EQUATIONS, AND IDENTITIES

At this point there is a big break because we move from expressions to *equations*. An *expression* is a bunch of letters and numbers connected by addition, subtraction, multiplication and division signs. It takes on different numerical values as the letters are assigned different values, but the important thing is that *an expression represents a number*. An *equation,* on the other hand, *expresses the equality between two expressions*. An expression therefore does not contain an equals sign, whereas an equation contains an equals sign connecting two expressions. For example,

$$\frac{2x^2 + 3x - 5}{x + 9} \text{ is an expression}$$

and

$$\frac{2x^2 + 3x - 5}{x + 9} = \frac{x^2 - 5}{2} \text{ is an equation.}$$

When working on an expression, you usually want to convert it from one form to some other more convenient form, but you always want to preserve the (numerical) value of the expression. For example, when you factor

$$x^2 + 3x \quad \text{into} \quad x(x + 3)$$

you are converting $x^2 + 3x$ into another expression that is equal to it in value, since, for any particular value of x, $x^2 + 3x$ and $x(x + 3)$ represent the same number.

Thus, *when working with an* expression, *you can do anything you like, provided that it doesn't change the value of the expression.* One example of this is the "golden rule of fractions," which says that you can multiply a fraction by any quantity (except zero), provided that you do it to both top and bottom. The reason that this is O.K. is that multiplying top and bottom by the same thing doesn't change the value of a fraction.

On the other hand, when you are working with an equation, the particular expressions on either side of the equals sign are usually of little importance to you, but what *is* of importance is the fact that whatever is on the left is equal to whatever is on the right.

Thus, *when working with an* equation, *you can do anything to the equation, provided that you do it to both sides.* This might be called the "golden rule of equations." Actually, there is one exception to this rule, namely, that you can't multiply or divide by zero, even if you do do it to both sides. Stated more precisely:

The Golden Rule of Equations

You can add anything to (or subtract anything from) an equation, *provided that you do it to both sides,* and you can multiply or divide by anything (except zero), *provided that you do it to both sides.*

There is one possible confusion about equals signs. I said above that an equation contains an equals sign but an expression does not. However, when replacing one expression by another of the same value, we often write them connected by an equals sign (to show that they have the same value). For example, when factoring $x^2 + 3x$, one writes:

$$x^2 + 3x = x(x + 3)$$

Hence, when we are working with an expression, equals signs do appear. But there is still a difference between

$$x^2 + 3x = x(x + 3)$$

which says that two expressions are equal because one is the factored form of the other, and

$$2x + 3 = x + 7$$

which is an honest-to-goodness equation. The difference is that $x^2 + 3x$ and $x(x + 3)$ have the same numerical value for *every* value of the variable x. In other words $x^2 + 3x$ and $x(x + 3)$ are simply two different forms of exactly the same expression. On the other hand,

$$2x + 3 \quad \text{and} \quad x + 7$$

are different expressions, which will be equal in value only for certain particular values of the variable x (in this case, for $x = 4$). Consequently, writing

$$x^2 + 3x = x(x + 3)$$

means that $x^2 + 3x$ and $x(x + 3)$ are equal for all values of x, whereas writing

$$2x + 3 = x + 7$$

means that $2x + 3$ and $x + 7$ are equal only when x takes on some particular value (usually a value we're trying to find).

Now,

$$x^2 + 3x = x(x + 3)$$

is called an *identity*, rather than an equation, to emphasize that it states that two expressions are always identical in value. If you're being proper you write $\equiv$ instead of $=$ in an identity. The name *equation* is reserved for something like

$$2x + 3 = x + 7$$

which is not an identity.

Summarizing:

An *expression* is a collection of algebraic symbols representing a number.

An *identity* expresses the equality of two expressions that holds for all values of the variable.

An *equation* expresses the equality of two expressions that holds only for certain values of the variable.

11.2   WHAT YOU DO WITH EXPRESSIONS, EQUATIONS, AND IDENTITIES

We've already done a good deal with expressions—we've factored them, added them, simplified them, and so on. All of these operations come under the heading of *manipulating expressions*, which means converting them from one form to another, without changing their value.

When faced with an identity, what you usually have to do is verify or prove it. *Verifying an identity* means using whatever manipulations you need to show that the expressions on either side of the equals sign are in fact the same.

Now, an equation contains two expressions that instead of being equal for all values of the variable are equal for only a few special values. So, in the case of an equation, what you want to know is what values of the variable, or *unknown*, make the expressions equal.

Finding these values is called *solving the equation*. The values of the variable that you find are called the *roots of the equation* or the *solutions to the equation*. They are said to *satisfy the equation*. The method you use to find the roots of an equation depends on what kind of equation it is; the next chapters describe methods used in the three most common kinds of equations: linear, quadratic, and simultaneous.

12   LINEAR EQUATIONS

12.1   SIMPLE LINEAR EQUATIONS

For the moment (the next four chapters to be precise), we will look at equations that have only one variable or unknown.

$$2x^2 - 9x = 5 + 2x$$

is an equation in one variable, or one unknown, and

$$4xy = 2x^2 - y^2 + 10$$

is an equation in two variables or two unknowns.

   This section is about linear equations in one unknown. Such equations are rigorously defined as being those that can be transformed (using simplification and the "golden rule of equations") into the form:

$$ax + b = 0 \qquad \text{where } a \text{ and } b \text{ are constants,}$$

$$a \neq 0, \text{ and}$$

$$x \text{ is the unknown.}$$

The problem with this definition is that it doesn't help you in practice, because when you actually are given a linear equation it usually isn't in this form. And by the time you have transformed it into this form and can recognize it, you've essentially solved the equation, and so it's a bit late. A looser definition for a linear equation, but one that will enable you to recognize one when you meet one, is as follows:

> An equation in one variable that contains the first but no higher power of the unknown, no fractions with the unknown in the denominator, and no roots of the unknown, is practically always a *linear equation*.

Therefore

$$2x - 5 = 6(x - 2)$$

and

$$\frac{2 - x}{3} = 4(x - 9) + \frac{1}{4}$$

are linear equations, while

$$1 - 2x = 4x^2 - 3x$$

$$\frac{1 + 2x}{3 - x^3} = 2x + 1$$

$$\sqrt{x} = 4 + x$$

are not. Linear equations are also called *first-degree equations* because they involve polynomials of the first degree.

Solving a linear equation means finding what value(s) of x (if any) make the two sides equal to one another. It turns out that there is always exactly one such value, and so a linear equation always has a *unique solution*. The solution is found by successively transforming the equation, always keeping the sides equal to one another, so that all the x's (or whatever the unknown is called) end up on one side of the equation, and eventually you get

$$x = \text{some number (or some number} = x)$$

This number is then the solution.

To transform an equation into this form, you follow the "golden rule of equations", which says that you may add or subtract anything on both sides of an equation and that you may multiply or divide both sides of an equation by anything except zero.

EXAMPLE: *Solve $x + 3 = 8$.*

Suppose we are trying to get the x's on the left, and the numbers on the right. Then all we need do is get rid of that 3 on the left—which we do by subtracting 3 from both sides of the equation.

$$x + 3 = 8$$

so

$$x + 3 - 3 = 8 - 3$$

The left-hand side now reduces to x (since $x + 3 - 3$ is $x + 0$ or x), and the right-hand side is 5 (since $8 - 3$ is 5), so the equation has become

$$x = 5$$

This is of the required form, so 5 is the solution to the original equation. In fact, we will usually refer to the final equation, $x = 5$, as the solution to the original equation.

CHECK: It is *always* a good idea to check that the solution you have found to an equation is in fact the right one. This is done by substituting the value you have found for x in each side of the equation and working out what the expressions on each side of the equation come to separately. Since we are looking for the value of x that makes the expressions on each side equal, if you get the same number when you substitute x into the left-hand side as when you substitute into the right-hand side, then you have the right x. In this case, when $x = 5$, the left-hand side $= x + 3 = 5 + 3 = 8$, and the right-hand side $= 8$; therefore 5 is the correct solution.

EXAMPLE: *Solve $2x + 3 = 9$.*

Here we will again try to collect the x's on the left side and the numbers on the right. First we will remove the 3 from the left-hand side by subtracting 3 from both sides, exactly as in the example above. If

$$2x + 3 = 9$$

then

$$2x + 3 - 3 = 9 - 3$$

The left-hand side reduces to $2x$ and the right-hand side is 6, so the equation now reads:

$$2x = 6$$

Now we want to get rid of the 2 which is multiplying the x. To do that we divide the equation by 2 (or, equivalently, multiply by $\frac{1}{2}$), which we can do provided that we do it to both sides. Therefore

$$2x = 6$$

divided by 2 gives

$$\frac{2x}{2} = \frac{6}{2}$$

The left-hand side reduces to x (since $\frac{\cancel{2}x}{\cancel{2}} = x$), and the right-hand side is 3. Therefore the equation now reads

$$x = 3$$

and so $x = 3$ is the solution to the original equation.

CHECK: If $x = 3$,

LHS = left-hand side = $2x + 3 = 2 \cdot 3 + 3 = 6 + 3 = 9$

RHS = right-hand side = 9

Therefore $x = 3$ is the correct solution.

EXAMPLE: *Solve $\frac{5}{6} = x - \frac{2}{3}$*

Here we will try and get all the x's on the right, since there are none on the left to start with. So we would like to remove the $-\frac{2}{3}$ on the right, which can be done by adding $\frac{2}{3}$ to both sides of the equation. If

$$\frac{5}{6} = x - \frac{2}{3}$$

then

$$\frac{5}{6} + \frac{2}{3} = x - \frac{2}{3} + \frac{2}{3}$$

Adding the fractions on the left-hand side of the equation gives $\frac{9}{6}$ or $\frac{3}{2}$, and the right-hand side reduces to x. Therefore the equation now reads

$$\frac{3}{2} = x$$

which is the same as

$$x = \frac{3}{2}$$

Therefore $x = \frac{3}{2}$ is the solution to the original equation.

CHECK: If $x = \frac{3}{2}$,

$$\text{LHS} = \text{left-hand side} = \tfrac{5}{6}$$

$$\text{RHS} = \text{right-hand side} = x - \tfrac{2}{3} = \tfrac{3}{2} - \tfrac{2}{3}$$

$$= \tfrac{9}{6} - \tfrac{4}{6} \qquad (\text{L.C.D.} = 6)$$

$$= \frac{9-4}{6}$$

$$= \tfrac{5}{6}$$

Therefore $x = \tfrac{3}{2}$ is the correct solution.

Alternative Method: Clearing of Fractions

In any equation involving fractions (be they numerical or algebraic), you can get rid of all the fractions right at the start by multiplying both sides of the equation by the L.C.D. of all the fractions occuring in the equation. Looking at

$$\frac{5}{6} = x - \frac{2}{3}$$

you see that the fractions occuring are $\tfrac{5}{6}$ and $\tfrac{2}{3}$, which have denominators of 6 and 3. Their L.C.D. is 6, and so we multiply both sides of the equation by 6 to get rid of the denominators. Now

$$\frac{5}{6} = x - \frac{2}{3}$$

multiplied by 6 gives

$$\left(\frac{5}{6}\right) \cdot 6 = \left(x - \frac{2}{3}\right) \cdot 6$$

The left-hand side becomes $\dfrac{5}{\cancel{6}} \cdot \cancel{6} = 5$. The right-hand side must be expanded using the distributive law:

$$\left(x - \frac{2}{3}\right) \cdot 6 = 6x - \frac{\overset{2}{\cancel{6}}}{1} \cdot \frac{2}{\underset{1}{\cancel{3}}} = 6x - 4$$

So the equation now reads

$$5 = 6x - 4$$

Thus, we have got rid of all the fractions, and are left with an equation that is exactly the same type as the last one, $2x + 3 = 9$. First we will get rid of the

−4 on the right by adding 4 to both sides; giving:

$$5 + 4 = 6x - 4 + 4$$

which reduces to

$$9 = 6x$$

To remove the 6 which is multiplying the x, divide both sides of the equation by 6 (or multiply by $\frac{1}{6}$, which comes to the same thing). If

$$9 = 6x$$

then

$$\frac{9}{6} = \frac{6x}{6}$$

Since $\dfrac{9}{6} = \dfrac{3}{2}$,   and $\dfrac{6x}{6} = x$, this reduces to

$$\frac{3}{2} = x$$

Therefore, as before, the solution is $x = \frac{3}{2}$.

Note on Clearing of Fractions in Equations.   In an equation involving fractions, use whichever of these two methods seems easiest to you; it really doesn't matter.

This method for solving equations by clearing of fractions underlines the difference between what you can do to expressions and what you can do to equations. In the example above we multiplied the equation by 6, which has the effect of removing all the denominators of all the fractions. If you were working with an expression you certainly couldn't multiply by 6, because that would change the value of the expression (it would multiply the value by 6), and so in general you can't possibly get rid of the denominators. The only thing you can do with a fractional expression is to multiply both top and bottom by 6, but that won't get rid of the denominator. Similarly, you can't add things to an expression without changing its value, but you can add things to an equation—provided that you do it to both sides.

EXAMPLE: *Solve for y*

$$4 = \frac{5}{4}y - \frac{1}{2}$$

The fact that the unknown in this equation is called y makes no difference—you solve for it exactly as you would for x.

We will get all the y's on the right, and so will first add $\frac{1}{2}$ to both sides:

$$4 = \frac{5}{4}y - \frac{1}{2}$$

$$4 + \frac{1}{2} = \frac{5}{4}y - \frac{1}{2} + \frac{1}{2}$$

which reduces to

$$\frac{8}{2} + \frac{1}{2} = \frac{5}{4}y$$

or

$$\frac{9}{2} = \frac{5}{4}y$$

To remove the $\frac{5}{4}$, the coefficient of y, we can either divide both sides by $\frac{5}{4}$, or multiply both sides by $\frac{4}{5}$, which comes to the same thing.

$$\frac{9}{2} = \frac{5}{4}y$$

gives

$$\frac{4}{5} \cdot \frac{9}{2} = \frac{4}{5} \cdot \frac{5}{4}y$$

Now since $\frac{\cancel{4}}{5} \cdot \frac{9}{\cancel{2}} = \frac{18}{5}$, and $\frac{\cancel{4}}{\cancel{5}} \cdot \frac{\cancel{5}}{\cancel{4}}y = y$, the equation reduces to

$$\frac{18}{5} = y$$

Therefore $y = \frac{18}{5}$ is the solution.

CHECK: If $y = \frac{18}{5}$

LHS $= 4$

RHS $= \frac{5}{4}y - \frac{1}{2} = \frac{5}{4} \cdot \frac{18}{5} - \frac{1}{2}$

$$= \frac{9}{2} - \frac{1}{2} = \frac{9-1}{2}$$

$$= \frac{8}{2} = 4$$

Therefore $y = \frac{18}{5}$ is correct.

EXAMPLE: *Solve 2a + 9 = 3a − 2 for a.*

Let's get the *a*'s on the right (because there are more there already) and the numbers on the left—although if you do it the opposite way it won't make any difference.

First we get rid of the −2 on the right by adding 2 to both sides:

$$2a + 9 = 3a − 2$$

giving

$$2a + 9 + 2 = 3a − 2 + 2$$

This reduces to

$$2a + 11 = 3a$$

Now remove the 2*a* on the left by subtracting 2*a* from each side:

$$2a + 11 − 2a = 3a − 2a$$

which, since $3a − 2a = a$, reduces to

$$11 = a$$

Therefore $a = 11$ is the solution.

CHECK: When $a = 11$;

$$\text{LHS} = 2a + 9 = 2 \cdot 11 + 9 = 22 + 9 = 31$$

$$\text{RHS} = 3a − 2 = 3 \cdot 11 − 2 = 33 − 2 = 31$$

Therefore $a = 11$ is correct.

EXAMPLE: *Solve 0.2u + 0.7 = 1.7 − 0.3u for u.*

We will get the *u*'s on the left. First add 0.3*u* to both sides giving

$$0.2u + 0.7 + 0.3u = 1.7 − 0.3u + 0.3u$$

And since $0.2u + 0.3u = 0.5u$, this reduces to

$$0.5u + 0.7 = 1.7$$

Then subtract 0.7 from both sides, giving

$$0.5u + 0.7 − 0.7 = 1.7 − 0.7$$

or

$$0.5u = 1.0$$

To remove the coefficient of 0.5, divide both sides by 0.5:

$$\frac{0.5u}{0.5} = \frac{1.0}{0.5}$$

which, since $\frac{0.5u}{0.5} = u$ and $\frac{1.0}{0.5} = 2.0$, reduces to

$$u = 2.0 \quad \text{or} \quad 2$$

Therefore $u = 2$ is the solution.

CHECK: When $u = 2$,

$$\text{LHS} = 0.2u + 0.7 = 0.2(2) + 0.7 = 0.4 + 0.7 = 1.1$$

$$\text{RHS} = 1.7 - 0.3u = 1.7 - 0.3(2) = 1.7 - 0.6 = 1.1$$

Therefore $u = 2$ is correct.

So far all the examples we have done have involved nothing except adding, subtracting, multiplying, and dividing both sides of the equation by some suitably chosen number.

Equations involving parentheses, however, often make it necessary to use the distributive law as well.

EXAMPLE: *Solve $2(x - 5) = 3x + 1$.*

With this equation, if we start trying to get all the x's on one side and all the numbers on the other, we will run into trouble. The parentheses are there in order to keep the $(x - 5)$ together, and so there seems to be a problem in trying to get the 5 on one side and the x on the other. However, if we use the distributive law on the left-hand side of the equation, we can remove the parentheses, and hence the problem. Remember that the distributive law says that $2(x - 5) = 2x - 10$. Therefore, an equation that starts as

$$2(x - 5) = 3x + 1$$

is equivalent to

$$2x - 10 = 3x + 1$$

and this is a type of equation that we have solved before. Subtract $3x$ from both sides, and we get:

$$2x - 10 - 3x = 3x + 1 - 3x$$

or

$$-x - 10 = 1$$

Then add 10 to both sides, giving:

$$-x - 10 + 10 = 1 + 10$$

or

$$-x = 11$$

To get rid of the minus sign in front of the x we have to multiply both sides of the equation by -1, giving:

$$(-1)(-x) = (-1)\,11$$

or

$$x = -11$$

Therefore $x = -11$ is the solution.

CHECK: When $x = -11$,

LHS $= 2(x - 5) = 2(-11 - 5) = 2(-16) = -32$

RHS $= 3x + 1 = 3(-11) + 1 = -33 + 1 = -32$

Therefore $x = -11$ is correct.

EXAMPLE: *Solve for t:*

$$\frac{1}{2}(t + 4) + 1 = 4 - (t - 1)$$

In order to be able to separate out the t's and the numbers, we first have to use the distributive law to get rid of the parentheses. Now:

$$\frac{1}{2}(t + 4) = \frac{1}{2} \cdot t + \frac{1}{2} \cdot 4 = \frac{t}{2} + 2 \quad \text{and} \quad -(t - 1) = -t + 1$$

Therefore our equation

$$\frac{1}{2}(t + 4) + 1 = 4 - (t - 1)$$

becomes

$$\frac{t}{2} + 2 + 1 = 4 - t + 1$$

or

$$\frac{t}{2} + 3 = 5 - t$$

Add t to both sides to remove the $-t$ on the right:

$$\frac{t}{2} + 3 + t = 5 - t + t$$

Since $\frac{t}{2} + t = \frac{1}{2} \cdot t + 1 \cdot t = \frac{3}{2}t$, this reduces to

$$\frac{3}{2}t + 3 = 5$$

Subtract 3 from both sides to get:

$$\frac{3}{2}t + 3 - 3 = 5 - 3$$

or

$$\frac{3}{2}t = 2$$

To get rid of the $\frac{3}{2}$ coefficient, multiply by $\frac{2}{3}$:

$$\frac{\cancel{2}}{\cancel{3}} \cdot \frac{\cancel{3}}{\cancel{2}}t = \frac{2}{3} \cdot 2$$

giving

$$t = \frac{4}{3}$$

So the solution is $t = \frac{4}{3}$.

CHECK: When $t = \frac{4}{3}$,

$$\text{LHS} = \frac{1}{2}(t + 4) + 1 = \frac{1}{2}\left(\frac{4}{3} + 4\right) + 1 = \frac{1}{2}\left(\frac{4}{3} + \frac{12}{3}\right) + 1$$

$$= \frac{1}{2} \cdot \frac{16}{3} + 1 = \frac{8}{3} + 1 = \frac{11}{3}$$

$$\text{RHS} = 4 - (t - 1) = 4 - \left(\frac{4}{3} - 1\right) = 4 - \frac{1}{3} = \frac{12}{3} - \frac{1}{3} = \frac{11}{3}$$

and $t = \frac{4}{3}$ is correct.

Alternative Method: Clearing of Fractions.

We could solve

$$\frac{1}{2}(t + 4) + 1 = 4 - (t - 1)$$

by first clearing of fractions, by multiplying both sides by 2:

$$2\left[\frac{1}{2}(t+4)+1\right] = 2[4-(t-1)]$$

and then using the distributive law to expand:

$$\not{2} \cdot \frac{1}{\not{2}}(t+4) + 2 \cdot 1 = 2 \cdot 4 - 2 \cdot (t-1)$$

or

$$(t+4) + 2 = 8 - 2(t-1)$$

(Note: It is *very* important—quite crucial in fact—that the square brackets are inserted when you multiply by the 2, to show that the 2 multiplies *everything* on both sides. Without them the equation would read $\not{2} \cdot \frac{1}{\not{2}}(t+4) + 1$ $= 2 \cdot 4 - (t-1)$ or $(t+4) + 1 = 8 - (t-1)$, which is quite different— and quite wrong—because the 1 and the $(t-1)$ have not been multiplied by 2.)

Now

$$(t+4) + 2 = 8 - 2(t-1)$$

is easily solved by using the distributive law to remove the parentheses, giving

$$t + 4 + 2 = 8 - 2t + 2$$

or

$$t + 6 = 10 - 2t$$

and then adding $2t$ to both sides:

$$t + 6 + 2t = 10 - 2t + 2t$$

or

$$3t + 6 = 10$$

Now subtract 6 from each side:

$$3t = 4$$

and divide both sides by 3:

$$\frac{3t}{3} = \frac{4}{3}$$

giving

$$t = \frac{4}{3}$$

EXAMPLE: *Solve*

$$\frac{5-x}{3} + 2 = 2(x-1) - 5(1+x)$$

The point of including this example is to show one solution in which fewer of the steps are included and which therefore looks shorter.

Use the distributive law:

$$\frac{5}{3} - \frac{x}{3} + 2 = 2x - 2 - 5 - 5x$$

$$\left(\text{Note: } \left(\frac{5-x}{3}\right) = \frac{1}{3}(5-x) = \frac{1}{3} \cdot 5 - \frac{1}{3} \cdot x = \frac{5}{3} - \frac{x}{3} \right)$$

Therefore

$$\frac{11}{3} - \frac{x}{3} = -3x - 7 \qquad \left(\frac{5}{3} + 2 = \frac{11}{3}\right)$$

Clear of fractions by multiplying everything by 3 (and it has to be *everything*):

$$11 - x = -9x - 21$$

Add $9x$ to both sides:

$$11 + 8x = -21$$

Subtract 11 from both sides:

$$8x = -32$$

Divide by 8 on both sides:

$$x = -\frac{32}{8} = -4$$

CHECK: When $x = -4$,

$$\text{LHS} = \frac{5-x}{3} + 2 = \frac{5-(-4)}{3} + 2 = \frac{5+4}{3} + 2 = \frac{9}{3} + 2$$

$$= 3 + 2 = 5$$

$$\text{RHS} = 2(x-1) - 5(1+x) = 2(-4-1) - 5(1-4)$$

$$= 2(-5) - 5(-3) = -10 + 15 = 5$$

Therefore $x = -4$ is correct.

PROBLEM SET   12.1

Solve the following equations:

1. $x + 11 = 8$

2. $\dfrac{8}{5} = \dfrac{2}{5} - y$

3. $0.2(x + 6) = 8x$

4. $x + 3 = 7$

5. $2x = 4x - 5$

6. $3p + 2 = 2p - 5$

7. $5(a - 1) = 1$

8. $2z - 1 = -3(z - 1)$

9. $2b + 5 = 7b - 2$

10. $10\left(z + \dfrac{1}{3}\right) = \dfrac{2}{8}$

11. $\dfrac{2x + 5}{5} - \dfrac{x}{5} = 0$

12. $\dfrac{2}{3}(x + 1) = 1 - (x + 1)$

13. $0.01(d - 0.01) = 1.01d - 1.0001$

14. $0.8(12 - h) = 0.4h$

15. $3a - 2 = \dfrac{5a + 2}{3}$

16. $y - 3 = \dfrac{4(y - 2)}{5}$

17. $-3(2x - 1) = -12(x + 1)$

18. $5 = \dfrac{x}{4} - \dfrac{x}{3}$

19. $2t + 7 = 3(t - 2)$

20. $\dfrac{s + 2}{6} + \dfrac{s - 3}{10} = \dfrac{s + 3}{6}$

21. $3(0.31x - 11) = 0.91x + 7$

22. $\dfrac{x + \frac{1}{2}}{0.1} = \dfrac{x + 0.1}{\frac{1}{2}}$

23. $w - 4 = \dfrac{w - 2}{3} + \dfrac{w + 3}{2}$

24. $2.2(y + 1) - 4.5y = 1.8y + 20$

25. $\dfrac{(3x + 1)}{2} = \dfrac{x}{3} + 2$

26. $5(0.04 - 2.71x) = \dfrac{1}{2}(3x + 0.52)$

27. $10(x - 2) - x + 2 - 2(x + 1) = 3(3 + x) - 9$

28. $-2[-5 - (x + 3)] = -1 - [x - (-10)]$

29. $4.7 - (0.3 + x) = -10\,[-1.1x - (0.1 - x)]$

30. $5(2b^2 - b) = 7 - 2(6 - 5b^2)$

31. $5(4p - 3) + 3(2p + 2) = 8p$

32. $\dfrac{2t - 4}{4} + 5 = t - (3 + t)$

33. $\dfrac{4}{5}(3m + 1) + m - \dfrac{3}{2} + \dfrac{2}{6}(m + 1) = 2$

34. $-3 = [-3(-5y + 1) + 2y] - 3[4 + (2y - 1) - y]$

12.2   DISGUISED LINEAR EQUATIONS: FRACTIONAL EQUATIONS

Suppose we have an equation like

$$\frac{(3 + x)}{x} = 2$$

This certainly doesn't look linear because it contains a fraction with an x in the denominator. But if you multiply both sides of the equation by x, to clear of fractions, you will see that it turns into a linear equation:

$$\frac{(3 + x)}{\cancel{x}} \cdot \cancel{x} = 2 \cdot x$$

or

$$3 + x = 2x$$

Since this is a perfectly ordinary linear equation, we can solve it by the usual methods. We find that $x = 3$ is the solution to

$$\frac{(3 + x)}{x} = 2$$

This is called a *fractional equation*, because it contains fractions with an x in the denominator. (In contrast, $\dfrac{3 + x}{5} = 2$ is not usually called a fractional equation because the denominator does not contain an x.)

In solving

$$\frac{(3 + x)}{x} = 2,$$

we first converted it to an equation containing no fractions; this equation turned out to be linear and something that we could easily solve. In general, the best way of dealing with a fractional equation is to convert it to one involving no fractions, and then look at the equation you've got and hope it fits into one of the types you already know how to solve.

Method for Attacking Fractional Equations

1. Find the L.C.D. of all the fractions occurring in the equation.
2. Multiply both sides of the equation by this L.C.D., using the distributive law, and remember to do any cancelling that can be done.
3. If possible, solve the equation that you get by one of the methods you already know.

EXAMPLE:　*Solve*

$$\frac{1}{p} + 3 = 2 - \frac{1}{p}$$

STEP 1.　　The only denominator occurring in the equation is p; therefore the L.C.D. $= p$.

STEP 2.   Multiply by p, using parentheses to make sure *everything* gets multiplied by p:

$$p\left(\frac{1}{p} + 3\right) = p\left(2 - \frac{1}{p}\right)$$

Use the distributive law:

$$\not{p} \cdot \frac{1}{\not{p}} + 3p = 2p - \not{p} \cdot \frac{1}{\not{p}}$$

So

$$1 + 3p = 2\mathrm{p} - 1$$

STEP 3.   This is a linear equation. To solve, subtract $2p$ from both sides:

$$1 + p = -1$$

Subtract 1 from both sides:

$$p = -2$$

Therefore the solution is $p = -2$.

Check: When $p = -2$,

$$\text{LHS} = \frac{1}{p} + 3 = -\frac{1}{2} + 3 = -\frac{1}{2} + \frac{6}{2} = \frac{5}{2}$$

$$\text{RHS} = 2 - \frac{1}{p} = 2 - \left(-\frac{1}{2}\right) = 2 + \frac{1}{2} = \frac{4}{2} + \frac{1}{2} = \frac{5}{2}$$

Therefore $p = -2$ is correct.

Note that in order to do the check we must go back to the *original* equation, and not the one that we got after clearing of fractions. If we were to substitute into

$$1 + 3p = 2 - p$$

we would only have checked that we had found the solution to

$$1 + 3p = 2 - p$$

which does not check that we had multiplied through correctly in the first place.

EXAMPLE: *Solve*

$$\frac{1}{5 + 2x} = \frac{1}{x - 3}$$

STEP 1.   The denominators are $(5 + 2x)$ and $(x - 3)$; therefore the L.C.D. is $(5 + 2x)(x - 3)$.

STEP 2.   Multiply through by $(5 + 2x)(x - 3)$:

$$\cancel{(5 + 2x)}\,(x - 3) \cdot \frac{1}{\cancel{(5 + 2x)}} = (5 + 2x)\,\cancel{(x - 3)} \cdot \frac{1}{\cancel{(x - 3)}}$$

$$x - 3 = 5 + 2x$$

STEP 3.   Solve by subtracting 5 and x from both sides:

$$-8 = x$$

or

$$x = -8$$

CHECK: When $x = -8$,

$$\text{LHS} = \frac{1}{5 + 2x} = \frac{1}{5 + 2(-8)} = \frac{1}{5 - 16} = -\frac{1}{11}$$

$$\text{RHS} = \frac{1}{x - 3} = \frac{1}{-8 - 3} = -\frac{1}{11}$$

EXAMPLE: *Solve*

$$\frac{3}{d - 2} + \frac{2}{d - 3} = \frac{1}{d - 2}$$

STEP 1.   The denominators are $(d - 2)$ and $(d - 3)$; therefore the L.C.D. $= (d - 2)(d - 3)$.

STEP 2.   Multiply through by $(d - 2)(d - 3)$:

$$(d - 2)(d - 3)\left(\frac{3}{d - 2} + \frac{2}{d - 3}\right) = (d - 2)(d - 3) \cdot \frac{1}{d - 2}$$

Using the distributive law:

$$\cancel{(d - 2)}\,(d - 3) \cdot \frac{3}{\cancel{(d - 2)}} + (d - 2)\,\cancel{(d - 3)} \cdot \frac{2}{\cancel{(d - 3)}}$$

$$= \cancel{(d - 2)}\,(d - 3) \cdot \frac{1}{\cancel{(d - 2)}}$$

Therefore

$$3(d - 3) + 2(d - 2) = d - 3$$

STEP 3.   This is a linear equation. Since it involves parentheses, we must first use the distributive law to remove them:

$$3d - 9 + 2d - 4 = d - 3$$

$$5d - 13 = d - 3$$

Subtract d and add 13 to both sides:

$4d = 10$

Divide by 4:

$$d = \frac{10}{4} = \frac{5}{2}$$

CHECK:

When $d = \dfrac{5}{2}$,

$$\text{LHS} = \frac{3}{d-2} + \frac{2}{d-3} = \frac{3}{\left(\frac{5}{2}-2\right)} + \frac{2}{\left(\frac{5}{2}-3\right)} = \frac{3}{\left(\frac{1}{2}\right)} + \frac{2}{\left(-\frac{1}{2}\right)}$$

$$= 3 \cdot \frac{2}{1} + 2 \cdot \left(-\frac{2}{1}\right) = 6 - 4 = 2$$

$$\text{RHS} = \frac{1}{d-2} = \frac{1}{\left(\frac{5}{2}-2\right)} = \frac{1}{\left(\frac{1}{2}\right)} = 2$$

Therefore $d = \dfrac{5}{2}$ is correct.

Now there is one serious thing that can go wrong with fractional equations. Watch this:

EXAMPLE: *Solve*

$$\frac{2}{x} + 1 - \frac{1}{2x} = \frac{3}{2x} - 1$$

STEP 1.   The denominators are x and $2x$, so the L.C.D. $= 2x$.

STEP 2.   Multiply through by $2x$:

$$2x\left(\frac{2}{x} + 1 - \frac{1}{2x}\right) = 2x\left(\frac{3}{2x} - 1\right)$$

Using the distributive law gives:

$$2x \cdot \frac{2}{x} + 2x \cdot 1 - 2x \cdot \frac{1}{2x} = 2x \cdot \frac{3}{2x} - 2x \cdot 1$$

or

$$4 + 2x - 1 = 3 - 2x$$

STEP 3.   This is a linear equation. To solve, add $2x$ and subtract 3 from both sides:

$$4x = 0$$

So

$$x = 0$$

CHECK: When $x = 0$,

$$\text{LHS} = \frac{2}{x} + 1 - \frac{1}{2x} = \frac{2}{0} + 1 - \frac{1}{2 \cdot 0} = ???$$

$$\text{RHS} = \frac{3}{2 \cdot 0} - 1 = ???$$

Unfortunately, division by zero is not defined.

Now we are in real trouble. The expressions on each side of the equation are undefined when $x = 0$, yet $x = 0$ is supposed to be the value making the two sides equal.

What has happened is this. $x = 0$ is indeed a solution to

$$4 + 2x - 1 = 3 - 2x,$$

but $x = 0$ is *not* a solution to the original equation,

$$\frac{2}{x} + 1 - \frac{1}{2x} = \frac{3}{2x} - 1$$

because the fractions in this equation are not even defined for $x = 0$. The reason for this curious discrepancy is that to get

$$4 + 2x - 1 = 3 - 2x$$

from

$$\frac{2}{x} + 1 - \frac{1}{2x} = \frac{3}{2x} - 1$$

we multiplied through by $2x$—which is zero when x is zero —and multiplying an equation by zero *always* causes trouble. (That's why the "golden rule of equations" forbids it.)

$x = 0$ is called an extraneous solution to the fractional equation. Since you should always check your solutions in the *original* equation, extraneous roots will identify themselves as such by making a denominator zero, and should be thrown out.

In this case, the only possible solution is extraneous, so the equation has no solution.

PROBLEM SET   12.2

Solve the following fractional equations:

1. $\dfrac{-1}{x + 3} = \dfrac{3}{x - 1}$

2. $\dfrac{-3}{x} + 1 = \dfrac{7}{2} + \dfrac{2}{x}$

3. $\dfrac{1}{2x} + \dfrac{1}{x} = \dfrac{1}{2}$

4. $\dfrac{4}{x-4} = \dfrac{2}{2x-5}$

5. $\dfrac{6}{y} = \dfrac{5}{y} + 1$

6. $\dfrac{a+3}{a-2} = \dfrac{8}{3}$

7. $\dfrac{2}{m} + 3 = \dfrac{8}{m}$

8. $0 = \dfrac{5}{2k+3} - \dfrac{2}{2k+1}$

9. $\dfrac{1}{5+2z} = \dfrac{1}{z-3}$

10. $\dfrac{1}{3m-4} = \dfrac{3}{2m+1}$

11. $\dfrac{-2}{2w-1} = \dfrac{4}{w+5}$

12. $\dfrac{3}{4x} - \dfrac{1}{2} = 7$

13. $\dfrac{3y-1}{2y+1} = -2\left(1 + \dfrac{3}{2y+1}\right)$

14. $\dfrac{5+x}{x+2} = \dfrac{2}{3}$

15. $\dfrac{4}{z} + \dfrac{3}{2z} = 1$

16. $\dfrac{1}{3} + \dfrac{3}{2x+1} = \dfrac{2}{5} - \dfrac{4}{2x+1}$

17. $\dfrac{2}{b-4} = 5\left(\dfrac{3}{b}\right)$

18. $\dfrac{s+1}{s-5} = \dfrac{4}{s-5} - 2$

19. $\dfrac{2}{p} = \dfrac{1}{p} - \dfrac{3}{8p-5}$

20. $\dfrac{5x+1}{3(x-2)} = 4$

21. $1 = \dfrac{4}{17} + \dfrac{1}{3d}$

22. $\dfrac{5y}{3y-1} = \dfrac{-2}{5}$

23. $\dfrac{1}{1 + \dfrac{1}{x}} = 4$

24. $\dfrac{1.2}{3x} - 1.1 = \dfrac{8.4}{1.2x}$

25. $\dfrac{\dfrac{5x+1}{3}}{\dfrac{4x-5}{2}} = 1$

26. $\dfrac{2}{3 + \dfrac{2}{x+1}} = -1$

27. $-7 = \dfrac{2}{a+1} - \dfrac{3}{2a+2}$

28. $\dfrac{12}{z+1} - 1 = \dfrac{4}{-2z-2}$

29. $0 = \dfrac{4w}{3w-1} - \dfrac{2}{3} + \dfrac{3w+2}{1-3w}$

30. $\dfrac{21}{t-3} - \dfrac{12}{2t-1} = \dfrac{4}{3-t}$

31. $\dfrac{1+2a}{2a+1} + 3a = 2$

12.3   MORE DISGUISED LINEAR EQUATIONS: LITERAL EQUATIONS

An equation in which some of the constants and coefficients are represented by letters is a *literal* equation. For example, if you write

$$2x + 3 = 9$$

as

$$ax + b = c$$

then you have a literal equation. Since solving

$$2x + 3 = 9$$

is extremely like solving

$$3x + 4 = 13$$

or

$$5x + 20 = 25$$

or even

$$-2x + 3 = -7$$

it would be much quicker if we could just solve the general case

$$ax + b = c$$

and then substitute values for a, b, and c to get the solution to any particular equation.

Before solving it, take one more look at the equation

$$ax + b = c$$

Since this now has *four* letters in it, you may be wondering why I am talking about it as though it was an equation in one variable that we can solve. The answer is that the letters a, b, c, and x are not on an equal footing: a, b, and c are *constants*, rather than variables, which means that they stand for fixed numbers, although we don't usually know what those numbers are. It is as though someone decided on values of a, b, and c (at least for the duration of the problem at hand), but then forgot to tell you what the values were. On the other hand, x is a *variable*. This means that its value is not fixed. Finding the value of x required by a certain problem is usually done by solving an equation.

Solving a literal equation does not mean finding a numerical value for x (or whatever the variable is called)—that would be impossible without knowing the numerical value of a, b, and c (or whatever the constants are called). It is possible, however, to find a formula or expression for x in terms of a, b, and c. This means that if you are given numerical values for a, b, and c, then by substituting those values into the formula or expression, you can find a numerical value for x.

Finding a formula for x in terms of a, b, and c is called *solving the equation for x*, and it is done by exactly the same method as was used on the linear equations in the section before. To make the parallel clear, I will give the solutions of $2x + 3 = 9$ and $ax + b = c$ alongside one another.

EXAMPLE:

Solve $2x + 3 = 9$.

Subtract 3 from both sides:

$2x + 3 - 3 = 9 - 3$

or

$2x = 6$

Divide both sides by 2:

$$\frac{2x}{2} = \frac{6}{2}$$

Therefore

$x = 3$

So the solution is

$x = 3$

CHECK:

When $x = 3$,

$$\text{LHS} = 2x + 3 = 2 \cdot 3 + 3$$
$$= 6 + 3 = 9$$

$\text{RHS} = 9$

Therefore $x = 3$ is correct.

Solve $ax + b = c$ for x.

Subtract b from both sides:

$ax + b - b = c - b$

or

$ax = c - b$

(notice that this can't be simplified)

Divide both sides by a:

$$\frac{ax}{a} = \frac{c - b}{a}$$

Therefore

$$x = \frac{c - b}{a}$$

(again, this can be simplified no further)

So the solution is

$$x = \frac{c - b}{a}$$

When $x = \frac{c - b}{a}$,

$$\text{LHS} = ax + b = \frac{\cancel{a}(c - b)}{\cancel{a}} + b$$
$$= c - b + b = c$$

$\text{RHS} = c$

Therefore $x = \frac{c - b}{a}$ is correct.

EXAMPLE:

Solve $ny + \dfrac{s}{t} = my - \dfrac{p}{q}$ for y.

Notice that before you can solve this equation you have to be told which quantity to solve for, or, in other words, which letter stands for the variable. In the previous example the variable was x; here it is y. The variable is often represented by one of the letters at the end of the alphabet, such as x, y, or z, while a, b, and c

usually stand for constants, and m, n, p, and q often stand for integers. But there's nothing logical or sacred about this—it's just convention, so you should be prepared to solve equations for any letter of the alphabet.

I will show the solution to

$$ny + \frac{s}{t} = my - \frac{p}{q} \quad \text{for} \quad y$$

alongside the solution to

$$4y + \frac{2}{3} = 7y - \frac{5}{2}$$

Solve

$$4y + \frac{2}{3} = 7y - \frac{5}{2}$$

Add $\frac{5}{2}$ to both sides:

$$4y + \frac{2}{3} + \frac{5}{2} = 7y$$

Therefore

$$4y + \frac{4 + 15}{6} = 7y$$

Therefore

$$4y + \frac{19}{6} = 7y$$

Subtract $4y$ from both sides:

$$\frac{19}{6} = 7y - 4y$$

$$\frac{19}{6} = 3y$$

Divide both sides by 3:

$$\frac{\frac{19}{6}}{3} = \frac{3y}{3}$$

Solve   (for y)

$$ny + \frac{s}{t} = my - \frac{p}{q}$$

Add $\frac{p}{q}$ to both sides:

$$ny + \frac{s}{t} + \frac{p}{q} = my$$

Therefore

$$ny + \frac{sq}{tq} + \frac{pt}{qt} = my$$

Therefore

$$ny + \frac{sq + pt}{tq} = my$$

Subtract ny from both sides:

$$\frac{sq + pt}{tq} = my - ny$$

$$\frac{sq + pt}{tq} = (m - n)\, y$$

$[my - ny = (m - n)y$ by the distributive law]

Divide both sides by $(m - n)$:

$$\frac{\frac{sq + pt}{tq}}{(m - n)} = \frac{(m - n)\, y}{(m - n)}$$

$$\frac{19}{6} \cdot \frac{1}{3} = y$$

$$\frac{19}{18} = y$$

Therefore the solution is

$$y = \frac{19}{18}$$

$$\frac{sq + pt}{tq} \cdot \frac{1}{(m - n)} = y$$

$$\frac{sq + pt}{tq(m - n)} = y$$

Therefore the solution is

$$y = \frac{sq + pt}{tq(m - n)}$$

Notice that the only new thing about the way we solve this literal equation is the factoring of $my - ny$ to get $(m - n)y$. The reason we do this is that we need to know what y is multiplied by before we can do the division.

The next example will be done without a numerical parallel. But any time you find a literal equation confusing, try replacing the constants by numbers, and see what you would do there. Then apply exactly the same procedure to the literal equation.

EXAMPLE: *Solve $s(a - p) + q = ta + r^2$ for a.*

This is an equation involving parentheses and is solved using the distributive law in the same way as $2(x - 5) = 3x + 1$ in the last section. To solve

$$s(a - p) + q = ta + r^2$$

use the distributive law to remove the parentheses:

$$sa - sp + q = ta + r^2$$

We are solving for a and so want to get all the terms involving a on one side, say the left, and all the other terms on the right. Add sp to both sides:

$$sa + q = ta + r^2 + sp$$

Subtract q from both sides:

$$sa = ta + r^2 + sp - q$$

Subtract ta from both sides:

$$sa - ta = r^2 + sp - q$$

Use the distributive law to factor out a so that you can see what a is being multiplied by:

$$(s - t)a = r^2 + sp - q$$

Divide both sides by $(s - t)$:

$$\frac{(s - t)a}{(s - t)} = \frac{r^2 + sp - q}{(s - t)}$$

So the solution is

$$a = \frac{r^2 + sp - q}{(s - t)}$$

EXAMPLE: *Solve*

$$\frac{x - t}{a} - \frac{x + t}{b} = 0 \quad \text{for } x$$

This is a parentheses problem in disguise because it could be written:

$$\frac{(x - t)}{a} - \frac{(x + t)}{b} = 0$$

or

$$\frac{1}{a}(x - t) - \frac{1}{b}(x + t) = 0$$

Therefore we will use the distributive law:

$$\frac{x}{a} - \frac{t}{a} - \frac{x}{b} - \frac{t}{b} = 0$$

Adding $\frac{t}{a}$ and then $\frac{t}{b}$ to both sides:

$$\frac{x}{a} - \frac{x}{b} = \frac{t}{a} + \frac{t}{b}$$

Use the distributive law on the left-hand side to factor out x:

$$x\left(\frac{1}{a} - \frac{1}{b}\right) = \frac{t}{a} + \frac{t}{b}$$

Dividing by $\left(\frac{1}{a} - \frac{1}{b}\right)$, the coefficient of x, looks awful:

$$\frac{x\left(\frac{1}{a} - \frac{1}{b}\right)}{\left(\frac{1}{a} - \frac{1}{b}\right)} = \frac{\left(\frac{t}{a} + \frac{t}{b}\right)}{\left(\frac{1}{a} - \frac{1}{b}\right)}$$

but gives us x:

$$x = \dfrac{\dfrac{t}{a} + \dfrac{t}{b}}{\dfrac{1}{a} - \dfrac{1}{b}}$$

This is the solution, which looks perfectly horrible at the moment. But that should make clear the need for the section on complex fractions, which showed how to simplify things like this. In case you don't remember, you have to combine the top and the bottom separately and then divide:

$$\dfrac{\dfrac{t}{a} + \dfrac{t}{b}}{\dfrac{1}{a} - \dfrac{1}{b}} = \dfrac{\dfrac{tb}{ab} + \dfrac{ta}{ab}}{\dfrac{b}{ab} - \dfrac{a}{ab}} = \dfrac{\dfrac{tb + ta}{ab}}{\dfrac{b - a}{ab}} = \dfrac{(tb + ta)}{\cancel{ab}} \cdot \dfrac{\cancel{ab}}{(b - a)} = \dfrac{tb + ta}{b - a}$$

Therefore the solution is

$$x = \dfrac{tb + ta}{b - a}$$

Alternatively, since

$$\dfrac{x - t}{a} - \dfrac{x + t}{b} = 0$$

contains fractions, we could get rid of them by multiplying through by the common denominator. Since the denominators appearing are a and b, we should multiply through by ab:

$$ab\left(\dfrac{x - t}{a} - \dfrac{x + t}{b}\right) = ab \cdot 0$$

Use the distributive law on the left, and the fact that $ab \cdot 0 = 0$ on the right:

$$\cancel{a}b\dfrac{(x - t)}{\cancel{a}} - a\cancel{b}\dfrac{(x + t)}{\cancel{b}} = 0$$

Therefore

$$b(x - t) - a(x + t) = 0$$

Now use the distributive law again:

$$bx - bt - ax - at = 0$$

Adding $bt + at$ to both sides, leaving only those terms containing x on the left:

$$bx - ax = bt + at$$

Factor the left side:

$(b - a)x = bt + at$

and divide by $(b - a)$:

$$x = \frac{bt + at}{b - a}$$

as before!

EXAMPLE: *Solve*

$$2 = \frac{A}{x + t} \quad for \quad x$$

This is a fractional equation, since we have to solve for x, and x is in the denominator. We will solve it alongside

$$2 = \frac{3}{x + 5}$$

and do it by the previous section's method for attacking fractional equations.

Solve	*Solve* *(for x)*
$2 = \dfrac{3}{x + 5}$	$2 = \dfrac{A}{x + t}$

STEP 1. L.C.D. $= x + 5$ L.C.D. $= x + t$

STEP 2. Multiplying both sides by L.C.D.:   Multiplying both sides by L.C.D.:

$$2(x + 5) = \cancel{(x + 5)} \cdot \frac{3}{\cancel{(x + 5)}} \qquad 2(x + t) = \cancel{(x + t)} \cdot \frac{A}{\cancel{(x + t)}}$$

So

$2(x + 5) = 3$ $2(x + t) = A$

Solve as a linear equation

STEP 3. First use the distributive law to get rid of parentheses: Use the distributive law:

$2x + 10 = 3$ $2x + 2t = A$

Subtract 10 from both sides: Subtract $2t$ from both sides:

$2x = -7$ $2x = A - 2t$

Divide by 2:   Divide by 2:

$x = -\dfrac{7}{2}$ $x = \dfrac{A - 2t}{2}$

CHECK: When   When

$x = -\dfrac{7}{2}$ $x = \dfrac{A - 2t}{2}$

LHS $= 2$ LHS $= 2$

RHS $= \dfrac{3}{-\dfrac{7}{2} + 5}$ RHS $= \dfrac{A}{\dfrac{A - 2t}{2} + t}$

$= \dfrac{3}{-\dfrac{7}{2} + \dfrac{10}{2}}$ $= \dfrac{A}{\dfrac{A - 2t}{2} + \dfrac{2t}{2}}$

$= \dfrac{3}{\dfrac{3}{2}}$ $= \dfrac{A}{\dfrac{A}{2}}$

$= \cancel{3} \cdot \dfrac{2}{\cancel{3}}$ $= \cancel{A} \cdot \dfrac{2}{\cancel{A}}$

$= 2$ $= 2$

Therefore $x = -\dfrac{7}{2}$ is correct.   Therefore $x = \dfrac{A - 2t}{2}$ is correct.

EXAMPLE: *Solve*

$$\frac{s}{a + by} - \frac{t}{c - dy} = 0 \quad for \ \ y$$

Another fractional literal equation!

STEP 1.   Denominators   are $(a + by)$ and $(c - dy)$;   therefore $(a + by)(c - dy)$ is the L.C.D.

STEP 2.   Multiply through by $(a + by)(c - dy)$:

$$(a + by)(c - dy) \left[\frac{s}{(a + by)} - \frac{t}{(c - dy)} \right] = (a + by)(c - dy) \cdot 0$$

By the distributive law,

$$(a + by)(c - dy) \cdot \frac{s}{(a + by)} - (a + by)(c - dy) \frac{t}{(c - dy)} = 0$$

which simplifies to

$$(c - dy)s - (a + by)t = 0$$

STEP 3.   This is a linear equation in y:

$$cs - dys - at - byt = 0 \qquad \text{(by the distributive law again)}$$

$$cs - at = dys + byt \qquad \text{(adding } dys + byt \text{ to both sides)}$$

$$cs - at = (ds + bt)y \qquad \text{(factoring out } y)$$

$$\frac{cs - at}{ds + bt} = y \qquad \text{(dividing both sides by } (ds + bt))$$

You can check this yourself.

PROBLEM SET   12.3

Solve the following literal equations for the variable in parentheses:

1. $a = \dfrac{Mr}{M + x}$ (M)

2. $aR = bR + c$ (R)

3. $y = mx + b$ (x)

4. $x + a = ax$ (x)

5. $A = \dfrac{1}{2}h(a + b)$ (a)

6. $f = \dfrac{ab}{a + b}$ (a)

7. $\dfrac{p}{p + q} = a$ (p)

8. $\dfrac{4m}{m - p} = \dfrac{6}{p}$ (m)

9. $\dfrac{4m}{m - p} = \dfrac{6}{p}$ (p)

10. $A = p + \dfrac{prn}{100}$ (p)

11. $\dfrac{pt}{1 + ptr} = A$ (t)

12. $a - b = \dfrac{nE}{E + nr}$ (n)

13. $a - b = \dfrac{nE}{E + nr}$ (E)

14. $s = \dfrac{a - rn}{1 - r}$ (r)

15. $a(x + 3) + b(x - 2) = c(x - 1)$ (x)

16. $E = \dfrac{Mr + t}{m + xm}$ (m)

17. $\dfrac{1}{a + db} - \dfrac{b}{a^2} = 2$ (d)

18. $at + (1 - a)t = 1$ *(t)*
19. $s(a - p) + q = ta + r^2$ *(a)*
20. $\dfrac{a + 2}{-x} + \dfrac{9}{4} = a^2 - 4$ *(x)*
21. $\dfrac{1}{3 + pb} - \dfrac{b}{a^2} = 2a$ *(p)*
22. $\dfrac{A}{p - t} = \dfrac{c}{4 + 2t}$ *(t)*
23. $\dfrac{s}{(a + by)} = \dfrac{t}{(c - dy)}$ *(y)*
24. $y = \dfrac{p - 2a}{p + b}$ *(p)*
25. $3(x - 2a) - 4(2x + 3a) = 5x$ *(x)*

26. $\dfrac{ax - b}{c} + \dfrac{d}{e} = \dfrac{f - gx}{h}$ *(x)*
27. $\dfrac{1}{p - q} - \dfrac{1}{p + q} = 0$ *(q)*
28. $\dfrac{r + s}{ax + b} = \dfrac{3b + s}{2x - r}$ *(x)*
29. $\dfrac{z(3 + 2a)}{4r + 1} = \dfrac{2}{r - 3}$ *(r)*
30. $\dfrac{a^2bx + bx + b^2}{b^2} - 1 - \dfrac{x}{b}$
$\quad = a^2$ *(x)*

CHAPTER 12 REVIEW

Solve the following equations:

1. $6Q - 7 = 9$
2. $\dfrac{1}{2}Y - \dfrac{1}{8} = \dfrac{1}{10}Y$
3. $3(a - 5) = 2(3 - a)$
4. $\dfrac{4}{3m - 1} = \dfrac{1}{m + 2}$
5. $\dfrac{2}{B} + 3 = \dfrac{-7}{B}$

Solve for R:

6. $\dfrac{R - 4}{Q - R} = 2$
7. $E = \dfrac{P - sR}{E + sR}$

Solve the following equations:

8. $\dfrac{1}{4}z + 2 - \dfrac{3}{2}z - z = z - 1$
9. $0.08m + 0.11(34 - m) = 2$
10. $\dfrac{w}{w + 3} - 7 = \dfrac{2}{w + 3}$

Solve for Y:

11. $Ax^2 + BYx + c = 0$
12. $w = \dfrac{z}{Y - z}$
13. $0.06(3 - Y) = 0.27 - 0.1(0.3 - 0.4Y)$

Solve the following equations:

14. $\dfrac{d}{7} + \dfrac{d}{14} + \dfrac{2d}{21} = \dfrac{5d}{42}$
15. $\dfrac{4 - Q}{2Q - 7} = \dfrac{6}{5}$
16. $\dfrac{s}{5} - \dfrac{(3 + s)}{7} = 2$

Solve for P:

17. $A = \dfrac{1}{R} B(Q - P)$ 18. $Q = \dfrac{HP}{H + QP}$ 19. $\dfrac{P - M}{M - 3P} = Q$

20. $a(P + a^2) = a(z - P) + b$

Solve the following equations:

21. $0.1\left(z + \dfrac{1}{0.3}\right) = \dfrac{0.2z}{0.5}$

22. $3(H - 2) - (4 + 6H) = 3H$

23. $0 = \dfrac{13}{2k - 1} - \dfrac{2}{3 - 2k}$

24. $\dfrac{1}{x} + 2 = \dfrac{2}{x}$

25. $\dfrac{2t}{t - 1} - 3 = \dfrac{7t}{1 - t}$

Solve for N:

26. $\dfrac{Q}{R + NQ} = \dfrac{P}{R - NP}$

27. $\dfrac{P - NS}{Q} = \dfrac{S}{P} + NS^2 - P$

28. $10 = \dfrac{N}{4} - \dfrac{5N}{7}$

Solve the following equations:

29. $\dfrac{3}{R} - \dfrac{0.6}{2R} + \dfrac{10}{3R} = 1$

30. $0.03(0.1W - 0.02) = 0.05W - 0.072$

31. $x - [1 - (x - 1)] = 3x$

32. $\dfrac{3}{r} = \dfrac{4}{3r} - 16$

33. $\dfrac{1}{3}(12P - 27) + 2\left(\dfrac{1}{4}P - 1\right) = 16P$

34. $\dfrac{-3}{2 - \dfrac{1}{3x - 1}} = 5$

Solve for S:

35. $\dfrac{R + S}{ax + B} = \dfrac{3B + S}{2x - r}$

36. $1 + R = \dfrac{a(1 - SN)}{1 - S}$

37. $3[1 - (S - 5)] = 2 - [(3S - 2) - 4(1 - S)]$

Solve the following equations:

38. $2 - 3\sqrt{5x - 1} = -7$

39. $5.2 - (0.3x - 0.01) = -0.2[-1.1 - (3 - x)]$

40. $\dfrac{x}{5} - \dfrac{1}{2}\left(x + \dfrac{1}{3}\right) = \dfrac{1}{3}\left(\dfrac{x}{2} + \dfrac{1}{6}\right)$

41. $\dfrac{5}{12}(3M + 2) + 2M = \dfrac{1}{3}(M - 6) - \dfrac{1}{2} + 2M$

42. $\dfrac{4}{b-2} + \dfrac{3}{b+3} - \dfrac{1}{b-2} = \dfrac{7}{b+3}$

43. $\dfrac{3}{4 - \dfrac{2}{x+1}} = -1$

44. $\dfrac{\dfrac{2Y-1}{3}}{\dfrac{3Y+1}{5}} = 2$

45. $-1 - (-10 - x) = 3(4 - x) - (-x - 3)$

Solve for C:

46. $\dfrac{1}{C-D} - \dfrac{3}{D+C} = 0$

47. $B - 3J = \dfrac{K - 23CJ}{2(1 + CJ)}$

48. $\dfrac{A^2BC}{D} + \dfrac{A^3D}{D^2} = \dfrac{C}{D} - 1$

49. Given that $T = \dfrac{A - B}{DB - EA}$ find the ratio of A to B.

Solve the following equations:

50. $\dfrac{3 + g}{1 + \dfrac{4g}{10}} = 5$

51. $\dfrac{2}{3}[1 - (x - 3)] - \dfrac{x}{5} = 3x + \dfrac{1}{10}$

52. $\dfrac{Q+1}{Q-4} - 2 - \dfrac{3}{2Q-8} = \dfrac{Q}{Q-4}$

53. $\dfrac{\dfrac{1-3x}{3}}{\dfrac{5-8x}{4}} = 1$

54. $\dfrac{3z-1}{2-z} = -5\left(\dfrac{3}{z-2} + 1\right)$

55. $\dfrac{2}{w+3} - 4 = \dfrac{7}{6w+18}$

56. $\sqrt{x+1} = 3$

57. Solve $T = \sqrt{\dfrac{R^3}{mG}}$ for $\left(\dfrac{m}{R^3}\right)$

58. Solve $l = \dfrac{u+v}{1 + \dfrac{uv}{c^2}}$ for u

59. Given that
$$\dfrac{b+c}{b} = \dfrac{c-a}{b-a}$$
show that $ac = b^2$.

60. Let $2kmx = c - x$
    (a) Solve for x.
    (b) Evaluate $x^2 + 1$ if $c = 10$, $k = 0.5$, and $m = 5$.

13 THE WORD PROBLEM PROBLEM

13.1 WHY WORD PROBLEMS NEED ALGEBRA

Most people learn mathematics in order to use it, and you are probably no exception. The chances are that the problems you want to solve will not start out as math problems, but will need math in their solution. Solving them consists of first translating the original problem into a math problem, and then doing the math problem. The original problem will usually be stated in words, as are most problems in chemistry and economics, for example. Often the hardest part of solving a word problem is translating it into a math problem, because there is no set or standard way of doing this. Hence word problems have sometimes got themselves a bad name as being the worst part of any math course, although once they are set up, solving them involves nothing more than the algebra you already know.

Whether or not they like math or think they can do math, everyone solves word problems as a part of everyday life. What to pay in a store, how long it will take to get somewhere, or what number of hours a week you will need a babysitter are all word problems that can be solved by pretty much everyday arithmetic; figuring out your income tax is an example of a problem that can be solved by slightly less than everyday arithmetic. But there are also problems that cannot be solved without algebra, and those will be the main concern of this section. Oddly enough, the type of problems that require algebra often don't look much different from those that require only arithmetic.

For example, consider the following problem: A man sets out to drive from Salisbury to Stonehenge. He drives for $\frac{3}{4}$ hour at an average speed of 40 miles per hour; then his car breaks down. He is unable to get a ride from a passing car, and so he has to walk the rest of the way to Stonehenge. This takes him $\frac{1}{2}$ hour, going at 4 miles per hour. How far is it from Salisbury to Stonehenge?

Solution: In order to do any problems involving a journey, you need to know the relation between the distance traveled, the time it took, and the speed traveled, namely,

$$\text{distance} = \text{speed} \cdot \text{time}$$

In this problem:

distance from Salisbury to Stonehenge

$$= \begin{pmatrix} \text{distance} \\ \text{traveled} \\ \text{by car} \end{pmatrix} + \begin{pmatrix} \text{distance} \\ \text{traveled} \\ \text{on foot} \end{pmatrix}$$

In the car the man traveled for $\frac{3}{4}$ hour at 40 miles per hour, so

$$\text{distance traveled by car} = 40 \cdot \frac{3}{4} = 30 \text{ miles}$$

On foot he traveled for $\frac{1}{2}$ hour at 4 miles per hour, so

$$\text{distance traveled on foot} = 4 \cdot \frac{1}{2} = 2 \text{ miles}$$

Therefore, the distance from Salisbury to Stonehenge $= 30 + 2 = 32$ miles.

Thus, to solve this problem nothing is needed except fairly straightforward arithmetic and a knowledge of how distance, speed, and time are related. However, let me now change the problem only very slightly, and let's look at it again:

Problem: The same man is going from Salisbury to Stonehenge, and his car again breaks down, forcing him to walk the rest of the way. His car still goes at 40 miles per hour, and he still walks at 4 miles per hour, but this time you are told that his journey takes a total of 2 hours and that the distance between Salisbury and Stonehenge is 32 miles. You now have to find the time for which he drove before the breakdown.

Solution: Solving this problem really involves finding two times: the time driving and the time walking. Fortunately, they are closely related to one another, because they add up to 2 hours (the total time of the journey), and hence, if both are measured in hours

$$\text{time walking} + \text{time driving} = 2$$

so

$$\text{time walking} = 2 - \text{time driving}$$

Now we know that the total distance (32 miles) is split up into two pieces, but we don't know how much time is spent on each part of the journey. So it seems that if we knew the time driving, we could find the time walking and hence the distances both driving and walking. And these distances would lead us back to the two times. As it stands, this argument seems circular and quite useless, but that's where the algebra comes in.

Suppose t stands for the time (in hours) he spends driving; t is a variable that stands for a quantity as yet unknown. Now we can find—in terms of t at least—all the distances mentioned in the circular-sounding argument.

$$\text{time walking} = 2 - t \text{ hours}$$

$$\text{distance traveled by car} = 40t \text{ miles} \qquad \text{(using distance}$$
$$= \text{speed} \cdot \text{time)}$$

$$\text{distance traveled by foot} = 4(2 - t) \text{ miles}$$

The only other thing we know is that these two distances must add up to the total distance from Salisbury to Stonehenge. Therefore,

$$40t + 4(2 - t) = 32$$

This is a perfectly straightforward linear equation which can be solved for t, and the t that we find will be the time we are looking for. To solve the equation first divide both sides by 4:

$$10t + (2 - t) = 8$$
$$10t + 2 - t = 8$$
$$9t = 8 - 2 = 6$$
$$t = \frac{6}{9} = \frac{2}{3} \text{ hour}$$

Therefore the driving time $= \frac{2}{3}$ hour.

This problem could not be done without algebra, because the unknown quantities (the times) enter into the calculation right at the start, and the only way to do that is to represent them by a letter. The only conceivable alternative is trial and error, but that is an extremely tedious method in all but the simplest problems. Hence it is worth spending some time on translating word problems into algebraic equations. Contrary to popular belief, this is actually not very hard. It is mostly a matter of reading very carefully so that you understand *precisely* what the word problem means—precisely enough to write an equation, in fact. But all the information you need to write the equation is contained in the problem—all you have to do is to look hard enough to find it there.

13.2   A METHOD FOR DEALING WITH WORD PROBLEMS

Unfortunately, there's no foolproof standard method for dealing with word problems, but following this outline should help:

1. Read the problem over, first quickly to get an idea of what it's about, and then *very* carefully. Make sure you have a very clear idea what every sentence means, and what every word is there for. In your mind's eye, get as vivid a picture as possible—it really helps to *see* what is going on in the problem!
2. Decide what the question is asking you to find. Let a variable represent it.
3. Draw a picture if appropriate. On it mark any lengths, distances, etc., that are given in the problem or that can be expressed in terms of the variable.
4. Write down *in words* an equation connecting the various quantities in the problem. This equation is usually hidden in, or implied by, the wording of the problem. If you can't see it, read through again carefully, and ask yourself what the problem is telling you about one quantity being equal to another.
5. List any formulas that might help.
6. Write expressions for all the other unknown quantities in terms of the variable.
7. Write the equation in symbols, by expressing the unknown quantities in terms of the variable and substituting any numerical values given in the problem.
8. Solve the equation.
9. Check that your solution is reasonable from a common sense point of view, and that it satisfies the conditions of the problem.

There are certain phrases that occur frequently in word problems and have standard translations into math. For example, "twice b" is "$2b$," "a increased by b" is "$a + b$," and "is," "was," "will be," and "becomes" all translate into "$=$."

Since there is no general formula for word problems, the rest of this chapter will be devoted to examples. Fortunately, word problems fall into several standard types, so this chapter will contain a couple of examples from eight of the most common ones.

13.3   NUMBER PROBLEMS

The point of these problems is to find a number with the properties specified in the question.

EXAMPLE: *Find a number whose double is eight more than the result of subtracting the original number from 25.*

STEP 2.   The problem is asking you to find a number, so let the number be x.

STEP 4.   The equation comes from the fact that:

Double of x = eight more than result of subtracting x from 25

STEP 6.   The double of x is $2x$.

The result of subtracting x from 25 is $(25 - x)$.

Eight more than the result of subtracting x from 25 is $8 + (25 - x)$.

STEP 7.   Substituting these expressions into the equations of Step 4:

$$2x = 8 + (25 - x)$$

STEP 8.   Solving this equation:

$$3x = 33$$

$$x = 11$$

STEP 9.   *Check:*

Double of x is 22.

Subtracting x from 25 gives 14, and eight more than 14 is 22.

Therefore, 11 is the answer to the problem.

EXAMPLE: *Find two consecutive integers whose sum is four more than the least integer.*

First remember that two integers are consecutive if they follow one another on the number line. For example, 5, 6 are consecutive integers as are $-7, -6$. Since 11 lies in between 10 and 12, they are not consecutive integers. However, 10 and 12 are consecutive even integers, since there is no even integer between them. I'll write out this example without labeling each step.

Let x = the small integer.

Then $x + 1$ = the large integer.

The word equation is

$$\frac{\text{small}}{\text{integer}} + \frac{\text{large}}{\text{integer}} = \text{four more than small integer}$$

Now four more than the small integer is $4 + x$, so the equation is

$$x + (x + 1) = 4 + x.$$

Solving gives

$$2x + 1 = 4 + x$$

$$x = 3.$$

So the small integer is 3, and the larger, being consecutive, is 4.

CHECK: Sum of the integers is $3 + 4 = 7$.
Four more than the smaller integer is $4 + 3 = 7$.
So we have the right solution.

PROBLEM SET   13.3

1. If 11 times a number is increased by 10, the result is 14 times the number, less 5. Find the number.

2. What number added to 40 is the same as five times the number?

3. If twice a number is increased by 6, the result is the same as decreasing four times the number by 14. Find the number.

4. If five times a number, minus 12, equals three times the number, increased by 10, find the number.

5. What number added to 15 is the same as half the number multiplied by eight?

6. Find three consecutive integers whose sum is 36.

7. Find three consecutive odd integers whose sum is 33.

8. Find five consecutive even integers whose sum is 60.

9. One number is five more than another, and the sum of the two is 71. Find the numbers.

10. Find three consecutive integers such that the product of the second and the third is eight more than the square of the first.

11. Find three consecutive odd integers such that twice the sum of the first and second equals four times the sum of the second and third.

12. Find two numbers whose difference is eight, such that the larger number is sixteen less than three times the smaller number.

13.4   AGE PROBLEMS

Age problems are very like number problems. You have to find a number, or in this case an age, with the properties specified in the question. The trick is to remember that if someone is x years old today, then in 3 years' time he or she will be $(x + 3)$ years old; and 2 years ago he or she was $(x - 2)$ years old; and so on.

EXAMPLE: *A suspicious bartender asks a customer whether he is of drinking age. The customer responds, "In four years, I shall be twice as old as I was last year." The bartender, who is no good at age problems, takes this display of mathematical sophistication as sufficient proof of the customer's maturity and serves him. Was the law broken?*

STEP 2.   Let $x =$ the customer's age.

STEP 4.   The word equation is

age in 4 years = twice age last year

STEP 6.   age in 4 years $= x + 4$

twice age last year $= 2(x - 1)$

STEP 7.   So the equation is

$x + 4 = 2(x - 1)$

STEP 8.   Solving gives $x = 6$, so the law was certainly broken!

STEP 9.   *Check:* In 4 years, the customer will be 10, and last year he was 5, which fits the problem.

EXAMPLE: *Some word problems have been around a long time. One word problem is so old that, 2 years ago, it was twice as old as it was 22 years ago. How old is it now?*

Let $x =$ the age of the word problem.

The word equation is

age 2 years ago = twice age 22 years ago

So

$$x - 2 = 2(x - 22)$$

Solving gives $x = 42$

So the word problem is 42 years old.

CHECK: Age 22 years ago is 20 and age 2 years ago is 40, so this age satisfies the problem.

PROBLEM SET   13.4

1. When Rob entered law school, he was one-fourth again as old as his brother Jeff, who was then entering college. Four years later, when Jeff graduated college, Rob was one-fifth again as old as Jeff. How old was Rob when he entered law school?

2. This problem is so old that in just eight years it will be four times as old as it was ten years ago. How old is the problem?

3. Michael and Janet are ten years apart in age. In 24 months, Michael will be twice as old as Janet. How old is Janet now?

4. A 33-year-old man has a 7-year-old daughter. In how many years will his age be double hers?

5. Doris is three years younger than Bob. Forty-one years ago she was two-thirds his age. How old is Bob now?

6. Don is five times as old as his daughter Susie. In five years he will be three times as old as Susie. How old is each now?

7. Laurie's grandfather is six times as old as she is. In six years he will be four times as old as she is. How old is Laurie?

8. When I was nine, my pride and joy was a 35-year-old silver dollar. However, as I grew older, the coin didn't seem as ancient, and by the time the coin was only twice as old as I was, I'd lost all interest in coin collecting. How old was the coin at that point?

9. Amy is six years younger than Sandee. Eighteen years ago, Amy was two-thirds Sandee's age. How old is Sandee now?

10. Car dealers are not well known for the directness of their language. A recent ad stated, "For a car so new that in five years it will only be twice as old as it was two years ago, this is a tremendous buy!!" How old is that car now?

11. Donna owns two sports cars, a Ferrari and a Porsche. She is six times as old as the Ferrari, which is two years older than the Porsche. In six years she will be four times as old as the Porsche. How old is Donna?

12. Earl is five years older than his favorite cousin. Thirteen years ago, he was twice his cousin's age. How old is each now?

13.5   GEOMETRY PROBLEMS

Remember the appendix of geometrical formulas at the end of the book!

EXAMPLE: *With the new interest in economy-size cars, parking lots find that they can reduce the size of their spaces and use the resulting area for other things. In one parking lot, each row consists of six spaces side by side. Each space is 12 feet long and 7 feet wide. By reducing the width of each space, they obtain 96 square feet at the end of each row for a garden. Find how much narrower each space is.*

STEP 2.   Let $x =$ the number of feet by which the width of each space is reduced.

STEP 3.   Pictures are *essential* here, so look at Figure 13.1.

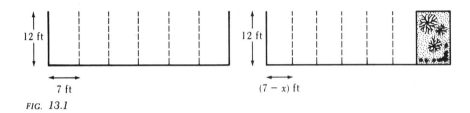

FIG. 13.1

STEP 4.   Since we are told that 96 square feet (ft²) are left over when the new spaces are made, this problem is about area (square feet are units of area). Therefore the area of the new row together with the garden is the same as the area of the old row.

$$\left(\begin{array}{c}\text{area of} \\ \text{new row}\end{array}\right) + \left(\begin{array}{c}\text{area of} \\ \text{garden}\end{array}\right) = \left(\begin{array}{c}\text{area of} \\ \text{old row}\end{array}\right)$$

STEP 5.   Since this problem is about areas, and the parking spaces are rectangular, we need the formula for the area of a rectangle:

$A = lw$

STEP 6.   In both the old and the new lots, there are six spaces to a row, so in each case we will find the area of one space and multiply by six.

area of one old space $= 12 \cdot 7 = 84$ ft²

So

area of old row $= 6 \cdot 84 = 504$ ft²

From the picture,

area of one new space $= 12(7 - x)$ ft²

So

area of new row $= 6 \cdot 12(7 - x) = 72(7 - x)$ ft²

STEP 7.
Substituting into the word equation, and using the fact that the area of the garden is 96 ft²

$72(7 - x) + 96 = 504$

STEP 8.
Dividing by 24:

$3(7 - x) + 4 = 21$

Solving for x:

$21 - 3x + 4 = 21$

$$x = \frac{4}{3} \text{ ft}$$

STEP 9.
Check:

area of one new space $= 12(7 - \frac{4}{3}) = 68$ ft²

area of new row $= 6 \cdot 68 = 408$ ft²

area of new row + garden $= 408 + 96 = 504$ ft², as it should.

EXAMPLE:
A living cell can be thought of as a sphere that absorbs oxygen through its surface. The speed at which the oxygen enters the cell is 10^{16} times its surface area in square centimeters (cm²). The cell uses up oxygen at a rate $3 \cdot 10^{19}$ times its volume in cubic centimeters (cm³). The cell grows until the ratio of the speed at which the oxygen is used up to the speed at which it enters the cell is 1. What is the largest possible radius the cell can have?

Let the largest possible radius $= x$ centimeters (cm). The equation that we need is that the ratio of the speed of oxygen consumption to the speed of oxygen entrance is 1.

$$\frac{\text{speed at which oxygen is used up}}{\text{speed at which oxygen enters cell}} = 1$$

Now,

speed oxygen is used up $= 3 \cdot 10^{19} \text{(volume of cell)}$

$$= 3 \cdot 10^{19} \left(\frac{4}{3} \pi x^3 \right)$$

and

speed oxygen enters cell $= 10^{16}$ (surface area of cell)

$$= 10^{16}(4\pi x^2)$$

So,

$$\frac{\cancel{3} \cdot 10^{19}\left(\frac{\cancel{4}}{3}\cancel{\pi}x^3\right)}{10^{16}\cancel{(4\pi x^2)}} = 1$$

which reduces to $10^3 x = 1$. So

$$x = \frac{1}{10^3} \text{ cm} \quad \text{or} \quad x = 10^{-3} \text{ cm}$$

Therefore the largest possible radius for the cell is 10^{-3} cm.

CHECK: When $x = 10^{-3}$,

$$\text{speed oxygen used up} = 3 \cdot 10^{19}\left[\frac{4}{3}\pi(10^{-3})^3\right]$$

$$= 4\pi 10^{19} \cdot 10^{-9}$$

$$= 4\pi 10^{10}$$

$$\text{speed oxygen enters cell} = 10^{16}[4\pi(10^{-3})^2]$$

$$= 4\pi 10^{16} \cdot 10^{-6}$$

$$= 4\pi 10^{10}$$

So

$$\frac{\text{speed oxygen used up}}{\text{speed oxygen enters cell}} = \frac{4\pi \ 10^{10}}{4\pi \ 10^{10}} = 1$$

PROBLEM SET   13.5

1. The length of a rectangle is 2 inches more than twice the width. If the perimeter of the rectangle is 76 inches, find the dimensions of the rectangle.

2. The width of a rectangle is 1 inch less than half the length. If the sum of the length and the width is 56 inches, find the dimensions of the rectangle.

3. The side of one square is 7 feet longer than the side of another. Find the sides of the squares if the sum of the two perimeters is 60 feet.

4. A piece of wire 72 inches long is to be used as a triangular Christmas

tree ornament. How long should the sides be if the triangle is to be isosceles and the base is to be 6 inches less than one of the equal sides?

5. Some movers attempt to carry a cube-shaped crate through a door, but they find that the crate is 1 foot too wide to fit through. They cut the crate in half and find that each half fits through the door with 18 inches to spare on each side. Find the volume of the crate before they cut it.

6. How many spherical raindrops of radius 0.2 inch would it take to fill a bucket with a volume of 100 cubic inches?

7. Complementary angles are two angles whose sum is 90°. If one angle is twice as big as another and the two are complementary, find the angles.

8. Supplementary angles are two angles whose sum is 180°. One of two supplementary angles is 10° more than two-thirds of the other. Find the angles.

9. A kite in the shape of a diamond has a total area of 240 square inches. If its longer diagonal has length 32 inches, find the length of the shorter diagonal.

10. A woman is bottling preserves in cylindrical jars of radius $1\frac{1}{2}$ inches. If she needs eight jars to hold 270 cubic inches of preserves, find the height of the jars. (Leave π in the answer.)

13.6   PERCENT PROBLEMS

To do percent problems you need to remember that "so many percent" means "so many hundredths," so the total number of percents making up a whole must be a hundred. Then "*A* is *p*% of *B*" means

$$A = \frac{p}{100} \cdot B$$

EXAMPLE: *In a certain national government, every employee is suspected of being corrupt, and all are under investigation. On the average, 99.5% of those investigated each week are never brought to trial, and the rest are prosecuted. If the courthouse contains 225 government employees per week on the average, how many employees are investigated each week?*

STEP 2.   Let x = number of employees investigated each week.

STEP 4.   The equation is going to come from the fact that:

number of people prosecuted per week = 225

STEP 5.   The only formula that is likely to be helpful is

$$A = \frac{p}{100} \cdot B$$

STEP 6.   If 99.5% of those investigated are not brought to trial, then $100\% - 99.5\% = 0.5\%$ of those investigated are prosecuted. So

number prosecuted per week $= 0.5\%$ of $x = \frac{0.5x}{100}$

STEP 7.   The equation is

$$\frac{0.5x}{100} = 225$$

STEP 8.   Solving gives

$x = 45{,}000$ investigated per week

STEP 9.   *Check:* The number of people investigated but not brought to trial each week is

$$99.5\% \text{ of } 45{,}000 = \frac{99.5}{100} \cdot 45{,}000 = 44{,}775$$

So the number of people brought to trial per week is $45{,}000 - 44{,}775 = 225$, and the answer checks.

EXAMPLE: *A magazine running a nationwide promotional sweepstakes sends entry blanks to everyone on its mailing list. 70% of those who receive the entry blanks throw them out immediately and, of the rest, only 10% enter the contest. The magazine receives 15,000 entries. How many people are on the mailing list?*

Let $x =$ the number of people on the mailing list.

The word equation here comes from

number of people entering contest = 15,000

Now,

number of people entering contest = 10% of those who didn't throw blanks away

number who threw blanks away   = 70% of people on mailing list

So,

number who didn't throw blanks away = 30% of people on
mailing list

$$= \frac{30}{100}x = 0.3x$$

Therefore

number entering contest = 10% of $0.3x$

$$= 0.1(0.3x)$$

Substituting into the word equation gives:

$0.1(0.3x) = 15,000$

Solving:

$$x = \frac{15,000}{(0.1)\,(0.3)}$$

$$= 500,000 \text{ people}$$

CHECK: number of blanks thrown out immediately = $0.7(500,000)$
$= 350,000$

number kept = $500,000 - 350,000 = 150,000$
number of entries = $0.1(150,000) = 15,000$, as expected.

PROBLEM SET   13.6

1. An air conditioner manufacturer automatically discounts the price of its units by 20% on the first day of fall. It further discounts the price of all remaining unsold units by 30% on the first day of winter. These successive discounts are equivalent to a single discount of what percent?

2. Sixty percent of Maine's electorate live in cities; the other 40% live in rural districts. If a candidate gets 60% of the city vote in Maine, what percent of the rural vote must he receive in order to have a majority (50% or more) of the votes in the state?

3. For some time, Gramer's department store made a healthy profit selling Idaho Digital (I.D.) calculators, the only brand on the market. Recently, however, Federal Analog (F.A.) has introduced an identical unit that is becoming increasingly popular with the public.   Each model costs Gramer's $200 apiece. They still have 1000 I.D.'s in stock which they have to sell at a 15% loss to sell them at all. If they can make a 10% profit on each F.A., how many F.A.'s must they sell to break even on calculators?

4. In order to protect U.S. automakers, the Federal Government places an import duty on all foreign cars brought into America. The effect of this duty is to raise the price of all foreign cars by $300. A local dealer, fearing that this price increase will cause an untimely slump in his sales, offers, for a limited time, a 10% rebate on all cars sold at the new higher price.

(a) Will a car that sold for $3000 before the import duty was imposed be more or less expensive after the duty and the rebate are applied? By what percent has the price changed?

(b) What percent rebate should the dealer give if it is just to cancel out the price increase due to the duty on this model?

5. Two hundred years ago, as it became more and more evident that this country would sever all ties with England, a large number of people who had moved here decided to return to their former home. This migration was so large, in fact, that by July 4, 1776, it had decreased the American population by 20%, and increased the population of the British Isles by 4%. Assuming that no one who left America went anywhere else, what percent of the British population was the American population before the migration?

6. Lumber companies reseed forests after they have cut down the mature trees that are used in making lumber. Twenty percent of the seeds planted never germinate, and 25% of all seedlings (young trees not ready to be cut for lumber) never become mature trees (people cut them for firewood, storms destroy them, etc.). If a paper company harvests 18,000 trees, how many seeds must they plant to ensure that they will be able to return to the same area several years later and harvest another 18,000 trees?

7. A factory along the Mississippi River draws in river water that contains 10% more pollutants than "clean" water (even clean water contains some pollutants). The water is used in a manufacturing process that causes the water to leave the plant containing 23% more pollutants than the water taken in. A municipal treatment facility downstream draws water in and removes 75% of all pollutants. How dirty is the water after going through the municipal treatment facility as compared to "clean" water?

8. "The bulk of the herbicide spraying program was addressed to the large inland forests of South Vietnam; of the total of about 25.9 million acres of such forests, 10.3% (6.5% of the total land area of South Vietnam) was subjected to one or more sprays." (From a National Academy of Sciences report, 1970.) What is the total land area of South Vietnam?

9. Labor negotiations between unions and management are a tricky business. In a recent contract dispute, the union asked for 60% more than the minimum amount necessary to keep them from striking. Management followed its usual practice and offered the union 20% less than it asked for.

(a) If the management offer was $6.40 per hour, what was the union's request?

(b) What was the minimum amount necessary to keep the workers from striking?

(c) By what percent did management's offer exceed the amount necessary to keep the workers from striking?

10. (a) Everyone who applies to be a spy for the ultrasecret D.I.O. is screened twice, first by a doctor, and later by a psychiatrist. If the doctors reject one applicant out of every 12, and the psychiatrists reject 40% of the applicants they see, what percent of the applicants make it past both screenings?

(b) If it costs $300 each time a doctor examines an applicant, and $1200 each time a psychiatrist examines one, would it be cheaper to run the screening process the other way around?

11. One day, as George Washington Carver was putting a fence around one of his peanut fields, he thought to himself, "It's too bad I'm not laying this same length of fence out in a circle, since it would enclose a lot more area." By what percent would the enclosed area be increased if the fence were laid out in a circle rather than a square? (Leave π in the answer.)

12. Two ships, the tanker MU and the supertanker Atlantis are about to pass through the Panama Canal. Before doing so, however, the Atlantis must reduce its load in order to draw less water, so it transfers 10% of its cargo to the MU. This increases the MU's cargo by 55%. Before the transfer, how many times heavier was the Atlantis' cargo than the MU's cargo?

13.7   MIXTURE PROBLEMS

Mixture problems are concerned with mixing things in such a way that the mixture comes out with certain specified characteristics. The problem is usually to find out what proportions are needed to give the required characteristic.

We often have to design a mixture that has a certain price, given the price of its constituents. In this case, we use the formula:

$$\begin{array}{ccc} \text{total cost} \\ \text{of mixture} \end{array} = \begin{array}{c} \text{total cost of} \\ \text{one constituent} \end{array} + \begin{array}{c} \text{total cost of} \\ \text{second constituent} \end{array}$$

and the fact that:

$$\text{total cost} = \frac{\text{price per unit}}{\text{quantity}} \times \text{quantity}$$

Another common kind of problem is to design a solution of a certain concentration or strength using two solutions of given concentrations. This is done using:

$$
\begin{array}{c}
\text{total quantity} \\
\text{of substance} \\
\text{in mixture}
\end{array}
=
\begin{array}{c}
\text{total quantity} \\
\text{of substance in} \\
\text{one solution}
\end{array}
+
\begin{array}{c}
\text{total quantity} \\
\text{of substance in} \\
\text{second solution}
\end{array}
$$

and:

$$
\begin{array}{c}
\text{total quantity} \\
\text{of substance} \\
\text{in solution}
\end{array}
= (\text{concentration})\,(\text{volume})
$$

Note: The concentration of a solution is the amount of the substance that has been dissolved per unit volume.

EXAMPLE: *An enterprising professor supplements his income by charging his students for their grades. He charges $60.00 per term for an A, $56.00 per term for a B+, and $55.00 per term for a B. If he were to sell only A's and B's to his class of 500, how many of each grade would he need to sell to make as much money as he would if he sold only B+'s?*

STEP 2.   We have to find out how many A's and B's he sells, and so it looks as though there are two unknowns in this problem. However, knowing that there are 500 students in the class tells you that you can find the number of B's given by subtracting the number of A's given from 500. So we'll say:

Let the number of A's given $= x$

Then the number of B's given $= 500 - x$

STEP 4.   The equation is

$$
\begin{pmatrix}
\text{revenue} \\
\text{from 500} \\
\text{B+'s}
\end{pmatrix}
=
\begin{pmatrix}
\text{revenue} \\
\text{from } x \\
\text{A's}
\end{pmatrix}
+
\begin{pmatrix}
\text{Revenue} \\
\text{from } (500 - x) \\
\text{B's}
\end{pmatrix}
$$

STEP 5.   We will use

$$
\text{total cost} = \left(\frac{\text{price per unit}}{\text{quantity}} \right) (\text{quantity})
$$

STEP 6.   Revenue from 500 B+'s is 500($56) = $28,000

Revenue from x A's is 60x$

Revenue from $(500 - x)$B's is $55(500 - x)$

STEP 7.   So the equation is

$$28,000 = 60x + 55(500 - x)$$

STEP 8.   Solving:

$$28,000 = 60x + 27,500 - 55x$$

$$500 = 5x$$

$$x = 100$$

Therefore 100 students get A's and 400 get B's.

STEP 9.   It is reasonable that there are more B's than A's because the rate he charges for B+'s is much closer to the B rate than to the A rate.

Check:

Revenue from A's = 100($60) = $6,000

Revenue from B's = 400($55) = $22,000

These do add up to $28,000, the revenue from the B+'s.

Next we'll do an example using the second pair of formulas at the start of this section.

EXAMPLE: *Commercial bleach contains 0.51 pound of sodium hypochlorite (the stuff that actually does the bleaching) per gallon. You wish to soak a tablecloth in a bucket of bleach to remove a stain. Suppose you know that if the bleach solution contains more than 0.03 pound of sodium hypochlorite per gallon, your tablecloth will rot. What is the largest volume of commercial bleach you can add to a gallon of water and be safe?*

We are looking for the amount of commercial bleach that we must add to 1 gallon of water to get a solution of the maximum strength, namely 0.03 lb of sodium hypochlorite per gallon. Let x = number of gals of commercial bleach that we add to 1 gal of water, giving $(1 + x)$ gals of soaking solution. See Figure 13.2 on page 252.

The equation we use is based on the fact that the total amount of sodium hypochlorite in the commercial bleach is unchanged by pouring the bleach into a bucket of water. Consequently:

amount of sodium
hypochlorite in   =   amount of sodium
soaking solution   hypochlorite in commercial
                        bleach

Now we will use the formula at the start of the section that says:

total quantity of sodium
hypochlorite in solution   =   (concentration) (volume)

We have $(1 + x)$ gallons of soaking solution containing 0.03 lb of sodium hypochlorite per gallon. Therefore,

amount of hypochlorite in the soaking solution $= (0.03)(1 + x)$ lb

Similarly, we have x gallons of commercial bleach containing 0.51 lb of sodium hypochlorite per gallon. Therefore,

amount of hypochlorite in commercial bleach $= 0.51x$ lb

Substituting into the equation gives

$$(0.03)(1 + x) = 0.51x$$

Solving:

$$0.03 + 0.03x = 0.51x$$

$$0.03 = 0.51x - 0.03x$$

$$0.03 = 0.48x$$

Therefore

$$x = \frac{0.03}{0.48} = \frac{1}{16} \text{ gallon}$$

Therefore we must add $\frac{1}{16}$ gal. of commmercial bleach to the bucket.

(*Note:* $\frac{1}{16}$ gal = 1 cup, which is about right according to the directions on most commercial bleach bottles.)

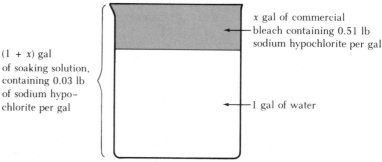

$(1 + x)$ gal
of soaking solution,
containing 0.03 lb
of sodium hypo–
chlorite per gal

x gal of commercial
bleach containing 0.51 lb
sodium hypochlorite per gal

1 gal of water

FIG. 13.2

CHECK: $\frac{1}{16}$ gallon of bleach contains $\frac{0.51}{16}$ lb hypochlorite.

1 gallon of water plus $\frac{1}{16}$ gallon of bleach gives $\frac{17}{16}$ gallons of solution containing $\frac{0.51}{16}$ lb hypochlorite.

Therefore the solution contains

$$\frac{\frac{0.51}{16}}{\frac{17}{16}} = \frac{0.51}{17} = 0.030 \text{ lb hypochlorite per gallon}$$

So the tablecloth won't rot!

PROBLEM SET   13.7

1. If 100 pounds of sea water contain 2.5 pounds of salt, how much water must be evaporated from 200 pounds of sea water in order that 12 pounds of the resulting solution will contain 1 pound of salt?

2. How much water must be added to 2.5 gallons of alcohol that is 90% pure to make a mixture that is 80% pure alcohol?

3. A mixture of silver and copper alloy weighs 128 pounds and contains 12 pounds of silver. How many pounds of silver must be added in order that 8 pounds of the resulting alloy will contain 1 pound of silver?

4. How much nonfat milk must be added to 50 gallons of milk containing 6% butterfat in order to obtain a mixture containing 4% butterfat?

5. How many pounds of coffee at $4.20 per pound should be mixed with 5 pounds worth $5.10 per pound to make a mixture worth $4.50 per pound?

6. Alloy A contains 40% copper and 60% zinc by weight; alloy B contains 20% copper and 80% zinc by weight. How many pounds of the first alloy should be combined with 50 pounds of the second to form a new copper-zinc alloy that contains 25% copper?

7. Ten gallons of chocolate ice cream containing 20% butterfat is mixed with 8 gallons of coffee ice cream contianing 14% butterfat to make mocha ice cream. What is the butterfat content of mocha ice cream?

8. A vineyard in France sells three types of champagne: dry, extra dry, and brut. Extra dry is a mixture of brut, which costs 55 francs per liter, and dry, which costs 47 francs per liter. If this vineyard wants to sell extra dry at 53 francs per liter, how many liters each of dry and brut are needed to make 800 liters of extra dry?

9. In the bookstore the other day, Carol was buying a magazine that cost $1.50. After the cashier rang up the sale she reached into her pocket to find she had left her wallet at home, and had only a handful of pennies, nickels, and dimes. She had twice as many pennies as she had nickels, and five times as many nickels as dimes. Her nickels and dimes together came to $1.05.

After she gave the cashier everything she had, how much did she have to borrow from the lady behind her to buy the magazine?

10. Argentieri's Greasy Spoon Cafe sells quarter-pound burgers, but unlike MacDonald's, Argentieri's are not all beef. In fact, Argentieri's burgers are a mixture of Beeftane (a soy product) and Crd-Brd-2 (a cellulose derivative). Beeftane costs 45¢ per pound and Crd-Brd-2 costs 21¢ per pound. If Argentieri wants to have the final product cost 25¢ per pound, how many pounds each of Beeftane and Crd-Brd-2 are needed to make 600 quarter-pound burgers?

11. A truant officer is licensed to round up students on the junior high and senior high school levels. He gets a commission of $4.20 per junior high student and $7.40 per senior high school student. In the first three weeks of October he rounds up a total of 12 junior high students and 10 senior high students. If the junior high school is on vacation the last week of the month and therefore has no truants, how many senior high school students must he snare to bring his average commission for the month up to $6.60 per truant?

12. According to the Bible, "...he (Judas) repented and took back the 30 silver pieces to the chief priests and elders" (Matthew 27:3). Assuming that the 30 coins were Greek drachmas and didrachmas (worth two drachmas) and that the total worth of the coins was 2.40 staters, how many of each coin were there? (1 stater = 20 drachmas.)

13.8 DISTANCE PROBLEMS

Most math books contain a mass of problems about distance, speeds, and times, which are all solved using the relation:

$$\boxed{\text{distance} = \text{speed} \cdot \text{time}}$$

or, equivalently,

$$\text{speed} = \frac{\text{distance}}{\text{time}} \quad \text{or} \quad \text{time} = \frac{\text{distance}}{\text{speed}}$$

My chief memory of these problems is that an abnormally large number of them were about trains, but perhaps that is because I was brought up on a textbook that must have been written in Victorian England in the heyday of the railway expansion. In any case, this type of problem actually embodies some useful ideas, and so you are going to get some of this type too. Both examples depend on your realizing that when one object meets or overtakes another, they are both at the same place at the same time—I know this is ridiculously obvious, but it is sometimes easy to forget it in the flurry of doing a problem.

EXAMPLE: *An overzealous student burns the candle at both ends. One night he lights the top and the bottom of a 12-inch candle at the same time. If the top flame burns at a speed of 1.5 inches per hour and the bottom burns at a speed of 2.1 inches per hour, how far from the top do the flames meet?*

STEP 2. Let x = the distance in inches from the top that the flames meet.

STEP 3. Total length of candle = 12 in., so if top burns down x in, bottom burns up $(12 - x)$ in. See Figure 13.3.

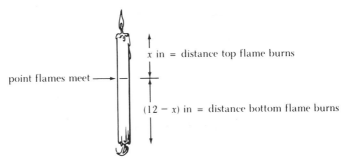

point flames meet ⟶

x in = distance top flame burns

$(12 - x)$ in = distance bottom flame burns

FIG. *13.3*

STEP 4. When the flames meet, they have both been burning for the same amount of time, so

$$\begin{pmatrix} \text{time top} \\ \text{flame has} \\ \text{been burning} \end{pmatrix} = \begin{pmatrix} \text{time bottom} \\ \text{flame has} \\ \text{been burning} \end{pmatrix}$$

STEP 5. Since the problem is about distances, speeds, and times, it seems likely that we'll need:

distance = speed · time

but since the equation above asks for time, the following form will probably be better:

$$\text{time} = \frac{\text{distance}}{\text{speed}}$$

STEP 6. Since the top flame burns at a speed of 1.5 in. per hr,

time for the top flame to burn x in. is $\dfrac{x}{1.5}$ hrs

Now the bottom flame burns at 2.1 in. per hr, so

time for the bottom flame to burn $(12 - x)$ in. is $\dfrac{12 - x}{2.1}$ hrs

STEP 7.   Substituting into the equation:

$$\frac{x}{1.5} = \frac{12 - x}{2.1}$$

STEP 8. $2.1x = 1.5(12 - x)$

$2.1x = 18 - 1.5x$

$3.6x = 18$

$x = 5$ in.

So the flames meet 5 inches from the top.

STEP 9.   This is reasonable because the top burns slower, so we would expect the meeting point to be less than halfway down the candle—i.e., less than 6 in.

Check:

time for top flame to burn 5 in. is $\dfrac{5}{1.5} = \dfrac{10}{3}$ hours

time for bottom flame to burn the other 7 in. is $\dfrac{7}{2.1} = \dfrac{10}{3}$ hours—O.K.

EXAMPLE: *(To preserve the train tradition.) One train leaves a station at noon and travels at 50 miles per hour (mph) toward the next city. One-and-a-half hours later an express train starts out from the same station, and travels at 80 mph along the same line. At what time does the express train overtake the slow train?*

We want to find the time at which the fast train catches up with the slow one; let this be x hours after the fast train has started. Our equation will come from the fact that the two trains have traveled the same distance when one overtakes the other.

$$\begin{pmatrix} \text{distance traveled by express} \\ \text{train when overtaking} \end{pmatrix} = \begin{pmatrix} \text{distance traveled by slow} \\ \text{train when overtaken} \end{pmatrix}$$

Now the express train has traveled for x hours at 80 mph when it overtakes the slow one, so

distance traveled by express train when overtaking $= 80x$ miles

The slow train started $1\frac{1}{2} = \frac{3}{2}$ hours earlier, and so has traveled for $(x + \frac{3}{2})$ hours at 50 mph by the time it is overtaken, so distance traveled by slow train when overtaken

$$= 50\left(x + \frac{3}{2}\right) \text{ miles}$$

Therefore the equation is

$$80x = 50\left(x + \frac{3}{2}\right)$$

Solving:

$$80x = 50x + 75$$

$$30x = 75$$

$$x = \frac{75}{30} = \frac{5}{2}$$

Therefore the express train travels for $\frac{5}{2} = 2.5$ hours before over-taking and the slow train travels $\frac{3}{2} + \frac{5}{2} = \frac{8}{2} = 4$ hours before being overtaken. Since the slow train started at noon, the trains meet at 4 p.m.

CHECK: At 4 p.m., the slow train has gone $50 \cdot 4 = 200$ miles, the express train has gone $80 \cdot \frac{5}{2} = 200$ miles; therefore they are at the same place—as would be expected.

Note: In this train problem the two trains started from the same place but at different times. In the candle problem we had two flames starting at the same time but from different places. Both problems wanted us to find where the two things met. In the train problem we did this by saying that they traveled for the same distance; in the candle problem by saying that they burned for the same time.

PROBLEM SET   13.8

Some problems in the Great Train Tradition:

1. A freight train and a passenger train leave Boston simultaneously for the 1050-mile trip to Chicago. If the freight train goes 20 miles per hour (mph) slower than the passenger train and takes 35 hours to make the trip, how far is the freight train from Chicago when the passenger train arrives there?

2. A train heading from Chicago to Santa Fe travels an hour and a quarter at 60 mph. It then picks up speed and continues westward at 85 mph for two and a half hours. How far has it traveled?

3. Two of the famous Japanese "bullet" trains approach one another

along straight parallel tracks. They start out at the same time from stations 180 miles apart. One train travels at 200 mph, the second train at 160 mph. How far has the faster train gone when they pass each other?

4. Two trains leave St. Louis headed in opposite directions, one going 40 mph faster than the other. After 5 hours, the trains are 300 miles apart. Find the speed of the slower train.

5. The New York-Boston "milk run" train, traveling at 30 mph, leaves New York 3 hours before the express train running along a parallel track at 75 mph. How long will it take the express train to overtake the "milk run" train? How far from New York will the trains be?

6. The track used by trains running from New York to Washington, D.C., is in much better shape than the track used by trains making the trip in the opposite direction. Thus, a round trip takes 16 hours, averaging 20 mph from Washington to New York and 30 mph for the return trip. How long does the New York-Washington leg take?

7. The B&M commuter trains travel from Acton into Boston at a speed of 25 mph during the morning rush hours. The return trip to Acton in the evening takes 10 minutes less, because the train travels 5 mph faster. How far is Acton from Boston?

8. A train leaves New York for Cleveland at the same time a train leaves Cleveland for New York, one train traveling 20 mph faster than the other; they meet in 10 hours. If the distance between Cleveland and New York is 600 miles, find the speed of the faster train.

9. A hitchhiker begins to walk along a straight road at the rate of 6 mph to the next town, which is 36 miles away. After 10 minutes, a car picks him up, and an hour and a quarter after being picked up he is in town. How fast was the car traveling?

10. The top runners in the Boston Marathon cover the 26-mile distance in 2 hours and 15 minutes. If they average 12 mph over the level part of the course, but only 8 mph in the infamous Heartbreak Hill area, how many of the 26 miles are level?

11. A skier leaves the lodge at the base of the mountain and immediately takes a 3000-foot gondola ride to the top of the mountain. She waits for 10 minutes and then skis down directly under the path of the gondola, returning to the lodge exactly 30 minutes after leaving it. If she skis four times as fast as the gondola moves, at what rate does she ski?

12. The East German Olympic champion crew left the boathouse and rowed upstream at the rate of 8 mph. It returned downstream at the rate of 16 mph. If the round trip takes 1½ hours, how far upstream did they row?

13.9   WORK PROBLEMS

Like the distance problems, work problems have a way of coming up in every math book, and of looking entirely useless. Indeed, I think they do verge on the redundant, but that may be because they are so often about filling swimming pools with one versus two pipes (something I've never worried about). Anyway, in spite of their possible uselessness, work problems provide very good examples of problems in which one has to add *rates*, rather than any other quantities. Most work problems revolve around the idea that:

$$\begin{array}{ccc} \text{rate of working of} \\ \text{A and B together} \end{array} = \begin{array}{c} \text{rate of working} \\ \text{of A alone} \end{array} + \begin{array}{c} \text{rate of working} \\ \text{of B alone} \end{array}$$

where

$$\text{rate of working} = \frac{\text{amount of work done}}{\text{time to do work}}$$

Rate of working represents the amount of work done in unit time. This equation tells you that the amount of work done in unit time by A and B together is equal to the amount of work done by A alone plus the amount of work done by B alone—which is obviously true unless A and B interfere with each other.

EXAMPLE: *(To preserve the swimming pool tradition.) A bathtub takes 10 minutes to fill using the hot tap only, and 15 minutes using the cold tap only. How long do both taps together take?*

First, please resist any impulses you have to say that both taps together take $10 + 15 = 25$ minutes. Both taps must fill the tub quicker than either tap alone!

STEP 2.   Let $x =$ the number of minutes to fill the tub using both taps.

STEP 4.   The work to be done in this problem is filling the tub, so we can say:

$$\begin{pmatrix} \text{rate at} \\ \text{which taps} \\ \text{fill the tub} \\ \text{together} \end{pmatrix} = \begin{pmatrix} \text{rate at} \\ \text{which hot} \\ \text{tap fills} \\ \text{the tub,} \end{pmatrix} + \begin{pmatrix} \text{rate at} \\ \text{which cold} \\ \text{tap fills} \\ \text{the tub} \end{pmatrix}$$

STEP 5.   We will need

$$\text{rate} = \frac{\text{amount of work}}{\text{time to do work}}$$

Here "amount of work" = 1, because 1 tub is to be filled.

STEP 6.   Rate at which hot tap fills tub

$$= \frac{\text{amount of work}}{\text{time}} = \frac{1 \text{ tub}}{10 \text{ min}} = \frac{1}{10} \text{ tub per min}$$

Rate at which cold tap fills tub

$$= \frac{1 \text{ tub}}{15 \text{ min}} = \frac{1}{15} \text{ tub per min}$$

Together, rate at which the two taps fill tub

$$= \frac{1 \text{ tub}}{x \text{ min}} = \frac{1}{x} \text{ tub per min}$$

STEP 7.   The equation becomes

$$\frac{1}{x} = \frac{1}{10} + \frac{1}{15}$$

STEP 8.   Solving:

$$\frac{1}{x} = \frac{3+2}{30} = \frac{5}{30} = \frac{1}{6}$$

so

$$x = 6 \text{ min}$$

STEP 9.   Being less than both 10 and 15, this is at least a reasonable answer.

Check: Use: total work done = rate · time.
In 6 min, hot tap fills $6 \cdot \frac{1}{10} = 0.6$ of tub and
cold tap fills $6 \cdot \frac{1}{15} = 0.4$ of tub. So at the end of 6 min, the tub is full.

EXAMPLE: *If a regular oil tanker splits open, it takes 30 minutes to cover 2 square miles of sea with oil. One day a supertanker collides with a regular tanker, both split open, and in 42 minutes cover a 7-square-mile bay with oil. How long would it have taken for the bay to have been covered with oil if the supertanker alone had split open?*

Let x = time (min) for the supertanker to cover the bay with oil.
Then

$$\begin{pmatrix} \text{rate at which} \\ \text{tankers spill} \\ \text{oil together} \end{pmatrix} = \begin{pmatrix} \text{rate at which} \\ \text{regular tanker} \\ \text{spills oil} \end{pmatrix} + \begin{pmatrix} \text{rate at which} \\ \text{supertanker} \\ \text{spills oil} \end{pmatrix}$$

Now, the tankers together cover 7 square miles (mi²) in 42 minutes, so

$$\text{rate together} = \frac{7}{42} = \frac{1}{6} \text{ mi}^2 \text{ per min}$$

The regular tanker covers 2 square miles in 30 minutes, so

$$\text{rate of regular tank} = \frac{2}{30} = \frac{1}{15} \text{ mi}^2 \text{ per min}$$

The supertanker covers 7 square miles in x min, so

$$\text{rate of supertanker} = \frac{7}{x} \text{ mi}^2 \text{ per min}$$

Therefore the equation reads

$$\frac{1}{6} = \frac{1}{15} + \frac{7}{x}$$

Solving:

$$\frac{7}{x} = \frac{1}{6} - \frac{1}{15} = \frac{5 - 2}{30} = \frac{3}{30}$$

$$\frac{7}{x} = \frac{1}{10}$$

$$x = 70$$

So it would be 70 minutes before the supertanker covered the bay with oil.

CHECK: We'll check that the entire bay does get covered in oil in 42 minutes if both tankers split. To do this use

total amount of oil spilt = rate · time

In 42 minutes, the regular tanker (whose rate is $\frac{1}{15}$ mi² per min) covers $(\frac{1}{15}) \cdot 42 = 2.8$ mi² with oil. During the same time the supertanker (whose rate is $\frac{7}{70} = \frac{1}{10}$ mi² per min) covers $(\frac{1}{10}) \cdot 42 = 4.2$ mi² with oil. Therefore in 42 minutes, both tankers together cover $2.8 + 4.2 = 7$ mi², that is, the whole bay.

PROBLEM SET 13.9

1. Two doctors are checking out the day's X-rays. Carol can do the job alone in 12 minutes, while Donna can do the job by herself in 9 minutes. How long does it take the two working together?

2. Ken can write all the boring word problems for this book in 10 hours. After he has worked 2 hours, Steve comes to help him and together they finish the problems in 3 more hours. How many hours would it have taken Steve working alone?

3. The United States possesses sufficient nuclear firepower to completely obliterate Europe in 3 hours. If the United States and the Soviet Union combine their efforts, Europe can be destroyed in 2 hours. How long would it take the Soviet Union alone to destroy Europe?

4. If Michael can clean an apartment in 3 hours, Ross can do it in 4 hours, and Earl in 5 hours, how long will it take them to clean the apartment working together?

5. Josh can set the type for one chapter of this book in 10 hours, while Alice takes 12 hours to do the same job. How long will it take to set type if both of them work together?

6. Suppose one pump can fill a pool in t_1 hours and a second pump can fill the pool in t_2 hours. How long will it take them together to fill three such pools? (*Note:* The answer will involve t_1 and t_2.)

7. Suppose one pump can fill a pool in 4 hours and a second pump can empty the full pool in 6 hours. If the pool is half full and both pumps are started, how long will it take for the pool to become full?

8. One photocopying machine can copy a 500-page paper in 12 minutes. A newer model of the same machine can copy the same paper in just 10 minutes. How long will it take the two machines working together to copy an 825-page paper?

9. One of the famous Vanderbilt mansions in Newport, Rhode Island, the Breakers, has a bathtub with four faucets—hot and cold for fresh water and hot and cold for salt water. If the salt water faucets together fill the tub in 12 minutes and the fresh water faucets together fill it in 8 minutes, how long will it take to fill the tub if all four faucets are used?

10. Still at the Breakers: The hot salt water faucet can fill the tub alone in 18 minutes. If the hot salt water faucet breaks, how long will it take the remaining three faucets to fill the tub?

11. Still at the Breakers: When the stopper is lifted, the filled tub empties in 10 minutes. If all four faucets are working and the stopper is lifted, how long will it take to fill two-thirds of the tub?

12. Still at the Breakers: The hot fresh water faucet fills the tub twice as fast as the cold fresh water faucet. How long does it take the hot fresh water faucet alone? (Remember from Problem 9 that they take 8 minutes together.)

13.10   PROPORTIONALITY PROBLEMS

If you are buying a certain kind of gasoline, the amount you pay depends on how much you buy. Buying twice as much will cost you twice as much; buying a third as much will cost you a third as much, and so on. The relationship between price and quantity is described by saying that the price is *proportional* to the quantity, and writing the equation

$$\text{price} = \text{price per gallon} \cdot \text{quantity}$$

which is of the form

$$\text{price} = \text{constant} \cdot \text{quantity} \qquad (\text{constant} = \text{price per gallon})$$

In general, *y is said to be proportional (or directly proportional) to x* if *x* and *y* are connected by an equation of the form:

$$\boxed{y = kx}$$

where k = some constant called the *constant of proportionality*. Here again, doubling x causes y to double, halving x causes y to halve, and so on.

Many quantities that occur frequently are proportional to one another. For example, if you are traveling at a steady speed, the distance traveled is proportional to the time spent traveling because

$$\text{distance} = \text{speed} \cdot \text{time} \qquad (\text{constant of proportionality} = \text{speed})$$

The number of legs in a room full of people is proportional to the number of people because

$$\text{number of legs} = 2 \cdot \text{number of people} \qquad (\text{constant proportionality} = 2)$$

Of course there are equally many things that are *not* proportional to one another. Height, for example, is certainly not proportional to age, and the grade you receive in a course is all too often not proportional to the work you put in. However, now I'd like you to look at an example of two quantities that, although not proportional, are related in a similar way—only "upside down."

Suppose you are going a fixed distance, say from San Francisco to Los Angeles. If you go twice as fast, it will take you half as long; if you go one-third as fast, it will take you three times as long, and so on. The time and the speed are said to be *inversely proportional* to one another, or to *vary inversely* with one another; they are connected by the equation:

$$\text{time (hours)} = \frac{404}{\text{speed (miles per hour)}} \qquad \begin{array}{l}(404 = \text{distance in} \\ \text{miles from S.F.} \\ \text{to L.A.})\end{array}$$

which is of the form:

$$\text{time} = \frac{\text{constant}}{\text{speed}}$$

In general, *y is said to be inversely proportional to x, or to vary inversely with x* if:

$$\boxed{\; y = \frac{k}{x} \quad}\; k = \text{constant}$$

Here doubling x causes y to halve, and halving x causes y to double.

Think now of the area of a circle. If the radius is doubled, the area is multiplied by four; if the radius is multiplied by one-third, the area is multiplied by one-ninth, and so. Here the area is said to *vary as the square* of the radius, which is expressed as

$$\text{area} = \pi(\text{radius})^2$$

which is of the form

$$A = \text{constant} \cdot (\text{radius})^2$$

In general, y is said *to vary as the square* of x if

$$y = kx^2 \qquad k = \text{constant}$$

y varies as the cube of x if

$$y = kx^3 \qquad k = \text{constant}$$

y varies inversely as the square of x if

$$y = \frac{k}{x^2} \qquad k = \text{constant}$$

y varies inversely as the fourth power of x if

$$y = \frac{k}{x^4} \qquad k = \text{constant}$$

Questions involving proportionality involve using these formulas.

EXAMPLE: *The number of new jokes a professional comedian needs to find each year is inversely proportional to the number of years he has been in the profession. If, after 5 years in the business, he needs to find 15 new jokes each year, how many will he need to find each year after 25 years?*

STEP 2. Let N = the number of jokes he will need to find after 25 years in the business.

STEP 4.
$$\text{number of new jokes} = \frac{k}{\text{years in business}}$$

STEPS 5 and 6. In this kind of problem we must be given either the constant of proportionality (which we aren't) or a complete set of data for the variables from which we can find the constant (this is the situation here). So now we find k using the fact that after 5 years he needs 15 jokes. This tells us that

$$15 = \frac{k}{5}$$

So

$$k = 75$$

STEP 7. Since, after 25 years, the number of jokes needed is N, substituting into

$$\text{number of new jokes} = \frac{k}{\text{years in business}}$$

gives

$$N = \frac{k}{25} = \frac{75}{25} \qquad (\text{since } k = 75)$$

STEP 8. So

$$N = 3$$

STEP 9.   The fact that the number of jokes is inversely proportional to the number of years means that as the number of years increases, the number of jokes should go down—which it does.

EXAMPLE: *The number of angels that can dance on the head of a pin is directly proportional to the square of the radius of the head of the pin. If 17 angels can dance on the head of a pin with a radius of 2 millimeters (mm), how many can dance on the head of a pin with radius 6 mm?*

Let n = the number of angels and r = the radius of the head of the pin. Then n is directly proportional to the square of r, so

$n = kr^2$

When $r = 2$, $n = 17$, so

$17 = k2^2 = 4k$

and therefore

$k = \dfrac{17}{4}$

Then if $r = 6$,

$n = k6^2 = \dfrac{17}{4} \cdot 36 = 153$

So 153 angels can dance on a 6 mm pinhead.

PROBLEM SET   13.10

1. In a bank, the annual interest received is directly proportional to the amount of money you put in your savings account. If the annual interest on $5500 is $247.50, what is the annual interest on $8000?

2. For meshed gears, the number of teeth and the number of revolutions vary inversely. A gear having 36 teeth drives another that has 48 teeth. If the first gear makes 200 revolutions, how many revolutions does the second gear make?

3. For pulleys connected by the same belt, diameters and revolutions vary inversely. If two pulleys connected by the same belt have diameters of 5 inches and 7 inches, and the larger pulley turns at 350 revolutions per minute (rpm), what is the turning speed of the smaller pulley?

4. For an enclosed gas at a constant temperature, Boyle's law states that pressure and volume are inversely proportional. If the volume is 21 cubic centimeters (cm³) when the pressure is 760 millimeters of mercury, what is the volume at a pressure of 400 millimeters of mercury?

5. After the brakes have been applied, the distance an automobile goes before stopping varies directly as the square of its speed. For an auto going 30 mph, the stopping distance is 54 feet. What is the stopping distance of an auto going 50 mph? (How does this compare to the National Safety Council's statement that drivers should allow one car length of stopping distance for every 10 mph of speed? An average car is about 16 feet long.)

6. The energy of a moving body varies directly as the square of its speed. If a body has an energy of 50 units at a speed of 5 feet per second, what is its speed when its energy is 450 units?

7. The intensity of light on a surface varies inversely as the square of the distance from a point source. If the distance from the source changes from 14 inches to 3.5 inches, how many times brighter will the intensity become?

8. The volume of a sphere varies as the cube of the radius. If sphere A has twice the volume of sphere B, what is the ratio of their radii?

A variable Q is said to vary jointly as x and y if Q varies directly as the product xy, that is, if $Q = k \cdot xy$.

9. The horsepower of a steam engine varies jointly as the average pressure in the cylinder and the speed of rotation. When the average pressure is 200 pounds per square inch (psi), the engine is turning at 375 revolutions per minute (rpm), the horsepower is 25. What is the horsepower when the average pressure is 300 psi and the engine is turning at 600 rpm?

10. The volume of a cone varies jointly as the altitude and the square of the radius. When the radius is 3 and the altitude is 8, the volume is 24π. What must the altitude be if the volume is 12π when the radius is 2?

11. Newton's law of gravitation states that the force F of attraction between two bodies varies jointly as their masses m_1 and m_2 and inversely as the square of the distance between them. Two bodies whose centers are 500 feet apart attract each other with a force of 10 pounds. What would be the force of attraction if their masses were tripled and the distance between their centers was doubled?

12. The safe load strength of a horizontal beam supported at both ends varies jointly as the breadth and the square of the depth, and inversely as the length of the beam. If a beam 20 feet long with breadth 4 inches and depth 8 inches can safely support 1600 pounds, what is the safe load for a beam 8 feet long with breadth 2 inches and depth 4 inches?

CHAPTER 13 REVIEW

1. At a certain university, there are 350 students enrolled in the introductory chemistry course. Of these, some are taking the course so that they can apply to medical school, some are taking it because it is required by their major, and the rest are taking it simply because they find the ideas interesting. If the number of pre-meds is three times the number of people taking the course for their major, and if there are 50 more people taking it for their major than there are taking it for general interest, find how many of each type of student there are. (Assume there is no overlap between the three groups.)

2. In five years time, a child will be three times as old as she was last year. How old is the child now?

3. Ellen and Ann sat in the bleachers at Fenway Park drinking beer and betting on balls and strikes. Ellen paid Ann a dollar for each strike and Ann paid Ellen a dollar for each ball. After 30 pitches Ann had won from Ellen $6 more than twice what Ellen had won from Ann. How many of the 30 pitches were strikes?

4. One spills one's milk less frequently as one grows older. In fact, the frequency of milk spilling is inversely proportional to age. If, at age 6, one spills one's milk 8 times per year, how frequently does one spill it at age 24?

5. A certain Hollywood personality spent a quarter of her life as a child movie star, a fifth as a bratty adolescent, a half as a sexy starlet, and has spent the last two years in obscurity. How old is she now?

6. A cheap beer-and-pizza place in a college town serves draft beer that has been diluted by water such that the final brew contains 20% tapwater. If the restaurant uses 20 kegs of undiluted beer each week, how much beer do they serve to students?

7. A physician treating a patient with hypertension (high blood pressure) prescribes two diuretics (substances that induce water loss): spironolactone and hydrochlorothiazide. Spironolactone costs 10¢ per 25 milligram (mg) tablet and hydrochlorothiazide is 3¢ per 25-mg tablet. The patient is told to take one spironolactone and two hydrochlorothiazide tablets each day. How many days supply of medication can he buy with $4?

8. The hypertension of the patient in Problem 7 persists and the doctor adds a prescription of 50 mg of hydralazine three times a day to the previous drugs. He tells the patient that $8.40 will cover a 30-day supply of all three drugs. How much is a 50-milligram tablet of hydralazine?

9. A glass packer is paid 8¢ for each article he packs, but is fined 40¢ for each one he breaks. One day near quitting time he realizes that he has packed 30 times as many articles as he has broken, and he quickly packs 13 more

without breakage. His net earnings for the day were $19.04. What was the total number of articles he packed successfully?

10. In Memorial Church, 310% more people come to the Easter service than on ordinary Sundays. If Reverend Cadence delivers the sermon, 40% of the congregation politely falls asleep with their eyes open. If the ordinary attendance is 50, how many people will be asleep in the pews if Reverend Cadence delivers the Easter sermon?

11. A company designs a circular dartboard such that the target area in the center is surrounded by a safety margin. Federal law requires that the area of the safety margin must be twice the area of the actual target in the center. If the total radius of the dartboard is 3 feet, what is the radius of the circular target in the center?

12. Suppose you have 36 coins, making a total of $5.05 in change. If $4.00 of this is in quarters, how many nickels do you have if the rest is all in dimes and nickels?

13. In a math text factory for making up word problems, each problem is tested before being published. On the average 96.5% of those tested are found to be boring and repetitive; the rest are thrown away. If the garbage can contains 560 interesting word problems a week on an average, how many word problems does the factory write per week?

14. An express train leaves New York at 3:00 p.m. and reaches Boston at 6:00 p.m. A slow train leaves Boston at 1:30 p.m. and arrives at New York at 6:00 p.m. If both trains travel at constant speeds, at what time do they meet?

15. If the wheat crop in the United States averages 680 million bushels per year over a 5-year period, what must be the average harvest during the next 2 years in order for the 7-year average to be 700 million bushels per year?

16. The sum of three consecutive odd numbers is 41 more than twice the smallest number. Find the largest of the integers.

17. A contractor estimated that a certain job could be done by 9 men in 8 hours, or by 16 boys in 9 hours. If 4 men and 4 boys started at 7:00 a.m., at what time did they finish if they took 1 hour off to eat?

18. In a protest against Star-Kist, three fish decide to commit suicide by drinking all the water in a 50-gallon tank. The sardine can drink 3 gallons per hour faster than the tuna, and the guppy can drink twice as fast as the sardine. If it takes 2 hours for the last fish to flop (i.e., until there is no water left in the tank), how fast (in gallons per hour) does the sardine drink his way to his dark, doleful death?

19. The rate of growth of Chris' snake is inversely proportional to the square of its age. If at 16 years old its rate of growth is 12 cubic inches per year, how fast will it be growing when it is 24 years old?

20. Two skiers race down a slope. Jill, the faster one, gives her brother Merrick a 100-yard lead. If Jill skis at 40 mph and Merrick at 35 mph, how far from Merrick's starting point does Jill overtake him? (Find the distance in *yards*.)

21. A nonprofit neighborhood health center is buying vitamins to sell to its patients. They find that, on the average, 2% of the bottles they order arrive broken and hence are useless.

(a) If the clinic has 4900 patients, how many bottles of vitamins should it order to have one unbroken bottle per patient?

(b) If the manufacturer charges $1.47 per bottle, what should the clinic charge to cover its costs, including broken bottles?

(c) An election puts a new director in charge of the clinic who insists that one-third of the clinic's revenue from sales must be profit. How much will the clinic now charge for a bottle of vitamins?

22. An obstetrician advises his patients to take Fero-gradumet® (an iron supplement) and vitamin C tablets while pregnant. If Fero-gradumet® costs $51.40 per thousand tablets, vitamin C costs $2.18 per hundred, and the prescription calls for twice as many vitamin C tablets per day as Fero-gradumet®, how many tablets of each kind should the patient get if she spends $142.50?

23. How soon after noon are the hands of a clock together again?

24. Two years ago, IBM made 70% of their profits in the United States and 30% in foreign countries. Last year their domestic profits declined by 4%, but their overall profit remained unchanged.

(a) What percent of their total profit came from domestic sales last year?

(b) By what percent did their foreign profits increase from 2 years ago to last year?

25. Two seagulls, Jonathan and Jenny, are racing between two cliffs 300 yards apart. They both start from the same cliff and head for the other. Now, if Jonathan takes off 4.5 seconds after Jenny and overtakes Jenny when she is nine-tenths of the way to the finishing line, what is Jonathan's average speed, if his time of flight from cliff to cliff is 15 seconds? What is Jenny's average speed?

26. Find the final age of Sam, an eccentric mathematician whose epitaph reads:
Sam spent one-sixth of his life as a child, one-twelfth as a youth, and one-seventh as a bachelor. Five years after his marriage, a son was born who died four years before his father at half his father's final age.

27. Now that she is over seventy, my great-aunt insists on demonstrating her youth on her birthday. Last year she hiked into the country at a rate of 3 mph and bicycled back at 24 mph. She had 6 hours at her disposal. How far into the country did she go?

28. A man made a 4000-mile journey in the United States. The first part of the journey was made before the announcement of the lowering of the speed limit due to the energy crisis; so he traveled then at an average speed of 60 mph. Then, when he made the second part of his journey, he had to cut down his average speed to 50 mph. Now, if the time taken for the second part of the journey was two-fifths of the time taken for the first part:

    (a) How many hours did first part of the trip take?

    (b) How many miles did he travel during the first part of the journey?

29. The larger of two numbers minus four times the smaller gives 43. If the two numbers average to 4, what are the numbers? (Hint: If the average is 4, what is the sum of the two numbers?)

30. An old, and very finicky, natural gas compressor in an oil field burns a fuel mixture that must be precisely two-thirds gasoline and one-third oil. After the cranky machine failed to operate, the head roustabout took a look at the fuel tank and discovered that some of the gas had evaporated, leaving 4 gallons of evenly mixed oil and gas. How much gas should the roustabout add to the tank to satisfy the compressor?

31. The "80-20" rule is frequently (and not always facetiously) used by engineers. It asserts that the first 80% of any project can be completed in the first 20% of the time, and that the last 20% of a project requires 80% of the time. If you were managing a project that had been 80% completed in 2 weeks, how long would you guess the total duration would be?

32. The Swindle Oil Company sells three types of gas: Super-Swindle, Regular-Swindle and Mini-Swindle. Regular-Swindle is a mixture of Super-Swindle ($1.48 per gallon) and Mini-Swindle ($1.42 per gallon). Suppose Swindle Oil wants to sell Regular-Swindle at $1.47 per gallon. How many gallons of Super- and Mini-Swindle are needed to make 900 gallons of Regular-Swindle?

33. You are building a house, and have 8 feet of wood to use for window ledges. Your blueprint plan calls for two windows, one rectangular, and one triangular, each 3 feet high, and energy considerations tell you that the total area of the two windows should be no more than 21 square feet. How should you cut up the wood for the window ledges to make the largest windows possible?

34. A pharmacist wants to make up 150 cubic centimeters (cm^3) of a solution containing $\frac{5}{3}$ grams per liter of hydrogen peroxide (H_2O_2). She has a standard solution (i.e., one already made up at some fixed concentration) of H_2O_2, and realizes that 100 cm^3 of the standard solution plus 50 cm^3 of water will give her what she needs. What was the concentration (in grams per liter) of the standard solution? (Note: 1 liter = 1000 cm^3.)

35. Suppose $1\frac{1}{2}$ men can do $1\frac{1}{2}$ jobs in $1\frac{1}{2}$ days. How long does it take one man to do one job?

36. The police department of a large city has recently come under fire for its failure to hire minorities. In fact, the federal government is withholding revenue-sharing funds until a recently implemented affirmative action programs brings the number of minority officers to a point where they comprise 12% of the force. If there are currently 1000 minority members in the 12,000-man department, and new officers are added from the civil service list at a rate of two minority to each white, how many new officers must be added before revenue-sharing funds start flowing again? (Assume there are no retirements.)

37. The demand, *D*, for a certain expensive sports car is related to the price, *p*, in dollars as follows:

$$D = 2.25(25{,}000 - p)$$

The supply equation relates the number of cars, *S*, that dealers are willing to supply, to the price *p:*

$$S = 3(p - 18{,}000)$$

What price will "clear the market"? That is, what is the price at which supply and demand will be equal? How many cars will be sold at this price?

38. Suppose the price, *P*, per egg in cents is related to weekly demand, *D*, and supply, *S*, by the following equations:

$$\frac{D}{10^8} + p = 10$$

$$S = \frac{1}{3} \cdot 10^9(p - 4)$$

Find the equilibrium price of eggs (i.e., when supply equals demand), and the quantity sold each week at this price.

39. Two nurses are responsible for taking the temperatures of all the patients on their floor. For 45 minutes before Rebecca came in, Rose took the temperature of 21 patients. Once Rebecca arrived, Rose took off for a coffee break and Rebecca worked alone for half an hour. When Rose got back, the two nurses worked together for an hour, taking 48 more temperatures. Assuming each nurse works at a constant rate, how many patients did they check in all?

40. Oil wells sometimes leak H_2S (hydrogen sulfide) gas while they are being drilled. The H_2S concentration generally varies inversely with the square of the distance from the drilling rig. If a concentration of $5 \cdot 10^3$ ppm (parts per million) is measured at 100 feet from the rig, what will the concentration be at 200 feet from the rig?

41. Safety requires that anyone working in an environment containing H_2S

at a concentration of 50 ppm or greater wear a gas mask. What is the minimum distance from the well described in Problem 40 at which such equipment is unnecessary?

42. A wildcatter figures that it costs him $10,000 to prepare a drilling site and then $15 a foot actually to drill the hole at a given location. If the wildcatter has just inherited $85,000 from a rich uncle, what's the deepest well he could drill?

43. A geologist has informed the wildcatter that she almost certainly knows where there is $250,000 worth of oil at a depth of 9000 feet. Assuming the wildcatter faces costs such as in the previous problem and wants to keep at least $50,000 in profits for himself, how much should he be willing to pay the geologist for the location of the oil?

44. A large chemical plant has applied to the Environmental Protection Agency for a permit to discharge diluted wastes into a stream running across its property. Engineering and environmental studies have concluded that the daily stream flow is 0.999 million gallons and that a waste concentration exceeding 100 parts per million would seriously affect the aquatic biological community. If the plant's discharge would contain 100,000 parts per million waste, what is the maximum daily volume of discharge that the EPA should allow the plant?

45. Thermal pollution is also a serious problem in American rivers. Water has a "heat content" equal to the product of its temperature and volume, and the temperature of a mixture can be found by dividing the sum of its heat contents by its total volume. A river has a flow of 100 million gallons per day and a normal temperature of 70°F. What is the maximum volume of water at 130°F that a nuclear power plant can be allowed to discharge into the river per day if a change in the river temperature of more than 8°F will wreck the balance of the aquatic ecosystem?

46. The first transcontinental railroad in this country was built by two crews working toward each other, meeting finally in Utah. Imagine a rail line being built in the opposite fashion: two crews starting in Utah and building away from each other. If the eastbound crew lays 12 miles more track each month than the westbound crew, and after 6 months the two crews are 648 miles apart, how fast is the westbound crew laying track?

47. A tree which grew squarely upright
    Broke in two and fell over one night.
    In the morning we found
    That the tip touched the ground
    At a distance just half the tree's height.

If this is true, see Figure 13.4 on page 274 and find how far up the tree the break occurred (i.e., what percentage is it of the total height of the tree)?

one-half the height of the tree

FIG. 13.4

48. How many apples are needed if four persons of six receive one-third, one-eighth, one-fourth, and one-fifth, respectively, of the total number, while the fifth receives ten apples, and one apple remains left for the sixth person? (From *Greek Anthology.*)

49. A powerful unvanquished excellent black snake which is 80 angulas in length, enters into a hole at the rate of $7\frac{1}{2}$ angulas in $\frac{5}{14}$ of a day, and in the course of $\frac{1}{4}$ of a day its tail grows $\frac{11}{4}$ of an angula. O ornament of arithmeticians, tell me by what time this serpent enters fully into the hole? (From *Mahāvira,* Hindu, c. 850.)

50. Of a collection of mango fruits, the king took $\frac{1}{6}$, the queen $\frac{1}{5}$ of the remainder, and the three chief princes $\frac{1}{4}$, $\frac{1}{3}$, and $\frac{1}{2}$ of the successive remainders, and the youngest child took the remaining three mangoes. O you who are clever in miscellaneous problems on fractions, give out the measure of that collection of mangoes. (From *Mahāvira,* Hindu, c. 850.)

14 QUADRATIC EQUATIONS

14.1   SETTING THE STAGE

Quadratic equations are as hard to define as linear equations. The rigorous definition is that a *quadratic equation* is one that can be transformed into the form

$$ax^2 + bx + c = 0 \qquad \text{where } a, b, \text{ and } c \text{ are constants,}$$
$$a \neq 0, \text{ and}$$
$$x \text{ is the unknown}$$

The difficulty with this is the same as before: Many quadratics don't come looking this way, and it is helpful to know what kind of equation you have from the start so that you know what to do with it. So here, for everyday use, is a loose definition:

An equation in one variable that contains the second but no higher powers of the unknown, no fractions with the unknown in the denominator and no roots of the unknown, is practically always a *quadratic equation*.

For example;

$$3x^2 = \frac{5}{2}x + 10$$

is a quadratic equation, and so is

$$(2x - 1)(5x + 3) = 9x$$

because when the left-hand side is multiplied out the equation becomes

$$10x^2 + x - 3 = 9x$$

On the other hand,

$$2x^2 - 9x = \frac{5}{x}$$

is not a quadratic equation because there is an x in the denominator. Nor is

$$(x^2 + 1)(x + 3) = x + 1$$

because when multiplied out it becomes

$$x^3 + 3x^2 + x + 3 = x + 1$$

Quadratic equations are also called *second-degree equations* because they involve polynomials of the second degree.

Unfortunately, quadratic equations cannot be solved the same way as linear equations. For example, suppose you have

$$x^2 - 3x + 2 = 0$$

and suppose you try to get this into the form

$$x = \text{some number}$$

by adding, subtracting, multiplying, or dividing things on both sides. Subtracting 2 from both sides of

$$x^2 - 3x + 2 = 0$$

gives

$$x^2 - 3x = -2$$

The trouble now is how to get just an x on the left-hand side. If you divide by x you get

$$x - 3 = -\frac{2}{x}$$

which can be rearranged into

$$x + \frac{2}{x} = 3$$

But neither way is much help because of the $\frac{2}{x}$. Dividing $x^2 - 3x = -2$ by 3 is even less help, since you still have an x^2 term:

$$\frac{x^2}{3} - x = -\frac{2}{3}$$

It is becoming increasingly clear that quadratics cannot be solved like linear equations, and we had better find some new methods specifically aimed at quadratics.

14.2   SOLVING QUADRATIC EQUATIONS BY FACTORING

One of the things we do know how to do with expressions like

$$x^2 - 3x + 2$$

is to factor them. You will be glad to know that the reason we spent so long on factoring was both because it is essential for adding algebraic fractions and because it gives us the easiest method for solving quadratic equations. This method depends on the following:

Useful Fact:

If the product of two numbers is zero, then one (or both) of the numbers is zero; that is,

$$\text{if } ab = 0 \quad \text{then } a = 0 \text{ or } b = 0 \text{ or both.}$$

If you aren't convinced of the truth of this Fact, think what happens if you multiply together numbers that are not zero. If you multiply two positive numbers or two negative ones, you will always get something positive—never zero. For example,

$$2 \cdot 7 = 14 \qquad (-3)(-2.5) = 7.5$$

$$\frac{1}{2} \cdot \frac{1}{8} = \frac{1}{16} \qquad (-0.2) \cdot (-0.1) = 0.02$$

Similarly, if you multiply a negative and a positive number, you will always get a negative number—never zero. For example,

$$2 \cdot \left(-\frac{1}{7}\right) = -\frac{2}{7} \qquad (-0.2) \cdot 4 = -0.8$$

So the only way the product of two numbers can be zero is if one of the numbers is zero to start with—which is just what the Fact says.

Note: This Useful Fact applies only when the product of the two numbers is *zero*. If their product is, say, 4, you cannot conclude that one of the numbers is 4 because both could be 2, or one could be 8 and one $\frac{1}{2}$, and so on.

Here is how a quadratic equation is solved:

If we factor the left-hand side of

$$x^2 - 3x + 2 = 0$$

we have

$$(x - 2)(x - 1) = 0$$

This is of the form

$$a \cdot b = 0 \qquad \text{with } a = (x - 2)$$
$$\text{and } b = (x - 1)$$

Therefore either $a = 0$ or $b = 0$ (or both). In other words, either $(x - 2) = 0$ or $(x - 1) = 0$ (or both). But if $(x - 2) = 0$, this means x satisfies the linear equation

$$x - 2 = 0$$

which has solution $x = 2$. And if $(x - 1) = 0$, then x satisfies the linear equation

$$x - 1 = 0$$

which has solution $x = 1$.

Therefore there are two solutions, namely, $x = 1$ and $x = 2$.

Check: When $x = 1$,

$$\text{LHS} = x^2 - 3x + 2 = 1^2 - 3 \cdot 1 + 2 = 1 - 3 + 2 = 0$$

$$\text{RHS} = 0$$

Therefore $x = 1$ is a solution.

When $x = 2$,

$$\text{LHS} = x^2 - 3x + 2 = 2^2 - 3 \cdot 2 + 2 = 4 - 6 + 2 = 0$$

$$\text{RHS} = 0$$

Therefore $x = 2$ is a solution.

To Solve Quadratic Equations by Factoring:

1. Get all the nonzero terms on the left-hand side of the equation, leaving a zero on the right.
2. Factor the left-hand side of the equation.
3. Set each factor of the left-hand side equal to zero, and solve the equations that you get.

EXAMPLE: *Solve $x^2 + 3x = 0$.*

x is a common factor on the left-hand side, and so this factors to:

$x(x + 3) = 0$

Since this means that the product of x and $(x + 3)$ is zero, one (or both) of them must be zero; therefore

either $x = 0$ or $x + 3 = 0$.

Now if $x + 3 = 0$ then $x = -3$.

Therefore the solutions are $x = 0$ and $x = -3$.

CHECK: When $x = 0$,

LHS $= x^2 + 3x = 0^2 + 3 \cdot 0 = 0$
RHS $= 0$

Therefore $x = 0$ is a solution.

When $x = -3$,
LHS $= (-3)^2 + 3(-3) = 9 - 9 = 0$
RHS $= 0$

Therefore $x = -3$ is a solution.

EXAMPLE: *Solve $6x^2 + x - 15 = 0$.*

Factoring the left-hand side gives

$(2x - 3)(3x + 5) = 0$

Therefore either $(2x - 3) = 0$ or $(3x + 5) = 0$.

If $2x - 3 = 0$	If $3x + 5 = 0$
then $2x = 3$	then $3x = -5$
or $x = \dfrac{3}{2}$	or $x = -\dfrac{5}{3}$

Therefore $x = \dfrac{3}{2}$ and $x = -\dfrac{5}{3}$ are the solutions.

CHECK: When $x = \dfrac{3}{2}$,

$$\text{LHS} = 6\left(\frac{3}{2}\right)^2 + \frac{3}{2} - 15 = 6 \cdot \frac{9}{4} + \frac{3}{2} - 15 = \frac{27}{2} + \frac{3}{2} - \frac{30}{2} = 0$$

RHS $= 0$

Therefore $x = \dfrac{3}{2}$ is a solution.

You can check $x = -\dfrac{5}{3}$!

EXAMPLE: *Solve $(6a - 5)(a + 1) = 10$.*

Here there is no zero on the right-hand side and we cannot use the factoring method until there is. Please notice that although we have

$$(6a - 5)(a + 1) = 10$$

we *cannot* deduce that $(6a - 5) = 10$ or $(a + 1) = 10$, because $(6a - 5)$ could be 2 and $(a + 1)$ could be 5, and so on. The Useful Fact is useful only when then there is a zero on the right!

In order to get

$$(6a - 5)(a + 1) = 10$$

into the usual form with a polynomial on the left and a zero on the right, we will have to multiply out on the left and then subtract 10 from both sides of the equation. This may seem like the long way round since we will then have to refactor, but it's the only way to get the 10 absorbed into the left-hand side. Multiplying out:

$$(6a - 5)(a + 1) = 10$$

gives

$$6a^2 + a - 5 = 10$$

so

$$6a^2 + a - 15 = 0$$

This is the same equation as in the example above, and so it factors to

$$(2a - 3)(3a + 5) = 0.$$

Therefore, the solutions are $a = \tfrac{3}{2}$ or $a = -\tfrac{5}{3}$.

EXAMPLE: *Solve $2x^2 = 2x + 24$.*

As in the last example, we must have everything on one side of the equation and a zero on the other before we factor. This is essential because the Useful Fact only tells us about two numbers *whose product is zero*—and no other situation. Therefore we subtract $2x + 24$ from both sides of this equation, giving

$$2x^2 - 2x - 24 = 0$$

Now take out the common factor of 2, and then factor as usual:

$$2(x - 4)(x + 3) = 0$$

An extension of the Useful Fact says that if the product of three numbers is zero, then at least one of them must be zero.

Now certainly $2 \neq 0$, (where $\neq$ means "is not equal to"); therefore,

either $x - 4 = 0$ or $x + 3 = 0$

so

$x = 4$ or $x = -3$

Therefore $x = 4$ and $x = -3$ are the solutions to $2x^2 = 2x + 24$.

Cancelling in Equations

To solve

$$2x^2 = 2x + 24$$

we might first have divided both sides of the equation by 2, or cancelled a 2 from both sides, leaving

$$x^2 = x + 12$$

We would then have proceeded as in the example above:

$$x^2 - x - 12 = 0$$
$$(x - 4)(x + 3) = 0$$

So either

$$(x - 4) = 0   \text{ or }   (x + 3) = 0$$

giving

$$x = 4   \text{ or }   x = -3$$

as before.

Thus, dividing both sides of the equation by 2 has exactly the same effect as throwing out the 2 after factoring (all the other factors led to a solution; the 2 did not because it cannot equal zero).

Suppose we try the same method on

$$x^2 = -3x$$

If we cancel an x from both sides, that is, divide both sides by x, we get

$$x = -3$$

But $x^2 = -3x$ should have two solutions because it is the same equation as $x^2 + 3x = 0$ whose solutions we earlier found to be $x = -3$ and $x = 0$. Therefore we have got one of the solutions to $x^2 = -3x$, but we seem to have lost the other one, namely $x = 0$. What has happened?

You may remember that when I introduced the "golden rule of equations" (that you can do anything you like to an equation, provided you do it to both sides), I said there was one exception—namely, that you can't multiply or divide by zero, even if you do it to both sides.

In the example above, we divided both sides by x, which can be zero. Dividing by zero is never a legal operation, and so it's not surprising that something went wrong (in this case we lost a solution to the equation).

In the first example, however, we were dividing by 2, which, not being zero, was perfectly legal, and so produced no problems. The moral is this:

You can cancel a factor from both sides of an equation (i.e., divide both sides of the equation by some quantity) only if you are sure that factor is not zero. In practice, this usually means that you can cancel only numerical factors and not an expression containing an x.

EXAMPLE: *Solve $t(2t - 1) = 3(2t - 1)$.*

Looking at the $(2t - 1)$ on both sides of the equation makes us think how nice it would be to simplify the equation by cancelling a $(2t - 1)$ from both sides. But this would be most illegal, since $(2t - 1)$ could be zero if $t = \frac{1}{2}$. Instead, move everything to the left side and take out the common factor of $(2t - 1)$:

$$t(2t - 1) - 3(2t - 1) = 0$$

$$(2t - 1)(t - 3) = 0$$

(You can also perfectly well multiply out on both sides of the original equation, collect terms on the left, and then factor, but it is much slower.)

Since

$$(2t - 1)(t - 3) = 0$$

then

$$(2t - 1) = 0 \quad \text{or} \quad (t - 3) = 0$$

so

$$t = \frac{1}{2} \quad \text{or} \quad t = 3$$

Notice that if you had cancelled the $(2t - 1)$ in the original equation, you would have been left with $t = 3$ and so would have lost the solution $t = \frac{1}{2}$.

How Many Solutions Should We Expect in a Quadratic Equation?

All the examples we have done so far have had two solutions, but they were chosen rather carefully, so we should look at a few more.

EXAMPLE: *Solve $t^2 = 1$.*

You can do this one by common sense: It is asking for the number(s) whose square is 1. There are two of them, 1 and -1, and therefore the solutions are

$$t = 1 \quad \text{and} \quad t = -1.$$

Alternatively, $t^2 = 1$ can be solved by factoring:

$$t^2 - 1 = 0$$

giving

$$(t - 1)(t + 1) = 0 \quad \text{(difference of squares)}$$

so $t = 1$ or $t = -1$ as before, and the equation has two solutions.

What about the next example?

EXAMPLE: *Solve $t^2 = -1$.*

Common sense says that you can't solve this one, because it asks you to find a number whose square is -1, and there is no such number (among the reals at least). If we try to solve the equation by factoring we get:

$$t^2 = -1$$

or

$$t^2 + 1 = 0$$

But $t^2 + 1$ is not factorable, and so we are stuck. Therefore the equation $t^2 = -1$ has no (real) solution.

EXAMPLE: *Solve*

$$\frac{y^2 + 1}{2} = y$$

It is very hard to factor something containing fractional coefficients, so we will first clear the equation of fractions by multiplying both sides by 2, giving

$$y^2 + 1 = 2y$$

or

$$y^2 - 2y + 1 = 0$$

Factoring:

$$(y - 1)^2 = 0$$

or

$$(y - 1)(y - 1) = 0$$

Therefore $y = 1$, and the equation has one solution.

It turns out that these are the only three possibilities and that all quadratic equations have either two, one, or no real roots. How to figure out how many solutions a given quadratic has will be dealt with in the next two sections, and in Section 15.1 we will see just why a quadratic equation has no more than two roots.

PROBLEM SET   14.2

Solve the following quadratic equations by factoring.

1. $x^2 + 3x = 0$
2. $a^2 - 6a + 8 = 0$
3. $-11q + 2q^2 + 5 = 0$
4. $12x^2 = 29x - 14$
5. $21 = t^2 - 4t$
6. $2f^2 + 5f = -2$
7. $6 + 6x^2 = -13x$
8. $0 = 9y^2 - 4$
9. $0.2y^2 = 0.1y + 0.6$
10. $z^2 - 3z - 54 = 0$
11. $m^2 = 11m - 24$
12. $-2 + 3x = -2x^2$

13. $6r^2 - 5r + 1 = 0$
14. $x^2 + a^2 + 2ax = 0$ for x
15. $\dfrac{k^2 - 12}{3} = \dfrac{k^2 - 4}{4}$
16. $y(y - b) = 2b(2y - 3b)$ for y
17. $x^2 - 5kx + 4k^2 = 0$
18. $\dfrac{t^2}{5} - \dfrac{t^2 - 10}{15} = 7 - \dfrac{50 + t^2}{25}$
19. $4v(v - 4) = v - 4$
20. $x^2 - 10x + 25 - 49a^2b^2 = 0$ for x
21. $(r + 1)^2 + (r + 2)^2 = 25$

22. $x^2 - (2a + b)x + 2ab = 0$ for x
23. $y(y + 0.2) = 0.1y + 0.02$
24. $\left(\dfrac{3x + 2}{2x + 1}\right)^2 - 3\left(\dfrac{3x + 2}{2x + 1}\right) + 2 = 0$

25. Given that $x = 1$ is one root to the equation $0 = x^3 - 5x^2 - x + 5$, find the other two roots to the equation.

26. Find k such that 3 is one of the roots of $x^2 + kx + 6 = 0$.
27. Find k such that the roots of $4x^2 + kx + 1 = 0$ are equal.

Find the average of the two roots of the following equations.

28. $0 = x^2 + 3x + 2$ 31. $3x^2 + 4x = 7$
29. $3 = -10x^2 - 17x$ 32. $0.06 - 0.7x = 3x^2$
30. $y + 6 = 2y^2$

Find quadratic equations that have the following as roots:

33. $-5, 7$ 36. 5 (double root)   39. $3, -\frac{2}{5}$
34. $3, -3$ 37. $\frac{3}{2}, -\frac{4}{5}$ 40. $4, \frac{5}{9}$
35. $\sqrt{2}, -\sqrt{12}$ 38. $0.01, 3$
                                             41. $\frac{a}{b}, -h$

14.3   SOLVING QUADRATIC EQUATIONS BY COMPLETING THE SQUARE

There are a great number of quadratic equations that cannot be factored and yet do have solutions. In fact, only those equations with rational solutions can be factored; those with irrational solutions must all be solved by some other method. One such method is completing the square, and using it you can solve any quadratic equation that has a solution. Factoring is still the quickest way, so we usually solve an equation by factoring if we can; if not, we use completing the square.

To introduce the method, suppose you want to solve:

$$x^2 - 4 = 0$$

This can be factored:

$$(x - 2)(x + 2) = 0$$

Therefore, either

$$x = 2 \quad \text{or} \quad x = -2$$

But instead of factoring you might have said:

$$x^2 = 4$$

and therefore

$$x = \pm \sqrt{4} \qquad \text{(where } x = \pm \sqrt{4} \text{ means } x = + \sqrt{4} \text{ or } x = - \sqrt{4})$$

Since $\sqrt{4} = 2$, we have $x = 2$ or $x = -2$ (as before). Notice that you need to take *both* the positive *and* the negative square root to get both solutions to the equation.

Suppose you now want to solve:

$$x^2 - 3 = 0$$

This does not factor, since it is not the difference of squares. However the second method certainly works:

$$x^2 = 3$$

$$x = \pm\sqrt{3} \quad \text{(here the solutions do not simplify any further)}$$

These are two examples of equations that have been solved not by factoring but by taking square roots. In some ways this is an obvious thing to try in solving a quadratic equation, because one of the things we want to get rid of in such an equation is the x^2 term. The problem is that in order to do this you must have the x's contained in a perfect square on one side of the equation.

For example,

$$(x - 3)^2 = 8$$

can be solved by taking square roots, giving

$$x - 3 = \pm\sqrt{8}$$

So

$$x - 3 = +\sqrt{8} \quad \text{or} \quad x - 3 = -\sqrt{8}$$

Then either

$$x = 3 + \sqrt{8} \quad \text{or} \quad x = 3 - \sqrt{8}$$

which can be written

$$x = 3 \pm \sqrt{8}$$

But now suppose we want to solve

$$x^2 + 2x - 2 = 0$$

This equation does not factor, and so the only hope we have is trying to get it into the same form as the example above, $(x - 3)^2 = 8$. If

$$x^2 + 2x - 2 = 0$$

then

$$x^2 + 2x = 2$$

Complete the square on the left-hand side by adding 1. To keep the equation balanced, we must add 1 to the right-hand side too:

$$x^2 + 2x + 1 = 2 + 1$$

Therefore

$$(x + 1)^2 = 3$$

Now the equation is in the form of the previous example, so

$$x + 1 = \pm \sqrt{3}$$

and either

$$x + 1 = \sqrt{3} \quad \text{or} \quad x + 1 = -\sqrt{3}$$

giving either

$$x = -1 + \sqrt{3} \quad \text{or} \quad x = -1 - \sqrt{3}$$

Therefore the solutions are

$$x = -1 \pm \sqrt{3}$$

To Solve Quadratic Equations by Completing the Square:

1. Put the constant term on the right of the equation and all terms containing x's on the left.
2. Divide through by the coefficient of x^2 (if it is not already 1).
3. Complete the square on the left, making sure that anything you add to the left side you add to the right also.
4. Take the square roots of both sides, remembering to take both the positive and the negative square roots of the right side.
5. Solve the two linear equations you get.

To convince you that this method really does work, we will first use it on an equation whose solutions we know already—i.e., one that factors.

EXAMPLE: *Solve $x^2 - x - 20 = 0$.*

Since $(x - 5)(x + 4) = x^2 - x - 20 = 0$, the solutions to this equation are $x = 5$ and $x = -4$. Doing it by completing the square:

$$x^2 - x - 20 = 0$$

STEP 1. $$x^2 - x = 20$$

STEP 3. $$x^2 - x + \frac{1}{4} = 20 + \frac{1}{4}$$

$$\left(x - \frac{1}{2}\right)^2 = \frac{81}{4}$$

STEP 4.

$$\left(x - \frac{1}{2}\right) = \pm \sqrt{\frac{81}{4}} = \pm \frac{9}{2}$$

STEP 5.

Either $x - \dfrac{1}{2} = \dfrac{9}{2}$ or $x - \dfrac{1}{2} = -\dfrac{9}{2}$

so

$$x = \frac{9}{2} + \frac{1}{2} = 5 \quad \text{or} \quad x = -\frac{9}{2} + \frac{1}{2} = -4$$

Therefore factoring and completing the square give the same answers.

EXAMPLE: *Solve $x^2 + x - 1 = 0$.*

This does not factor, so we have to use completing the square.

$$x^2 + x - 1 = 0$$

STEP 1.

$$x^2 + x = 1$$

STEP 3.

$$x^2 + x + \frac{1}{4} = 1 + \frac{1}{4}$$

$$\left(x + \frac{1}{2}\right)^2 = \frac{5}{4}$$

STEP 4.

$$x + \frac{1}{2} = \pm \sqrt{\frac{5}{4}} = \pm \frac{\sqrt{5}}{2}$$

STEP 5.

In order to solve these equations, you don't have to write down both separately. Solving them together gives

$$x = -\frac{1}{2} \pm \frac{\sqrt{5}}{2}$$

or

$$x = \frac{-1 \pm \sqrt{5}}{2}$$

EXAMPLE: *Solve $3x^2 - 2x - 7 = 0$.*

This is the first example where the coefficient of x^2 is not 1.

STEP 1.

$$3x^2 - 2x = 7$$

STEP 2.
$$x^2 - \frac{2}{3}x = \frac{7}{3}$$

STEP 3.
$$x^2 - \frac{2}{3}x + \frac{1}{9} = \frac{7}{3} + \frac{1}{9}$$

$$\left(x - \frac{1}{3}\right)^2 = \frac{22}{9}$$

STEP 4.
$$x - \frac{1}{3} = \pm \sqrt{\frac{22}{9}} = \pm \frac{\sqrt{22}}{3}$$

STEP 5.
$$x = \frac{1}{3} \pm \frac{\sqrt{22}}{3} = \frac{1 \pm \sqrt{22}}{3}$$

How Different Numbers of Solutions Appear Using Completing the Square

You remember that in the section on factoring quadratic equations we showed some equations that had two roots, some that had one root, and some that had no (real) roots. The same three things happen when you use completing the square. It is instructive to see exactly what it is in the process that goes wrong when there are no roots, and how you sometimes get one root instead of two.

Let us first look at an equation that does have two roots:

$$x^2 - 2x - 2 = 0$$

Using completing the square:

$$x^2 - 2x = 2$$

$$x^2 - 2x + 1 = 2 + 1$$

$$(x - 1)^2 = 3 \qquad \text{(this is step 3)}$$

$$x - 1 = \pm \sqrt{3}$$

$$x = 1 \pm \sqrt{3}$$

so there are two roots:

$$x = 1 + \sqrt{3} \quad \text{and} \quad x = 1 - \sqrt{3}$$

The existence of two different solutions comes from the two square roots of 3, the positive one ($\sqrt{3}$) and the negative one ($-\sqrt{3}$).

Now let us look at

$$x^2 - 2x + 1 = 0$$

Using completing the square:

$$x^2 - 2x = -1$$

$$x^2 - 2x + 1 = -1 + 1$$

$$(x - 1)^2 = 0 \qquad \text{(step 3)}$$

Zero has only one square root—itself—so

$$x - 1 = 0$$

$$x = 1$$

In this example there is only one root because we have to take the square root of zero—the only number with exactly one square root.

At this stage you might be able to guess what happens in equations that have no roots. Realize that an equation with two roots got them because we had to take the square root of some number, such as 3, which had two square roots, a positive and a negative one. An equation that had only one root ended up that way because we had to take the square root of zero—the only number with exactly one square root. Doesn't it seem logical, then, that equations that have no (real) roots will be that way because in trying to solve them we came to a point where we have to take the square root of a number that has no (real) square root (in other words, a negative number)?

Let us look at:

$$x^2 - 2x + 2 = 0$$

Using completing the square:

$$x^2 - 2x = -2$$

$$x^2 - 2x + 1 = -2 + 1$$

$$(x - 1)^2 = -1 \qquad \text{(step 3)}$$

The next stage would be taking the square roots of both sides, but since there is a negative number on the right-hand side this can't be done because it would lead to

$$x - 1 = \pm \sqrt{-1}$$

and $\sqrt{-1}$ is not a real number. Therefore we have to stop at

$$(x - 1)^2 = -1$$

and say that the equation cannot be solved.

The number that occurs on the right-hand side of a quadratic equation at step 3 of the completing-the-square method tells us a great deal about the equation—namely, whether the equation has two, one, or no (real) roots. To summarize the three examples above:

$x^2 - 2x - 2 = 0$ can be rewritten

$$(x - 1)^2 = \text{\textcircled{3}}\, positive \text{ so the equation has } two \text{ real roots}$$

$x^2 - 2x + 1 = 0$ can be rewritten

$$(x - 1)^2 = \text{\textcircled{0}}\, zero \text{ so the equation has } one \text{ real root}$$

$x^2 - 2x + 2 = 0$ can be rewritten

$$(x - 1)^2 = \text{\textcircled{-1}}\, negative \text{ so the equation has } no \text{ real roots}$$

PROBLEM SET   14.3

Solve the following quadratic equations by completing the square.

1. $y^2 - 8y + 12 = 0$
2. $3r^2 + 6r + 1 = 0$
3. $x^2 + 10x - 2 = 0$
4. $x^2 + 6x = 16$
5. $6s = s^2 - 17$
6. $p^2 - 2 = 7p$
7. $x^2 + 3x = 2$
8. $5 + 4x = x^2$
9. $5b^2 = 1 + 6b$
10. $25 + 9y^2 = 30y$
11. $2z^2 - 14z = 4$
12. $3m + 3 = 5m^2$
13. $-7y + y^2 - 25 = 0$
14. $5x + 7 = x^2$
15. $-2 + 2x^2 - x = 0$

16. $9 = t^2 + 4t$
17. $x^2 - 0.06x = 0.9991$
18. $y^2 = 1.9y - 0.8$
19. $0.2w^2 - 0.76w + 0.7 = 0$
20. $-2 - 6y + y^2 = 0$
21. $d^2 + \frac{9}{2}d - \frac{5}{2} = 0$
22. $x^2 - 7 + x = 2$
23. $x^2 + 7x = 2x - 1$
24. $p^2 - 4 = 3p - 2$
25. $x(x + 3) = 23 - 4x$
26. $(y + 2)(y - 3) = 12$
27. $(s + 1)(s + 2) - 3 = 2s - 1$
28. $2x(5 + 2x) = 15(1 - x)$
29. $(x + 3)(2x - 1) - 3(x - 2) = 5$
30. $ax^2 + bx + c = 0$

14.4   THE QUADRATIC FORMULA

After solving a few quadratic equations by completing the square, you should begin to realize that you do pretty much the same thing in every case. This is fortunate, for it enables us to derive a formula for finding the solutions to a quadratic equation in terms of the coefficients of the equation. The idea is to solve a perfectly general quadratic equation such as

$$ax^2 + bx + c = 0 \qquad (a, b, c \text{ are constants})$$

by completing the square and so find a formula for the solutions in terms of a, b, and c. This way, the completing the square is done once and for all, and to solve a specific equation all we have to do is substitute the relevant values of *a*, *b*, and *c*.

> *To solve $ax^2 + bx + c = 0$ by completing the square:*

STEP 1.   Put the constant term on the right:

$$ax^2 + bx = -c$$

STEP 2.   Divide through by the coefficient of x^2:

$$x^2 + \frac{b}{a}x = -\frac{c}{a}$$

STEP 3.   Now, to complete the square on the left-hand side. The coefficient of *x* is $\frac{b}{a}$; therefore we add $\left(\frac{b}{2a}\right)^2 = \frac{b^2}{4a^2}$ to both sides:

$$x^2 + \frac{b}{a}x + \frac{b^2}{4a^2} = \frac{b^2}{4a^2} - \frac{c}{a}$$

$$\left(x + \frac{b}{2a}\right)^2 = \frac{b^2 - 4ac}{4a^2}$$

STEP 4.   Taking square roots:

$$x + \frac{b}{2a} = \pm\sqrt{\frac{b^2 - 4ac}{4a^2}} = \frac{\pm\sqrt{b^2 - 4ac}}{2a}$$

STEP 5.   Therefore, either

$$x + \frac{b}{2a} = \frac{\sqrt{b^2 - 4ac}}{2a} \qquad \text{or} \qquad x + \frac{b}{2a} = \frac{-\sqrt{b^2 - 4ac}}{2a}$$

$$x = -\frac{b}{2a} + \frac{\sqrt{b^2 - 4ac}}{2a} \qquad\qquad x = -\frac{b}{2a} - \frac{\sqrt{b^2 - 4ac}}{2a}$$

$$= \frac{-b + \sqrt{b^2 - 4ac}}{2a} \qquad\qquad = \frac{-b - \sqrt{b^2 - 4ac}}{2a}$$

Therefore the solutions to $ax^2 + bx + c = 0$ are

$$x = \frac{-b + \sqrt{b^2 - 4ac}}{2a} \qquad \text{or} \qquad x = \frac{-b - \sqrt{b^2 - 4ac}}{2a}$$

which is often written:

$$\boxed{x = \frac{-b \pm \sqrt{b^2 - 4ac}}{2a}}$$

This is called the *quadratic formula*.

EXAMPLE: *Solve $2x^2 + 9x - 18 = 0$ using the quadratic formula.*

In this example $a = 2$, $b = 9$, and $c = -18$. Therefore the solutions are

$$x = \frac{-b \pm \sqrt{b^2 - 4ac}}{2a}$$

$$= \frac{-9 \pm \sqrt{81 - 4 \cdot 2 \cdot (-18)}}{4}$$

$$= \frac{-9 \pm \sqrt{81 + 144}}{4}$$

$$= \frac{-9 \pm \sqrt{225}}{4}$$

$$= \frac{-9 \pm 15}{4}$$

So one solution is

$$x = \frac{-9 + 15}{4} = \frac{6}{4} = \frac{3}{2},$$

and the other is

$$x = \frac{-9 - 15}{4} = \frac{-24}{4} = -6$$

Therefore the solutions are $x = \dfrac{3}{2}$ and $x = -6$

CHECK: These solutions can be checked by substitution. Notice that

$$2x^2 + 9x - 18 = (2x - 3)(x + 6)$$

so that we could have got these solutions by factoring. Section 15.1 will tell you that any equation that has rational roots can be factored, so the quadratic formula or completing the square is only absolutely essential for an equation with irrational roots.

EXAMPLE: *Solve $3x^2 = 2x + 7$ using the quadratic formula.*

First we must rewrite the equation in the form $ax^2 + bx + c = 0$:

$$3x^2 - 2x - 7 = 0$$

so $a = 3$, $b = -2$, $c = -7$. Therefore the solutions are

$$x = \frac{-(-2) \pm \sqrt{(-2)^2 - 4 \cdot 3 \cdot (-7)}}{2 \cdot 3}$$

$$= \frac{2 \pm \sqrt{4 + 84}}{6}$$

$$= \frac{2 \pm 2\sqrt{22}}{6} \qquad (\sqrt{88} = \sqrt{4 \cdot 22} = \sqrt{4} \cdot \sqrt{22} = 2\sqrt{22})$$

$$= \frac{2(1 \pm \sqrt{22})}{6}$$

$$= \frac{1 \pm \sqrt{22}}{3}$$

CHECK: This example was done by completing the square in the previous section, so you can check that we got the same answer both ways; otherwise you have to substitute (which turns out to be pretty hard work).

Different Numbers of Solutions and the Quadratic Formula; The Discriminant

Remember the discussion in the last section of what made some quadratic equations have two roots, some one, and some none? There we saw that the number of real roots to a quadratic equation is determined by the sign of the number on the right-hand side of the equation when the left-hand side is written as a perfect square. This is the number whose square root we have to take during the completing-the-square method. If it is positive we get two roots, if zero we get one, if negative we get none.

In deriving the quadratic formula, we arrive at:

$$\left(x + \frac{b}{2a}\right)^2 = \frac{b^2 - 4ac}{4a^2}$$

The number whose square root we have to take is

$$\frac{b^2 - 4ac}{4a^2}$$

and so whether the equation has two, one, or no roots depends on the sign of this quantity. But since its denominator is $4a^2$, which is always positive, the sign of $\dfrac{b^2 - 4ac}{4a^2}$ is always the same as the sign of $b^2 - 4ac$.

Therefore:

If $b^2 - 4ac$ is positive, the equation has two real roots.
If $b^2 - 4ac$ is zero, the equation has one real root.
If $b^2 - 4ac$ is negative, the equation has no real roots.

The expression $b^2 - 4ac$ is called the *discriminant*.

EXAMPLE: *Without finding them, figure out how many (real) roots $x^2 + x + 1 = 0$ has. How about $x^2 - x + 1 = 0$? and $x^2 + x - 1 = 0$?*

First, $x^2 + x + 1 = 0$ has $a = b = c = 1$. The discriminant is

$$b^2 - 4ac = 1^2 - 4 \cdot 1 \cdot 1 = -3$$

Therefore there are no real roots.

Next, $x^2 - x + 1 = 0$ has $a = c = 1$, $b = -1$. The discriminant is

$$b^2 - 4ac = (-1)^2 - 4 \cdot 1 \cdot 1 = -3$$

Therefore there are no real roots.

Finally, $x^2 + x - 1 = 0$ has $a = b = 1$, $c = -1$. The discriminant is

$$b^2 - 4ac = 1^2 - 4 \cdot 1(-1) = 5$$

Therefore there are two real roots.

EXAMPLE: *For what value of h will $3x^2 + 4x + h = 0$ have only one root?*

To do this problem, we have to find the value of h making the discriminant zero. Now:

$$\text{discriminant} = 4^2 - 4 \cdot 3 \cdot h = 16 - 12h$$

So the equation has one root if

$$16 - 12h = 0 \quad \text{or} \quad h = \frac{4}{3}$$

PROBLEM SET 14.4

Solve the following quadratic equations by using the quadratic formula. (Note: Some have no real roots.)

1. $x^2 + 8x - 20 = 0$
2. $6x^2 - 7x = 5$
3. $5w - 7 + 3w^2 = 0$
4. $\frac{1}{2} - \frac{7}{4}x = x^2$

5. $5 = 5x^2 - x + 14$

6. $5x = 7 - x^2$

7. $5c^2 - 7c = 1$

8. $t^2 + 4t = 9$

9. $2x^2 - 3x + 1 = 2x - 1$

10. $(y - 2)(y - 3) = 14$

11. $4x(x + 1) = 2.75$

12. $(x + 1)^2 - (x + 1) = 20$

13. $(x + 2)\,4x + 3(x + 2) = 0$

14. $4 = q^2 - \sqrt{2}q$

15. $3a(a - 5) = 4a^2 + 10a - 20$

16. $(x + 2)(x + 4) = 1$

17. $(x + 1)(3x - 2) = 2x - 7$

18. $x^2 + 1 = -4x$

19. $(x - 3)^2 + (x + 3)^2 = 18$

20. $0.3x^2 + 0.62 = 0.24x$

21. $\frac{1}{3} - \frac{7}{4}x = x^2$

22. $2\left(\dfrac{x + 2}{3x + 1}\right)^2 + 4\left(\dfrac{x + 2}{3x + 1}\right) + 5 = 0$

23. $(y + 2)(2y + 5) = y + 2$

24. $\left(\dfrac{x + 1}{3} + 2\right)\left(\dfrac{x + 1}{3} + 4\right) = -1$

Solve the following quadratic equations for the variable in parentheses.

25. $x^2 + 4bx + b^2 = 0$ (x)

26. $a^2 + ap + q = 0$ (a)

27. $s = \frac{1}{2}at^2$ (t)

28. $2\pi f^2 c = \dfrac{df}{1 + 3c}$ (f)

29. $hf = w + \dfrac{mv^2}{2}$ (v)

30. $A = 2\pi R_1^2 - R_2^2$ (R_2)

31. $m - (w + d)\,2wn^2 = 0$ (n)

32. $s - s_0 = v_0 t + \frac{1}{2}gt^2$ (t)

33. $\dfrac{y^2 + 3ay}{2} = \dfrac{y + 1}{3c}$ (y)

34. $d(3a + 2d) - 2ad - 3a$
$= a - d(2 + 3a)$ (d)

For each of the equations in Problems 35–39, (a) find the discriminant; (b) say how many real roots the equation has; and (c) find all the real roots.

35. $3x^2 - 5x + 2 = 0$

36. $(3 \cdot 10^2)x^2 + (6 \cdot 10)x = 4$

37. $2x^2 + 7x + 15 = 0$

38. $4x^2 + 3 = 4\sqrt{3}x$

39. $2x^2 - (x + 3)(x - 1) = 5$

40. Show that $x^2 + x + 1 = 0$ has no real roots.

41. Find the values of b for which $x^2 + bx + 1 = 0$ has real roots.

Find the sum and the product of the two roots for each of the following:

42. $15x^2 + 23x - \frac{1}{3} = 0$

43. $y^2 - 1 = 0$

44. $-7.2 + 7z^2 - \sqrt{3}z = \sqrt{2}z$

45. $2x(x - 1) = 3(x + 4)$

46. $\sqrt{2}y^2 + \sqrt{2} = 7y$

47. $-0.3p^2 + 0.02p = -(0.63 + 0.22p)$

48. $(x - \frac{2}{3})^2 = \frac{4}{5}(x - \frac{1}{5})$

In Problems 49–50, show that if r_1 and r_2 are the two roots of the quadratic equation $ax^2 + bx + c = 0$, then

49. $r_1 + r_2 = \dfrac{-b}{a}$

50. $r_1 \cdot r_2 = \dfrac{c}{a}$

51. Find the relation between a, b, and c if one root of the equation $ax^2 + bx + c = 0$ is three times the other.

52. Show that the sum of the solutions to $ax^2 + bx + c = 0$ and the solutions to $ax^2 - bx + c = 0$ is 0.

14.5   COMPLEX NUMBERS

This section is about what happens when you take the square root of a negative number. Since there is no such thing as a real number whose square is negative, if we're considering only real numbers, we have to say that $\sqrt{-1}$ and $\sqrt{-4}$ don't exist. However, the mathematicians of the eighteenth century were not happy with this and so, being creative people, they decided to invent them. What they did was to define a new number, i, to be the square root of -1:

$$i = \sqrt{-1} \quad \text{and} \quad i^2 = -1$$

The square root of any negative number can now be expressed in terms of i. For example,

$$\sqrt{-4} = \sqrt{4 \cdot (-1)} = \sqrt{4} \cdot \sqrt{-1} = 2i$$
$$\sqrt{-3} = \sqrt{3 \cdot (-1)} = \sqrt{3} \cdot \sqrt{-1} = \sqrt{3}i$$

Now we will use i to define complex numbers. A *complex number* is any number of the form $a + bi$, where a and b are real. For example, $2 + 3i$, $1 - 5i$, $1.2 - 3i$, $\pi - 0.7i$ are complex numbers. $2i$ is also a complex number because you can think of it as $0 + 2i$. The real numbers are included in the complex numbers because, for example, you can write $3 = 3 + 0i$ or $\pi = \pi + 0i$. In a complex number such as $2 + 3i$, the 2 is called the *real part*, and $3i$ the *imaginary part*. A number such as $7i$, which has no real part, is called *purely imaginary*.

In spite of their name, complex numbers aren't hard to deal with, and they can be added, subtracted, multiplied, and divided just like real ones.

Adding and Subtracting Complex Numbers

To add $2 + 7i$ and $6 - 4i$, *add the real and the imaginary parts separately:*

$$(2 + 7i) + (6 - 4i) = (2 + 6) + (7i - 4i)$$
$$= (2 + 6) + (7 - 4)i$$
$$= 8 + 3i$$

To subtract two complex numbers, *subtract the real and the imaginary parts separately.* For example, subtracting $3 - 2i$ from $5 + 12i$ looks like this:

$$(5 + 12i) - (3 - 2i) = (5 - 3) + (12i + 2i) \qquad \text{[because}$$
$$-(-2i) = 2i\,]$$

$$= 2 + 14i$$

Multiplying Complex Numbers

To multiply $(2 + 3i)$ by $(4 - 5i)$, use the distributive law and the fact that $i^2 = -1$:

$$(2 + 3i)(4 - 5i) = 2 \cdot 4 - 2 \cdot 5i + 3i \cdot 4 - 3i \cdot 5i$$

$$= 8 - 10i + 12i - 15i^2$$

$$= 8 + 2i + 15 \qquad (15i^2 = -15)$$

$$= 23 + 2i$$

Higher powers of i can be simplified using the fact that $i^2 = -1$. For example,

$$i^3 = i^2 \cdot i = (-1)i = -i$$
$$i^4 = i^2 \cdot i^2 = (-1) \cdot (-1) = 1$$

Thus you can get rid of any terms in i^2 or higher powers of i and so get any product into the form $a + bi$.

Dividing Complex Numbers

Suppose we have to divide

$$\frac{19 + 17i}{3 + 4i}$$

As it stands, this fraction is not of the form $a + bi$ and so doesn't look like a complex number. However, it can be made to look like one by multiplying the fraction top and bottom by $(3 - 4i)$:

$$\frac{19 + 17i}{3 + 4i} = \frac{(19 + 17i)}{(3 + 4i)} \cdot \frac{(3 - 4i)}{(3 - 4i)} = \frac{19 \cdot 3 - 19 \cdot 4i + 17i \cdot 3 - (17 \cdot 4)i^2}{3^2 - (4i)^2}$$

$$= \frac{125 - 25i}{9 + 16} \qquad \begin{array}{l} \text{[since } (4i)^2 = 16(-1) = -16 \\ \text{and } -(17 \cdot 4)i^2 = -68i^2 = 68] \end{array}$$

$$= \frac{25(5 - i)}{25}$$

$$= 5 - i$$

Multiplying by $(3 - 4i)$ is a really clever trick because it makes the bottom of the fraction real and so helps you to get the fraction into the usual form. This trick always works, so: *To divide by a + bi, express the division as a fraction and multiply top and bottom by a − bi, called the* conjugate *of a + bi.*

EXAMPLE: *Divide* $\dfrac{1 - i}{-4 - 2i}$

The conjugate of $-4 - 2i$ is $-4 + 2i$, so multiply top and bottom by this:

$$\frac{1 - i}{-4 - 2i} = \frac{(1 - i)}{(-4 - 2i)} \cdot \frac{(-4 + 2i)}{(-4 + 2i)} = \frac{-4 + 2i + 4i - 2i^2}{(-4)^2 - (2i)^2}$$

$$= \frac{-2 + 6i}{16 + 4} \qquad [\text{since } -2i^2 = 2$$
$$\text{and } (2i)^2 = -4\,]$$

$$= \frac{\cancel{2}(-1 + 3i)}{\cancel{20}\,10}$$

$$= \frac{-1}{10} + \frac{3i}{10}$$

Finding the Complex Roots of a Quadratic Equation

The reason that we're so interested in complex numbers is that they enable us to find a solution for *any* quadratic equation. So far, any quadratic can be solved using the quadratic formula unless it gives a negative number under the square root sign.

However, using i we can take the square roots of any number, even negative ones, and so the quadratic formula will give us a solution to any equation.

EXAMPLE: *Solve $x^2 - 4x + 5 = 0$.*

The quadratic formula tells us that

$$x = \frac{-(-4) \pm \sqrt{(-4)^2 - 4(1)(5)}}{2} = \frac{4 \pm \sqrt{-4}}{2}$$

Without complex numbers we'd have to say that this equation has no real solutions because we couldn't find the square root of -4. However, allowing i we have

$$\sqrt{-4} = 2i$$

so

$$x = \frac{4 \pm \sqrt{-4}}{2} = \frac{4 \pm 2i}{2} = \frac{2(2 \pm i)}{2} = 2 \pm i$$

Therefore, the equation $x^2 - 4x + 5 = 0$ has two complex roots, $2 + i$ and $2 - i$.

EXAMPLE: *Solve $t^2 - 2t + 4 = 0$.*

The quadratic formula gives

$$t = \frac{-(-2) \pm \sqrt{(-2)^2 - 4(1)(4)}}{2} = \frac{2 \pm \sqrt{-12}}{2}$$

Again, running into the square root of a negative number tells us that there are no real roots.

But if we say

$$\sqrt{-12} = \sqrt{4 \cdot 3 \cdot (-1)} = 2\sqrt{3}i$$ (See Chapter 6 for simplification of radicals)

then we can write

$$t = \frac{2 \pm 2\sqrt{3}i}{2} = \frac{2(1 \pm \sqrt{3}i)}{2} = 1 \pm \sqrt{3}i$$

So the equation $t^2 - 2t + 4 = 0$ has roots $1 + \sqrt{3}i$ and $1 - \sqrt{3}i$.

You may have noticed that $2 + i$ and $2 - i$, the roots of the first equation, are conjugates of one another, as are $1 + \sqrt{3}i$ and $1 - \sqrt{3}i$, the roots of the second. This is not an accident—in fact, whenever a quadratic equation (with real coefficients) has a complex root, its conjugate will also be a solution to the equation.

The reason is that you get a complex root only when the discriminant $(b^2 - 4ac)$ is negative. Then $\sqrt{b^2 - 4ac}$ is purely imaginary, that is, of the form ki, with k real. So the quadratic formula says:

real⟶ $x = \dfrac{\boxed{-b} \pm \boxed{\sqrt{b^2 - 4ac}}}{\boxed{2a}}$ ⟵imaginary / real

Since the "$\pm$" only applies to the imaginary part, the two possible values of x are conjugates of one another.

Summarizing: The quadratic formula can always be used to find the roots of $ax^2 + bx + c = 0$, and, extending the results of the last section:

If $b^2 - 4ac$ is positive, the equation has two real roots.
If $b^2 - 4ac$ is zero, the equation has one real root.
If $b^2 - 4ac$ is negative, the equation has two complex roots, which are conjugates of one another.

PROBLEM SET   14.5

Perform the indicated operations and simplify.

1. $(3 + 2i) + (5 - 3i)$
2. $(4 - i) + 7i$
3. $(\tfrac{4}{3} - 6i) + 2$
4. $(\tfrac{1}{2} + 7i) - (5 + 2i)$
5. $(\tfrac{6}{5} + \tfrac{3}{4}i) + (\tfrac{1}{6} - \tfrac{1}{3}i)$
6. $(-6 - 2i) + (\tfrac{1}{2} + 2i)$
7. $(1.02 - 0.34i) - (0.35 - 0.41i)$
8. $6i(1 - i)$
9. $(3 + 2i)^2$
10. $3i(2 - 3i)^2$
11. $\dfrac{2 + 4i}{1 - i}$
12. $\dfrac{1}{8 - 2i}$
13. $\dfrac{2i}{9 - \tfrac{3}{2}i}$
14. $(3 + i)(3 - i)$

15. $\dfrac{2}{4 - 7i}$
16. $(\sqrt{2} - 3\sqrt{2}i) + (\sqrt{8} + \sqrt{8}i)$
17. i^3
18. $(2i)^4$
19. $\left(\dfrac{a}{2} + \dfrac{b}{4}i\right) + \left(-\dfrac{a}{2} - \dfrac{i}{2}\right)$
20. $(a + b) - (\tfrac{2}{3} - 11i)$
21. $(1 + i)^3$
22. $\dfrac{(1 - i)(2 + i)}{3 - 7i}$
23. $(a + bi)(a - bi)$
24. $\dfrac{2i + 1}{i + 1} - \dfrac{3i}{2 - i}$
25. $\left(\dfrac{a}{a^2 + b^2} - \dfrac{b}{a^2 + b^2}i\right)(a + bi)$

Find all of the roots (real and imaginary) to the following quadratic equations.

26. $2x^2 - 2x + 5 = 0$
27. $0.2y^2 = -3y + 0.5$
28. $3r^2 + 3(r + 5) = 2 - r$
29. $(c - 1)(c + 2) = 14 - c$
30. $t(1 - t) + 2t(3 + 2t) = 5 - t(t + 1)$
31. $\dfrac{8}{3} - \dfrac{1}{6}t = t\left(\dfrac{1 - 3t}{2}\right)$

Perform the indicated operations and simplify.

32. $\dfrac{(2 - 3i)(8 + i)}{(\tfrac{2}{3} + 4i)(1 + 2i)}$

33. $\left[\dfrac{a + 2}{3} - (a - 3)i\right]\left(\dfrac{2}{3} + \dfrac{a}{3} + (a - 3)i\right)$

34. $\dfrac{3i}{2 - i} + i$

35. $(\sqrt{2} + 3i)(1 - \sqrt{3}i)$

36. $2\left(\dfrac{2 - i}{\sqrt{3} + i}\right) + 6i$

37. $\dfrac{1}{2 - i} + (4 + 2\sqrt{3})(\sqrt{2} - i)$

38. $(1 + 2i)(2 - 3i)(i - \tfrac{2}{3})$

39. $\dfrac{1.4 + 6.2i}{2.1 - 2.2i}$

40. $\dfrac{x}{y + zi}$

41. $\dfrac{-\sqrt{-9}(3 + 2i)}{4 - i}$

42. i^{31}

43. i^n for n an integer [Hint: There are four cases to consider.]

Find the complex conjugate of the following.

44. $(3 + 4i) - 2 - 3i$

45. $(\frac{1}{2} - \frac{3}{8}i)(2 + \frac{1}{4}i)$

46. $\left[\dfrac{\dfrac{1 + 3i}{4}}{i(3 - i)^2}\right]$

47. $\dfrac{(1 + i) + 2i(3 - 4i)}{i}$

A bar over a complex number represents the conjugate of the number. For example, $\overline{4 + 3i} = 4 - 3i$. Using this, state whether the following are true or false. Justify your answers.

48. $\overline{(a + bi) + (c + di)} = \overline{(a + bi)} + \overline{(c + di)}$

49. $\overline{(a + bi)(c + di)} = \overline{(a + bi)} \cdot \overline{(c + di)}$

50. $\overline{(a + bi) - (c + di)} = [\overline{(a + bi) - (c + di)}]$

51. $\overline{\left(\dfrac{a + bi}{c + di}\right)} = \dfrac{\overline{(a + bi)}}{\overline{(c + di)}}$

Find all of the roots (real and imaginary) to the following quadratic equations.

52. $3z^2 + \dfrac{4}{5}z\left(1 - \dfrac{z}{4}\right) = z + 1$

53. $0 = 4a + (a + 2)(1 - 2a)$

54. $x^2 + \frac{1}{2} = 0$

55. $d - 7 = 7d^2$

56. $0.08\,[d + 2(0.03 - d)(-0.2d)] = 0$

57. $4(x + 1)^2 + 5(x + 1) = -2$

58. $12\left(\dfrac{x - 1}{2x}\right) = 4\left(\dfrac{x - 1}{2x}\right)^2 + 9$

59. $2\left(\dfrac{0.3x}{4}\right)^2 + \left(\dfrac{0.3x}{4}\right) + 3 = 0$

Find quadratic equations with real coefficients that have the following as one of their roots.

60. $3 + i$

61. $4 - 2i$

62. $2i$

63. $\dfrac{2i}{5}$

64. $\dfrac{4 + 3i}{2}$

CHAPTER 14 REVIEW

Solve for x.

1. $x^2 - 5x - 14 = 0$

2. $2x^2 + x - 6 = 0$

3. $2x^2 = 162$

4. $-13x = 6x\left(x + \dfrac{1}{x}\right)$

5. $\dfrac{x^2 + 1}{2\sqrt{2}} = x$

6. $3x^2 + 5x = 1$

7. $\dfrac{2x^2 + 3}{5} = \dfrac{x^2 - x}{2} - \dfrac{x^2}{10}$

8. $x = \dfrac{x^2}{2}$

9. $(2x + 1)^2 = 4(2x + 1 - 1)$

10. $36x^4 + 7x^2 - 4 = 0$

11. $\dfrac{2x^2 + 7x}{5} = \dfrac{x^2 + 8x}{6} + \dfrac{1}{15}$

12. $(2x^2 + 2x + 1)^2 - 2(2x^2 + 2x + 1) - 8 = 0$

13. $x^2 - 0.2x + 0.05 = 0$ 17. $b^2x^2 - 1 = 2bx^2 - x^2$

14. $x^2 + 2xy + y^2 = 0$ 18. $x^2 - (r - t)x + rs + st - rt - s^2 = 0$

15. $4(a^2 - x^2) = b^2 + c^2$

16. $u^2 = 4xu + k$ 19. $x = -\dfrac{(2bx^2 + c)}{2c + b}$

20. If the roots of $ax^2 + bx + c = 0$ are r_1 and r_2, what are the roots of $ax^2 + bkx + ck^2 = 0$?

21. Write an equation with integral coefficients with roots

$$\frac{1 + 2i}{3} \quad \text{and} \quad \frac{1 - 2i}{3}$$

22. Solve for k so that one root of $x^2 - kx - \frac{1}{2} = 0$ is two units less than the other.

23. Write an equation with roots $\sqrt{5} - 1$ and $1 + \sqrt{5}$. Check that they are roots.

24. Find an equation with real coefficients and one root equal to

$$\frac{1}{\sqrt{2}}(-1 + i)$$

25. Find all roots of $x^3 = -1$. Remember that $x = -1$ is one root.

Perform the indicated operations.

26. $\left(\dfrac{\sqrt{3}}{2} + \dfrac{1}{2}i\right)^3$ 29. $\left(\dfrac{1 - i}{1 + 3i}\right)\left(\dfrac{2 + 3i}{-1 + 4i}\right)$

27. $(ap + bpi)(a - bi)$

28. $\left(\dfrac{3}{2 - 5i}\right)\left(\dfrac{2}{-2 - 5i}\right)$ 30. $\left(\dfrac{\sqrt{3}}{2} + \dfrac{1}{2}i\right)^6$

31. The area of a square equals one-half its perimeter less one unit. How long is each side?

32. The sum of the squares of two consecutive integers is 85. What are the integers?

33. Seven times a certain integer is seven less than the square of the next consecutive integer. Find the number.

34. The square of the admiral's age is twenty times the difference between his age and the age of his house (400 years). How old is the admiral?

35. However you cut it, the admiral likes cake (coffee cake, that is). If he eats half of a circular birthday cake, gives half of what's left to his wife, and finds that the top of the last piece (which he feeds to the goldfish) has an area of 0.0025π square meters, what was the radius of the cake?

15   YET MORE EQUATIONS

15.1   POLYNOMIAL EQUATIONS: WHAT TO EXPECT

In the last few sections we spent a great deal of time solving linear and quadratic equations. It turns out that it is also useful to be able to solve equations involving polynomials of higher degree. The *degree of such a polynomial equation* is defined as the highest power of the unknown occurring in the equation. For example, a linear equation is a first-degree equation; a quadratic equation is second-degree; $4x^3 - 6x = 7x^2$ is a third-degree, or cubic, equation; $x^2(x^9 - 12x) = 1 + 2x^9(x^7 - 5)$ multiplies out to $x^{11} - 12x^3 = 1 + 2x^{16} - 10x^9$ and so has degree 16.

In the next section we will go into the question of how polynomial equations are actually solved. But before doing that we will look back over quadratics and see what general results there can be more widely applied.

First, *what is the relation between solutions and factors?*

Let's suppose we're solving a quadratic equation. Imagine that everything has been put on the left-hand side so the equation is in the form:

$$p = 0$$

where p is a quadratic polynomial. Then all the quadratics you have solved (Section 14.2) should tell you that:

Whenever p has a factor of $(x - k)$ the equation has a solution $x = k$.

Whenever p has a factor of $(lx - k)$ the equation has a solution $x = \dfrac{k}{l}$.

(where k and l are integers)

For example, $x^2 - 3x + 2 = 0$ factored into

$$(x - 2)(x - 1) = 0$$

and had solutions

$$x = 2 \quad \text{and} \quad x = 1$$

Also, $6x^2 + x - 15 = 0$ factored into

$$(2x - 3)(3x + 5) = 0$$

which can be thought of as

$$(2x - 3)[3x - (-5)] = 0$$

and had solutions

$$x = \frac{3}{2} \quad \text{and} \quad x = -\frac{5}{3}$$

Again, $x^2 + 3x = 0$ factored into

$$x(x + 3) = 0$$

which can be thought of as

$$(x - 0)[x - (-3)] = 0$$

and had solutions

$$x = 0 \quad \text{and} \quad x = -3$$

It is also true that:

If the equation has solution of $x = k$ then p has a factor of $(x - k)$.

If the equation has a factor of $x = \dfrac{k}{l}$ then p has a factor of $(lx - k)$

(where k and l are integers)

The results in these two boxes are together called the *factor theorem*, and they are equally true when p is a polynomial of degree higher than two.

EXAMPLE: *Construct an equation with solutions $x = 3$ and $x = -5$.*

An equation with solution $x = 3$ must have a factor of $(x - 3)$. Because -5 is also a solution, the equation must have a factor of $[x - (-5)] = (x + 5)$ as well. The equation

$$(x - 3)(x + 5) = 0$$

or

$$x^2 + 2x - 15 = 0$$

has both these factors, and therefore fills the bill.

Actually,

$$2(x - 3)(x + 5) = 0 \quad \text{or} \quad 2x^2 + 4x - 30 = 0$$

has exactly the same solutions, as does

$$3(x - 3)(x + 5) = 0 \quad \text{or} \quad 3x^2 + 6x - 45 = 0$$

and any number of other equations. If you don't mind having solutions besides the 3 and -5, you can multiply

$$x^2 + 2x - 15 = 0$$

by any (nonzero) factor you like. For example,

$$x^3 + 2x^2 - 15x = 0$$

has 0, 3, and -5 as solutions.

Second, *how many solutions should you expect from an equation, and what does that have to do with factors?*

All the linear equations that we solved had exactly one solution, or root. The quadratics had two, one, or zero, assuming that we are interested only in real roots.

The reason for this is as follows. Remember what happens when you try and factor a quadratic. Suppose you take out any numerical common factors first; after that, the rest either factors into two things of the form $(x - k)$ or $(lx - k)$ or doesn't factor at all. The factor theorem says that if you have factors like $(x - k)$ or $(lx - k)$, then you have a solution, so, with two such factors you will either have two solutions (if the factors are different) or one (if they're the same). If the quadratic won't factor, or if you can only factor out a constant, it doesn't necessarily mean that there are no real roots, as completing the square and the quadratic formula remind you. But those methods will again lead to at most two real roots, and possibly one or none, so a quadratic equation can never have more than two real roots.

Therefore, a linear, or first-degree equation, has at most one root; and a quadratic, or second-degree equation, has at most two roots. And, in general, it is true that:

The number of solutions to a polynomial equation is at most equal to its degree.

Note: This result is still true even if we allow complex solutions.

PROBLEM SET 15.1

What is the degree of the following equations?
1. $x^5 + 7x^2 - 2x^{19} = x^7$
2. $(t^2)^2 + t^3 - 2t - 3 = t^3$
3. $0 = p^4(p^4 + p^5) - p^5(p^4 + p^5)$
4. $(3x + 2)^2 - (4 - x)^3 = x^2$
5. $y^3 \left[\dfrac{1}{2} - y(y^2 - 1) - \dfrac{y}{2} \right] + 4 = 0$

What is the degree of the polynomial equation obtained after clearing the fractions?

6. $\dfrac{2}{x} + \dfrac{3}{x^2} + \dfrac{4}{x^3} + 1 = 0$

7. $0 = \dfrac{y + 3}{2 - y} + \dfrac{y - 3}{2 + y}$

8. $\dfrac{5}{a^2 + a^3} + a = a - \dfrac{1}{a^3 - a^2}$

9. What is the degree of $ax^2 + 2x + 8xy^3 + y^2a = 25$
   (a) As an equation for x regarding y and a as constants?
   (b) As an equation for y regarding x and a as constants?
   (c) As an equation for a regarding x and y as constants?
10. What is the degree of $R^2(Rq + T)^2 - 17RT^2 + T(3R + q^2)^2 = 0$
   (a) As an equation for R regarding q and T as constants?
   (b) As an equation for q regarding R and T as constants?
   (c) As an equation for T regarding R and q as constants?

Find the roots of the following equations by inspection.

11. $(x - 1)(x - 17) = 0$

12. $(y - 1)(2 - 3y)(y + 1) = 0$

13. $(p - \pi)(p + \pi)(p - \sqrt{2}) = 0$

14. $0 = z(z - 10^{-3})(z + 10^{-4})$

15. $z^2(z - 1)^2(z - i) = 0$

16. $0 = (0.11b - 2.2)(0.03b + 0.6)\left(0.05b - \frac{1}{2}\right)\left(\frac{b}{2} + 1.13\right)$

Find all the values for the unknown making the following expressions 0.

17. $(z - 7)(z + 1)$

18. $q(2q + 3)(3q - \frac{1}{3})^2(0.01 - 0.001q)$

Find an equation that has the following solutions.

19. $1, -1, \frac{1}{2}, 2$

20. $i, \dfrac{-1}{i}, 0$

21. Find an equation with two different roots that add up to 6.

22. Find an equation with two (nonreal) complex roots that add up to 6.

23. Find an expression that becomes 0 when $x = 0$ or $x = 5$ is substituted into it.

Find a quadratic equation that has

24. Two real distinct roots

25. Two real distinct irrational roots

26. One real root

27. No real roots

28. Two distinct complex roots

29. One complex root

True or false? Why?

30. A linear equation with real coefficients has a real root.

31. A linear equation with (nonreal) complex coefficients has a (nonreal) complex root.

32. A quadratic equation with real coefficients has real roots.

33. A quadratic equation with (nonreal) complex coefficients always has (nonreal) complex roots.

34. A cubic equation with real coefficients and two real roots has a third real root.

15.2 POLYNOMIAL EQUATIONS: HOW TO SOLVE THEM

Now we come to the question of which of the methods used for linear and quadratic equations might be useful for equations of higher degree. The answer is not many—except for factoring. Consequently, which higher-degree equations you can solve depends on which higher-degree polynomials you can factor (Section 9.10). We also know (from the last section) that each different linear factor will give us a root, and that the total number of solutions to an equation will be no greater than its degree.

EXAMPLE: *Solve $x^3 + 2x^2 = 3x$.*

To solve this by factoring, first put everything on the left, because eventually we will want to use the Useful Fact, namely that if $ab = 0$, then either $a = 0$ or $b = 0$ or both, and this works only with a zero on the right. Then:

$$x^3 + 2x^2 - 3x = 0$$

There is a common factor of x:

$$x(x^2 + 2x - 3) = 0$$

Please note that you cannot cancel the x, because it could be zero. The quadratic inside the parentheses factors again, so we get:

$$x(x + 3)(x - 1) = 0$$

If the product of three things is zero, at least one of them must be zero, and therefore either $x = 0$ or $x + 3 = 0$ or $x - 1 = 0$

So either

$$x = 0 \quad \text{or} \quad x = -3 \quad \text{or} \quad x = 1$$

These are the solutions to the equation, which here has the maximum number possible for a cubic.

EXAMPLE: *Solve $9x^4 - 37x^2 + 4 = 0$.*

This can be factored as though it were a quadratic because it contains only x^2 and x^4 terms, and so can be written:

$$9(x^2)^2 - 37(x^2) + 4 = 0$$

Using the systematic method, this factors to

$$(9x^2 - 1)(x^2 - 4) = 0$$

But each of those factors is a difference of squares, and factors again:

$$(3x + 1)(3x - 1)(x + 2)(x - 2) = 0$$

Hence either

$$3x + 1 = 0 \quad \text{or } 3x - 1 = 0 \quad \text{or} \quad x + 2 = 0 \quad \text{or} \quad x - 2 = 0$$

giving

$$x = -\frac{1}{3} \quad \text{or} \quad x = \frac{1}{3} \quad \text{or} \quad x = -2 \quad \text{or} \quad x = 2$$

Therefore the solutions are $x = \pm\frac{1}{3}$, $x = \pm 2$.

EXAMPLE: *Solve $3t + 10 = 2t^2 + t^3$.*

Rewrite as

$$t^3 + 2t^2 - 3t - 10 = 0$$

Then, using the method for guessing linear factors (Section 9.10), look for a factor of the form $(t - k)$, where k is one of the factors of 10, namely $\pm 10, \pm 5, \pm 2, \pm 1$. By the factor theorem, if $(t - k)$ is a factor of the polynomial, then $t = k$ is a solution to the equation. Therefore $t = \pm 10, \pm 5, \pm 2, \pm 1$ are possible solutions to the equation.

The quickest way to find out which (if any) are solutions is to try them out by substitution:

$$t = 1: \qquad t^3 + 2t^2 - 3t - 10 = 1 + 2 - 3 - 10 = -10$$

$$t = -1: \quad t^3 + 2t^2 - 3t - 10 = -1 + 2 + 3 - 10 = -6$$

$$t = 2: \qquad t^3 + 2t^2 - 3t - 10 = 8 + 8 - 6 - 10 = 0$$

Therefore $t = 2$ is a solution and $(t - 2)$ a factor.

Before we go looking for any other linear factors, we divide $(t - 2)$ into $t^3 + 2t^2 - 3t - 10$ and get $t^2 + 4t + 5$.

Therefore the equation factors to

$$(t - 2)(t^2 + 4t + 5) = 0$$

So either

$$t - 2 = 0 \quad \text{or} \quad t^2 + 4t + 5 = 0$$

Now $t - 2 = 0$ gives the solution $t = 2$ that we already knew, and $t^2 + 4t + 5 = 0$ is a quadratic, which can be solved the usual way. However, since discriminant $= 4^2 - 4 \cdot 1 \cdot 5 = -4$, it has no real solutions, though we could certainly find complex roots.

Therefore $t = 2$ is the only real solution to $3t + 10 = 2t^2 + t^3$.

PROBLEM SET   15.2

Find all the real roots to the following equations.

1. $x^4 + 4x^3 + 4x^2 = 0$
2. $4z^4 - 37z^2 = -9$
3. $y^6 - 9y^3 + 8 = 0$
4. $13t^2 = 9t^4 + 4$
5. $c^3 - c^2 - 14c + 24 = 0$
6. $3r^4 + 2r^2 - 1 = 0$
7. $0 = y^4 - 29y^2 + 100$
8. $36 = 4x^4$
9. $a^3 - 6 = 6a^2 - 11a$
10. $13b^2 - 36b = -36 + 4b^3 - b^4$
11. $s^4 - 3s^2 = -2$
12. $24 = 10k + 3k^2 - k^3$
13. $4(b - 1) = (b - 1)b^2$
14. $10y^4 = 7y^2 - 1$
15. $x^6 - 1 = 0$
16. $12t^4 = 5t^8 + 7$

17. $(x - 1)^3 + 5(x - 1) = -4(x - 1)^2 - 2$
18. $(z^2 + 9)^2 - 36z^2 = 0$
19. $-19y^2 - 16y^4 = -13y + 3 - 4y^5 - 17y^3$
20. $11(x + 2) - 1 = 5 + 6(x + 2)^2 - (x + 2)^3$

Find all of the complex roots to the following equations.

21. $x^4 - 2x^2 = -1$
22. $t(t^2 + 2t) + 2(8 - 4t) = -2t^2 + 24$
23. $8 - r^3 = 0$
24. $(r + 1)^4 - 3(r + 1)^2 + 2 = 0$
25. $13\left(\dfrac{2s - 1}{s + 1}\right)^2 = 9\left(\dfrac{2s - 1}{s + 1}\right)^4 + 4$
26. $-0.1t(1 - t^2) = -0.2(t^2 - 1)$
27. $r(1 + r^2) - 2r = -3r(\frac{1}{3} + r) + 4$
28. $x^3 - 4\left[2x\left(1 - \dfrac{x}{2}\right)\right] = 32$
29. $4(2a)^4 - 11(2a)^2 = 3$
30. $0.4x^2(x^2 - 9) = 0.1x^2 - 0.9$

15.3   FRACTIONAL EQUATIONS

Just as it is possible to have a fractional equation that is really a linear equation in disguise, and that becomes a linear equation when you clear it of fractions, so it is possible to have a fractional equation that is really a quadratic or higher-degree polynomial equation in disguise, and that shows its true colors when you clear it of fractions. For example,

$$2x - 13 = \frac{7}{x}$$

certainly does not look like a quadratic equation. But if you multiply both sides by x to clear it of fractions, it turns into

$$2x^2 - 13x = 7$$

which is an entirely ordinary quadratic. This can be solved by the usual methods—in this case, factoring works:

$$2x^2 - 13x = 7$$

$$2x^2 - 13x - 7 = 0$$

$$(2x + 1)(x - 7) = 0$$

$$x = -\frac{1}{2} \quad \text{or} \quad x = 7$$

For such equations it is useful to use the *method for attacking fractional equations* described in Section 12.2, which involves finding the least common denominator (L. C. D.), multiplying every term in the equation by it to clear of fractions, and then solving the equation you get.

EXAMPLE: *Solve*

$$\frac{4(t - 5)}{(3t + 5)} + \frac{1}{(t + 5)} = \frac{t - 1}{(t + 5)}$$

The denominators occurring are $(3t + 5)$ and $(t + 5)$, and so the L. C. D. is $(3t + 5)(t + 5)$. Multiplying through by this:

$$\cancel{(3t + 5)}(t + 5) \cdot \frac{4(t - 5)}{\cancel{(3t + 5)}} + (3t + 5)\cancel{(t + 5)} \cdot \frac{1}{\cancel{(t + 5)}}$$

$$= (3t + 5)\cancel{(t + 5)} \cdot \frac{(t - 1)}{\cancel{(t + 5)}}$$

So

$$4(t + 5)(t - 5) + (3t + 5) = (3t + 5)(t - 1)$$

This is clearly a quadratic, and we'll hope it can be solved by factoring or completing the square. But before we can do either, we must get the equation into the usual form with a quadratic polynomial on the left and a zero on the right. This means multiplying out all the products and collecting like terms:

$$4(t + 5)(t - 5) + (3t + 5) = (3t + 5)(t - 1)$$

$$4t^2 - 100 + 3t + 5 = 3t^2 + 2t - 5$$

$$t^2 + t - 90 = 0$$

This factors (fortunately):

$(t - 9)(t + 10) = 0$

giving

$t = 9$ or $t = -10$

Of course, you should check these solutions in the original equation.

EXAMPLE: *Solve*

$$\frac{32}{\pi r} = \frac{r^2}{2\pi^4} \quad for\ r$$

Multiply both sides by $2\pi^4 r$:

$$2\pi^{\overset{3}{4}}r \cdot \frac{32}{\cancel{\pi r}} = \cancel{2\pi^4 r} \cdot \frac{r^2}{\cancel{2\pi^4}}$$

$$64\pi^{3|} = r^3$$

A number has only one real cube root. Therefore

$r = 4\pi$

Now things can go just as wrong with the solutions to fractional equations here as they did in Section 12.2.

EXAMPLE: *Solve*

$$\frac{x + 1}{x - 1} = \frac{x}{3} + \frac{2}{x - 1}$$

Multiply both sides by $3(x - 1)$:

$$3(\cancel{x - 1}) \cdot \frac{x + 1}{\cancel{x - 1}} = \cancel{3}(x - 1) \cdot \frac{x}{\cancel{3}} + 3(\cancel{x - 1}) \cdot \frac{2}{(\cancel{x - 1})}$$

$$3(x + 1) = x(x - 1) + 6$$

$$3x + 3 = x^2 - x + 6$$

$$x^2 - 4x + 3 = 0$$

$$(x - 1)(x - 3) = 0$$

Therefore solutions are

$x = 1$ or $x = 3.$

But a strange thing has happened in this example: If you check the solution $x = 3$, everything goes fine:

$$\text{LHS} = \frac{x+1}{x-1} = \frac{3+1}{3-1} = \frac{4}{2} = 2$$

$$\text{RHS} = \frac{x}{3} + \frac{2}{x-1} = \frac{3}{3} + \frac{2}{3-1} = 1 + 1 = 2$$

So $x = 3$ is a solution. But when you check $x = 1$, there's trouble:

$$\text{LHS} = \frac{x+1}{x-1} = \frac{1+1}{1-1} = \frac{2}{0} \quad \text{(undefined)}$$

$$\text{RHS} = \frac{x}{3} + \frac{2}{x-1} = \frac{1}{3} + \frac{2}{1-1} = \frac{1}{3} + \frac{2}{0} \quad \text{(undefined)}$$

So one of the numbers we have come up with as a solution is the *one* number for which the equation makes no sense—neither side of the equation is even defined, so they can't possibly be equal. This clearly means we can't consider $x = 1$ as a solution, and so the only answer is $x = 3$.

In general, it is important to check solutions to fractional equations and to discard any that make a denominator zero.

EXAMPLE: *Solve*

$$\frac{y-3}{2-3y} + \frac{5y-1}{y+2} = \frac{88}{3y^2 + 4y - 4}$$

The denominators are $(2 - 3y)$, $(y + 2)$, and

$$3y^2 + 4y - 4 = (3y - 2)(y + 2)$$

Since $(2 - 3y) = -(3y - 2)$, the L.C.D. is $(3y - 2)(y + 2)$.

Multiplying through:

$$\overset{(-1)}{\cancel{(3y-2)}}(y+2) \cdot \frac{(y-3)}{\cancel{(2-3y)}} + (3y-2)\cancel{(y+2)} \cdot \frac{(5y-1)}{\cancel{(y+2)}}$$

$$= \cancel{(3y-2)}\cancel{(y+2)} \cdot \frac{88}{\cancel{(3y^2+4y-4)}}$$

gives

$$-(y+2)(y-3) + (3y-2)(5y-1) = 88$$

$$-y^2 + y + 6 + 15y^2 - 13y + 2 = 88$$

$$14y^2 - 12y - 80 = 0$$

$$2(7y - 20)(y + 2) = 0$$

so

$(7y - 20)(y + 2) = 0$

therefore either

$7y - 20 = 0$ or $y + 2 = 0$

which means

$y = \dfrac{20}{7}$ or $y = -2$

So $y = \frac{20}{7}$ and $y = -2$ are possible solutions. But since $y = -2$ makes the denominator $y + 2$ zero, the only solution is $y = \frac{20}{7}$.

PROBLEM SET 15.3

Solve the following fractional equations for all real roots.

1. $x + \dfrac{x}{2} - \dfrac{1}{x} = 4 + \dfrac{1}{2x}$

2. $\dfrac{1 - z - z^2}{z} = \dfrac{z}{9}$

3. $\dfrac{4}{t + 1} - \dfrac{1}{t} - 1 = 0$

4. $\dfrac{y + 2}{2y + 5} = y + 2$

5. $\dfrac{21}{w + 2} - \dfrac{1}{w - 4} = 2$

6. $\dfrac{x + 3}{3} = \dfrac{4}{x - 1}$

7. $\dfrac{s + 2}{s^2 - s - 6} = 3 - \dfrac{4}{s - 3}$

8. $0 = \dfrac{1}{6c^2} - \dfrac{3}{4c^2} + \dfrac{7}{3}$

9. $a - 1 + \dfrac{2}{1 - a} = 0$

10. $x + \dfrac{1}{x - 1} = \dfrac{9}{2}$

11. $\dfrac{w - 1}{2} + \dfrac{2}{w - 1} = \dfrac{5}{2}$

12. $\dfrac{6}{2x + 3} = x + 1$

13. $\dfrac{3z + 1}{5z + 4} - \dfrac{z + 3}{4z + 5} = 0$

14. $\dfrac{2}{3}(5 - x) - 1 = \dfrac{2}{4 - x}$

15. $\dfrac{3p - 1}{p - 1} - \dfrac{1 + 2p}{1 + p} = 1$

16. $\dfrac{2}{b - 1} + 1 = \dfrac{3b - 2}{4b + 5}$

17. $\dfrac{r - 2}{3r + 1} - \dfrac{2(r + 4)}{r - 4} = 0$

18. $\dfrac{k}{k + 1} - \dfrac{2}{3k + 1} = 0$

19. $3\left(\dfrac{x + 1}{x - 1}\right) = \dfrac{2}{x - 4} + \dfrac{6}{x - 1}$

20. $\dfrac{(-9 - 2x + 3x^2)}{4x^2 - 16} = \dfrac{2}{3}$

21. $\dfrac{\dfrac{1}{z} - z}{1 - \dfrac{2}{z + 1}} = \dfrac{3(1 - 2z)}{2}$

22. $\dfrac{2 + \dfrac{1}{r - 1}}{r + 1 - \dfrac{3}{2r}} = \dfrac{r - 4}{r + 1 - \dfrac{3}{2r}} + \dfrac{2r}{r - 1}$

23. $\dfrac{3t + 2}{t - 2} + \dfrac{3}{t + 2} = \dfrac{4t}{t - 2}$

24. $3 - \left(\dfrac{3 - x^2}{11 + x}\right) = 2(x + 2) + 1 + 3\left(\dfrac{2x - 1}{x + 11}\right)$

Solve the following equations for the variable in parentheses.

25. $4\pi f c = \dfrac{1 + f}{3\pi f L}$ (f)

26. $\dfrac{y^2}{b^2} + \dfrac{6ay}{b} + 9a^2 = 0$ (b)

27. $\dfrac{1}{p} - \dfrac{1}{p + x} = \dfrac{1}{q - x} - \dfrac{1}{q}$ (x)

28. $\dfrac{a(x - 2)}{b - 2a} = \dfrac{b + 2a}{a(x + 2)}$ (x)

29. $\dfrac{1 - p^2}{ap + b} - \dfrac{bp + a}{ab} = 0$ (p)

30. $1 - \dfrac{y}{x + y} = \dfrac{y}{x - y}$ (y)

31. $\dfrac{a - 2}{s - a} = \dfrac{a}{s^2 - a}$ (s)

32. $(m - p)^2 - 2(m - p) - 15 = 0$ (m)

15.4   RADICAL EQUATIONS

Just as a linear or quadratic equation can be disguised in a fractional equation, so they can be disguised in a radical equation—meaning one involving a root of the unknown. For example,

$$4\sqrt{x} - 1 = 4x \qquad \sqrt{x^2 - a^2} = 3a^2 \qquad \dfrac{1}{\sqrt{t + 1}} = \sqrt{t}$$

are radical equations. If you can figure out how to get rid of the radical sign, these equations can often be turned into quite ordinary and respectable equations. However, getting rid of the radical often entails squaring, which means that you have to be terrifically careful about checking your solutions—as you will see in the last example in this section.

EXAMPLE: *Solve*

$$\sqrt{t^2 + 1} - \dfrac{1}{\sqrt{t^2 + 1}} = 0$$

Clear of fractions by multiplying by $\sqrt{t^2 + 1}$:

$$\sqrt{t^2 + 1} \cdot \sqrt{t^2 + 1} - \sqrt{\cancel{t^2 + 1}} \cdot \frac{1}{\sqrt{\cancel{t^2 + 1}}} = \sqrt{t^2 + 1} \cdot 0$$

so

$$t^2 + 1 - 1 = 0$$

giving $t^2 = 0$

so $\qquad t = 0$

Notice that no squaring was needed here.

CHECK: When $t = 0$,

$$\text{LHS} = \sqrt{t^2 + 1} - \frac{1}{\sqrt{t^2 + 1}} = \sqrt{1} - \frac{1}{\sqrt{1}} = 1 - 1 = 0$$

$$\text{RHS} = 0$$

So the solution is $t = 0$.

EXAMPLE: *Solve for u:* $\quad \sqrt{u^2 + 9a^2} = a + u \quad$ *where a is positive (not zero).*

In order to be able to solve for u, we must get rid of that square root. Therefore we will square both sides of the equation, giving

$$(\sqrt{u^2 + 9a^2})^2 = (a + u)^2$$

or

$$u^2 + 9a^2 = a^2 + 2au + u^2$$

Therefore

$$9a^2 = a^2 + 2au$$

or

$$8a^2 = 2au$$

so

$$u = \frac{8a^2}{2a} = 4a$$

CHECK: When $u = 4a$,

$$\text{LHS} = \sqrt{u^2 + 9a^2} = \sqrt{(4a)^2 + 9a^2} = \sqrt{16a^2 + 9a^2} = \sqrt{25a^2}$$

$$= 5a \quad (a \text{ is positive})$$

$RHS = a + u = a + 4a = 5a$

So the solution is $u = 4a$.

EXAMPLE: *Solve $4p = 4\sqrt{p} - 1$.*

If we square both sides of this equation to try and get rid of the radical, we have:

$(4p)^2 = (4\sqrt{p} - 1)^2$

or

$16p^2 = 16p - 8\sqrt{p} + 1$

which isn't much help because it still has a radical in it. Is there any way we could have avoided the $8\sqrt{p}$? Since it came up as the middle term from multiplying out $(4\sqrt{p} - 1)^2$, if the 1 had not been there (i.e., if we had just been squaring $4\sqrt{p}$), all would have been well because we would have got $(4\sqrt{p})^2 = 16p$. So, obviously, the thing to do is to move the 1 to the other side of the equation before squaring:

$4p + 1 = 4\sqrt{p}$

so

$(4p + 1)^2 = (4\sqrt{p})^2$

$16p^2 + 8p + 1 = 16p$

so

$16p^2 - 8p + 1 = 0$

$(4p - 1)^2 = 0$

Therefore the solution is $p = \dfrac{1}{4}$.

CHECK: When $p = \frac{1}{4}$:

$LHS = 4p = 4 \cdot \dfrac{1}{4} = 1$

$RHS = 4\sqrt{p} - 1 = 4\sqrt{\dfrac{1}{4}} - 1 = 4 \cdot \dfrac{1}{2} - 1 = 1$

So $p = \frac{1}{4}$ is a correct solution.

EXAMPLE: *Solve $\sqrt{2x - 5} = 1 + \sqrt{x - 3}$.*

Squaring both sides to try and remove the radicals, we see that

$$(\sqrt{2x-5})^2 = (1 + \sqrt{x-3})^2$$
$$2x - 5 = 1 + 2\sqrt{x-3} + (x-3)$$

The $\sqrt{x-3}$ term is the cross (or middle) term from multiplying out $(1 + \sqrt{x-3})^2$, but this time no amount of moving the 1 to the other side would help because there is a radical there too. However, at this stage, there is only one radical left, so if we move everything else to one side and square again, we will get rid of the last radical. If

$$2x - 5 = 1 + 2\sqrt{x-3} + (x-3)$$

then

$$x - 3 = 2\sqrt{x-3}$$

Squaring both sides gives:

$$(x-3)^2 = (2\sqrt{x-3})^2$$

or

$$x^2 - 6x + 9 = 4(x-3)$$

or

$$x^2 - 10x + 21 = 0$$

which factors to

$$(x-3)(x-7) = 0$$

therefore

$$x = 3 \quad \text{or} \quad x = 7$$

CHECK:

When $x = 7$:

LHS $= \sqrt{2x-5} = \sqrt{2 \cdot 7 - 5} = \sqrt{14 - 5} = \sqrt{9} = 3$

RHS $= 1 + \sqrt{x-3} = 1 + \sqrt{7-3} = 1 + \sqrt{4} = 1 + 2 = 3$

So $x = 7$ is a solution.

When $x = 3$:

LHS $= \sqrt{2x-5} = \sqrt{2 \cdot 3 - 5} = \sqrt{6-5} = \sqrt{1} = 1$

RHS $= 1 + \sqrt{x-3} = 1 + \sqrt{3-3} = 1 + \sqrt{0} = 1$

So $x = 3$ is a solution.

If you are wondering why I have been so careful about checking the solutions to these equations, just look at the next one:

EXAMPLE: *Solve* $2\sqrt{x} + 3\sqrt{x-5} = 0$.

Suppose we rewrite this as

$$2\sqrt{x} = -3\sqrt{x-5}$$

If we square it in this form we get no cross terms:

$$(2\sqrt{x})^2 = (-3\sqrt{x-5})^2$$

and the equation becomes

$$4x = 9(x-5)$$

that is,

$$4x = 9x - 45$$

giving

$$x = 9$$

CHECK: Substituting into the original equation: When $x = 9$,

$$\text{LHS} = 2\sqrt{x} + 3\sqrt{x-5} = 2\sqrt{9} + 3\sqrt{9-5} = 2 \cdot 3 + 3 \cdot 2 = 12$$

$$\text{RHS} = 0$$

So $x = 9$ does *not* seem to be a solution because it makes the left-hand side of the equation 12, while the right-hand side is 0.

What has happened?
The problem is this: When we squared

$$2\sqrt{x} = -3\sqrt{x-5}$$

we got

$$4x = 9(x-5)$$

If we were to go backwards by taking square roots again, we get

$$2\sqrt{x} = \pm 3\sqrt{x-5}$$

This means that if x satisfies $4x = 9(x-5)$, then x satisfies either $2\sqrt{x} = -3\sqrt{x-5}$ (our equation) or $2\sqrt{x} = +3\sqrt{x-5}$. In other words, the solution to $4x = 9(x-5)$ may satisfy our original equation or it may satisfy another equation. And as it turns out, the $x = 9$ we got does not satisfy our original equation, but satisfies $2\sqrt{x} = +3\sqrt{x-5}$ instead. So we have to throw it away and say $2\sqrt{x} + 3\sqrt{x-5} = 0$ has no solution at all.

If you look again at

$$2\sqrt{x} + 3\sqrt{x-5} = 0$$

and remember that $\sqrt{x}$ and $\sqrt{x-5}$ are defined to be the positive square roots of x and $x-5$, you will see that it is impossible for the left-hand side of this equation to be zero unless both $\sqrt{x}$ and $\sqrt{x-5}$ are zero—which is impossible for the same value of x. Therefore, it's not surprising that this equation has no solutions.

As in the example above, squaring both sides of an equation may introduce extra roots that are not solutions to the original equation. These are called *extraneous roots*. For example, squaring $x = 2$ gives $x^2 = 4$, which has roots $x = 2$ *and* $x = -2$ and so squaring has added the root $x = -2$. Solving an equation containing radicals usually involves squaring, and the possibility of extra roots. Therefore, the solutions you end up with may, in fact, not be solutions to the original equation, and the only way to find out if they are is to check. Hence, checking is vitally important!

PROBLEM SET   15.4

Solve each of the following equations. Be sure to check your answers.

1. $\sqrt{2x-1} - \sqrt{x-1} = 1$

2. $x - 5\sqrt{x} + 6 = 0$

3. $x + \sqrt{x} = 6$

4. $2x - \sqrt{2x-3} = 3$

5. $x = \sqrt{5x-4} + 2$

6. $\sqrt{x^2+1} - \dfrac{1}{\sqrt{x^2+1}} = 0$

7. $\sqrt{2a-1} = 8 - a$

8. $\sqrt{3p} = \dfrac{2}{\sqrt{9p+1}}$

9. $\sqrt{r-3} + 1 = \sqrt{2r-5}$

10. $\sqrt{\dfrac{3t^2+4}{t-1}} = 4$

11. $\dfrac{1}{\sqrt{x}} + 2\sqrt{x} = \sqrt{2x+1}$

12. $\sqrt{s+1} = 2\sqrt{1-s} + 1$

13. $\sqrt{1+3z} = 3 + \sqrt{z}$

14. $a + 3 + \sqrt{a+5} = 0$

15. $\dfrac{t+1}{\sqrt{t-2}} = \sqrt{2t-3}$

16. $2\left(k - 1 + \sqrt{\frac{5}{4}k - \frac{9}{4}}\right) = k + 1$

17. $\sqrt{\frac{4}{3}w + 2} = \frac{5}{3} + w$

18. $\sqrt{b+5} = \dfrac{4}{\sqrt{b+2}}$

19. $\sqrt{2v-1}\,(1 + \sqrt{v}) = 0$

20. $\sqrt{2r+12} = 2(r-1)$

Solve the following literal equations for z. Checking your answers can be difficult here unless you have values for the other letters.

21. $\sqrt{az+b} = z - a$

22. $\sqrt{pz + r}\ \sqrt{pz - r} = p + r$

23. $\sqrt{\dfrac{sz + t}{a}} = z$

24. $\dfrac{2\sqrt{zx}}{\sqrt{z + x}} = \sqrt{x + z}$

25. $\dfrac{1}{\sqrt{1 - az}} = \sqrt{\dfrac{az + 1}{2az}}$

15.5   YET MORE WORD PROBLEMS

You may have noticed how all the word problems in Chapter 13 led to linear equations. This, of course, was not an accident—they were chosen to do so—and you should realize that a word problem can just as well lead to a quadratic, or to any other kind of equation. Such problems fall into many of the same types as the problems of Chapter 13, and are solved in the same way.

Number Problems   (see Section 13.3)

EXAMPLE: *Find two consecutive integers such that if twice the square of the smaller is subtracted from 27 times the larger, the result is 117.*

(by the method of Section 13.2)

STEP 2.   We have to find two integers, but since they are consecutive we know that the larger is exactly one bigger than the smaller. Let the smaller integer $= x$. Then the larger integer $= x + 1$.

STEP 4.   The problem tells us that

27 (larger integer) $- 2$ (square of smaller integer) $= 117$

STEP 6.   Now

27 (larger integer) $= 27(x + 1)$

and

2(square of smaller integer) $= 2x^2$

STEP 7.   The equation becomes

$27(x + 1) - 2x^2 = 117$

STEP 8. This is a quadratic equation which (we hope) can be solved by factoring:

$$27x + 27 - 2x^2 = 117$$

$$2x^2 - 27x + 90 = 0$$

$$(2x - 15)(x - 6) = 0$$

giving

$$x = \frac{15}{2} \quad \text{or} \quad x = 6$$

So there seem to be two possible answers for the smaller number, namely $\frac{15}{2}$ and 6. But we are looking for an *integer*, which $\frac{15}{2}$ is certainly not.

Therefore the smaller integer must be 6, making the larger 7.

STEP 9. *Check:* 27 (larger integer) = 189

2(square of the smaller) = 72

The difference of 189 and 72 is 117, as required.

Extraneous Solutions

But what has gone wrong that we got $\frac{15}{2}$ as an answer for the smaller integer? This extra—and wrong—solution to the word problem arises from a rather delicate point. In deriving the quadratic equation for x, namely $27(x + 1) - 2x^2 = 117$, it is clear that the x which is the answer to this problem *must* satisfy this equation. However, we did not show that any x which satisfies this equation is also a solution to the word problem. In fact this is not true, as you can see from the fact that $\frac{15}{2}$ satisfies the equation but not the word problem. Numbers that are solutions to the equation but not the word problem are called *extraneous solutions* to the equation, and they should just be ignored when you are listing solutions to the original problem. Such a situation never arises with word problems leading to linear equations, because a linear equation has only one solution, which therefore must be the one that is also a solution to the problem.

Geometry Problems   (see Section 13.5)

EXAMPLE: *A candy manufacturer of dubious reputation makes a rectangular chocolate bar with a total volume of 6 cubic inches. The advertising department says that if the bar is made longer, the* *public will think that it contains more chocolate. Cashing in on this idea, the manufacturer decides to make a bar of the same volume but 2 inches longer, 1 inch narrower, and of the same $\frac{1}{2}$ inch height. What was the length of the original bar?*

(This time I won't number the steps.)

Let the length of the old bar $= x$ inches. Then the length of the new bar $= (x + 2)$ inches.

It's helpful to draw a picture, so look at Figure 15.1.

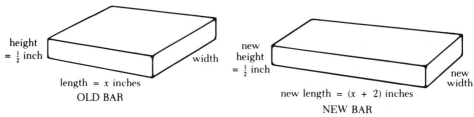

height = $\frac{1}{2}$ inch

width

length = x inches
OLD BAR

new height = $\frac{1}{2}$ inch

new width

new length = $(x + 2)$ inches
NEW BAR

FIG. 15.1

Our equation is going to come from the fact that

volume of new bar $= 6$ cubic inches (in^3)

So

new length $\cdot$ new width $\cdot$ new height $= 6$

Now the only thing we don't know here is the new width. However, we do know that

volume of old bar $= 6$

old length $\cdot$ old width $\cdot$ old height $= 6$

So

$$\text{old width} = \frac{6}{\text{old length} \cdot \text{old height}} = \frac{6}{x \cdot \frac{1}{2}} = \frac{12}{x} \text{ inches}$$

The new width is 1 inch less, so

$$\text{new width} = \left(\frac{12}{x} - 1\right) \text{ inches}$$

Therefore the equation,

new length $\cdot$ new width $\cdot$ new height $= 6$

becomes

$$(x + 2)\left(\frac{12}{x} - 1\right) \cdot \frac{1}{2} = 6$$

Solve by multiplying out:

$$6 + \frac{12}{x} - \frac{x}{2} - 1 = 6$$

and multiplying through by $2x$:

$$12x + 24 - x^2 - 2x = 12x$$

giving

$$x^2 + 2x - 24 = 0$$

which factors to

$$(x + 6)(x - 4) = 0$$

so

$$x = -6 \quad \text{or} \quad x = 4$$

Now you can't have a candy bar of length -6, so $x = -6$ is an extraneous solution. Therefore the original bar is 4 inches long.

CHECK: If the old bar is 4 in long and $\frac{1}{2}$ in high, then it must be 3 in wide if its volume is to be 6 in³ (because $6 = 4 \cdot 3 \cdot \frac{1}{2}$). The new bar is 2 in longer, 1 in narrower and the same height, so its dimensions are 6 in by 2 in by $\frac{1}{2}$ in. Therefore the volume of the new bar is $6 \cdot 2 \cdot \frac{1}{2} = 6$ in³, which checks.

Distance Problems   (see Section 13.8)

As before, all distance problems depend on the formula

$$\text{distance} = \text{speed} \cdot \text{time}$$

EXAMPLE: *The Boston Marathon is 26 miles long. If a runner averages 2 mph less than the winner, and finishes 39 minutes behind him, what is the winner's average speed?*

(numbering the steps of Section 13.2)

STEP 2.   Let the winner's average speed be x mph. Then the other runner's average speed is $(x - 2)$ mph.

STEP 4.   Since the speeds are given in miles per hour, we must change the 39 minutes to $\frac{39}{60} = \frac{13}{20}$ hours. Our equation will come from the fact that if a runner finishes 39 minutes or $\frac{13}{20}$ of an hour behind the winner, then the runner's time is $\frac{13}{20}$ hours longer than the winner's, so

$$\left(\begin{array}{c}\text{runner's}\\\text{time (hours)}\end{array}\right) = \left(\begin{array}{c}\text{winner's}\\\text{time (hours)}\end{array}\right) + \frac{13}{20}$$

STEP 5.   Now we have to write each of these times in terms of x. We do this using the formula

$$\text{time} = \frac{\text{distance}}{\text{speed}}$$

STEP 6.   Each person goes 26 miles. The winner's speed is x mph, so

$$\text{winner's time} = \frac{26}{x} \text{ hours}$$

The other runner's speed is $(x-2)$ mph, so

$$\text{other runner's time} = \frac{26}{(x-2)} \text{ hours}$$

STEP 7.   Therefore the equation tells us:

$$\frac{26}{(x-2)} = \frac{26}{x} + \frac{13}{20}$$

STEP 8.   Cancelling 13 from the numerators to simplify:

$$\frac{2}{(x-2)} = \frac{2}{x} + \frac{1}{20}$$

Multiplying through by the L.C.D., which is $20x(x-2)$:

$$20x(x-2)\frac{2}{(x-2)} = 20x(x-2)\cdot\frac{2}{x} + 20x(x-2)\cdot\frac{1}{20}$$

gives

$$40x = 40(x-2) + x(x-2)$$

or

$$x^2 - 2x - 80 = 0$$

So

$$(x+8)(x-10) = 0$$

giving

$$x = -8 \quad \text{or} \quad x = 10$$

We are looking for a speed, which can't very well be negative, so $x = -8$ is an extraneous solution.

Therefore, the winner's average speed is 10 mph.

CHECK: The winner's time is $\frac{26}{10} = \frac{13}{5}$ hours; the other runner's speed is 8 mph, so his time is $\frac{26}{8} = \frac{13}{4}$ hours. The other runner therefore finishes $\frac{13}{4} - \frac{13}{5} = \frac{13}{20}$ hours behind the winner.

There's another type of distance problem that often gives rise to a quadratic equation: the *current problem*. These are usually about boats that sometimes go upstream against the current and sometimes go downstream with the current.

There's one important thing you have to realize to be able to do current problems, which is this: If a boat goes 5 mph in still water, then in moving upstream against a current of 1 mph, it will go 4 mph.

To see why this is, think of the boat and the current working separately. Suppose that in an hour the boat runs freely (without the current) upstream for 5 miles, and then the current takes it 1 mile back downstream again. Thus the boat has ended up going 4 miles upstream by the end of the hour, and therefore its speed upstream against the current is 4 mph. If, however, the boat goes downstream, its speed will be 6 mph. In this case the boat will move 5 miles downstream on its own in an hour, and the current will take it 1 mile further. Therefore it will go 6 miles downstream in an hour, and have a speed of 6 mph.

This kind of reasoning applies to any situation in which one thing is moving relative to something that is itself moving, such as boats and currents, planes and airstreams, and the following example.

EXAMPLE: *In a sustained (and often successful) attempt to harass his mother, a child runs up and down the escalators in a department store. His mother observes that her child takes 1 minute to make a round trip up and down a 40-foot escalator which travels at 30 feet per minute downwards. Assuming the child can run the same speed up and down stationary stairs, find that speed.*

Let the child's speed on stationary stairs be x ft per min. Then, just as with the boat going upstream, the child's speed running up the escalator is $(x - 30)$ ft per min, and his speed running down the escalator is $(x + 30)$ ft per min.

Our equation comes from the fact that the whole trip (up and down the escalator) takes 1 minute, so

$$\begin{pmatrix} \text{time (min)} \\ \text{to go 40 ft} \\ \text{up} \end{pmatrix} + \begin{pmatrix} \text{time (min)} \\ \text{to go 40 ft} \\ \text{down} \end{pmatrix} = 1$$

We use

$$\text{time} = \frac{\text{distance}}{\text{speed}}$$

to say that if the child goes 40 feet up the escalator at $(x - 30)$ ft per min, then

$$\text{Time to go up} = \frac{40}{(x - 30)} \text{ min}$$

Similarly, the child goes 40 feet down the escalator at $(x + 30)$ ft per min so

$$\text{Time to go down} = \frac{40}{(x + 30)} \text{ min}$$

So the equation reads

$$\frac{40}{(x - 30)} + \frac{40}{(x + 30)} = 1$$

Multiplying through by $(x - 30)(x + 30)$ gives

$$(\cancel{x - 30})(x + 30)\frac{40}{(\cancel{x - 30})} + (x - 30)(\cancel{x + 30})\frac{40}{(\cancel{x + 30})}$$

$$= (x - 30)(x + 30)$$

or

$$40(x + 30) + 40(x - 30) = (x - 30)(x + 30)$$

$$40x + 1200 + 40x - 1200 = x^2 - 900$$

$$x^2 - 80x - 900 = 0$$

$$(x - 90)(x + 10) = 0$$

So

$$x = 90 \quad \text{or} \quad x = -10$$

A negative speed makes no sense, so $x = -10$ is an extraneous solution.

Therefore the child's speed is 90 ft per min

CHECK: The child's speed going up is 60 ft per min, so the time to go 40 feet up is $\frac{40}{60} = \frac{2}{3}$ min.

The speed going down is 120 ft. per min, so the time to go down is $\frac{40}{120} = \frac{1}{3}$ min.

Therefore the total time is $\frac{2}{3} + \frac{1}{3} = 1$ min.

(*Note:* 90 ft per min is only slightly over 1 mph, so this kid has a way to go before he's a competitive runner!)

PROBLEM SET   15.5

1. Two consecutive integers multiply to 462. What are the integers?

2. Two consecutive even integers multiply to 168. What is the smaller of the two?

3. The base of a triangle is 4 feet less than the altitude (or height), and the area of the triangle is 48 square feet. Find the length of the base.

4. Bamilla throws two dice. The difference in face value between the two dice is three; and the product of the numbers thrown is equal to twice the sum of the two numbers. What numbers did Bamilla throw?

5. Theophilus spent 30 years of his life writing two books. The first, *The Origins of Human Thought*, was 350 pages, while the second, *Corrections to the Origins of Human Thought*, was 800 pages. Fortunately, Theophilus averaged 25 pages per year more writing the second book. How long did the first book take?

6. XB-112 is a Martian general, determined to take over control of that planet by brute force. Given enough time, he could manage such a takeover. XB-113, his younger brother, is a politician, also determined to take over the planet. XB-113's method is more effective; he could conquer Mars in 5 fewer years than his brother. Rather than compete, however, the brothers decide to work together, so that the conquest will take only 6 years. How long would it have taken XB-112 alone?

7. You are planning a high-rise office building on a site of 300,000 square feet. The city has a law saying that all new buildings must be able to provide off-street parking for all employees assuming each employee drives his own car to work. Therefore, part of the site must be made into a parking lot. The problem is that if you make the building taller you have to make the base larger (or it will fall down), but at the same time, more people can work in a bigger building and so you need more space for the parking lot. Given the following data:

Number of employees in each story $= 50$

Space needed to park one car $= 50$ square feet

Size of base of building of x stories is $5x^2$ square feet

Find the maximum number of stories your building can be.

8. A hare and a tortoise are racing between two trees 150 feet apart. The tortoise has a headstart of 30 seconds, but the hare catches up in 5 seconds. The hare then runs all the way to the finish line, turns around, and runs back to meet the tortoise again 10 seconds later (the hare has now run 15 seconds in all). What is the speed of each, assuming they remain constant throughout the race?

9. The giant albatross migrates seasonally from Guam to Kwajalein and back over the central Pacific, where the trade winds blow at a constant 15 mph. If the islands are 2250 miles apart, and the trip takes 100 hours longer against the winds than with the winds, how fast does the albatross fly in calm air?

10. When Sam and Susy reached retirement, they were finally able to buy the house of their dreams. It had a 40-by-50 foot rectangular yard. Susy cultivated $\frac{2}{3}$ of the area of the yard in flowers, which she grew in an even border around the central court. Every morning Sam walked the dog once around the central grass area inside the flowers. What distance did he cover in his morning stroll?

11. Two numbers differ by 4. The reciprocal of the smaller exceeds the reciprocal of the larger by $\frac{1}{3}$. What is the larger of the two numbers?

12. At 4:32 the Silver Streak Express left Chicago with $100,000 bound for the First National Bank of Springfield, 100 miles away. An hour later the Dalton Gang left Chicago riding after the Express. If the Gang moves at 5 mph faster than the train, and manages to catch up with the train just as it reaches Springfield, how fast was the train moving?

13. Marge Arane and "Sugar" Kane were selected as finalists in a biscuit baking contest. Each had to bake 72 biscuits for the final judging round. Unfortunately, the contest only had one oven, so Marge had to wait for Sugar to finish her baking before she could begin. Each batch of biscuits took 20 minutes. If each batch Marge made contained two more biscuits than each of Sugar's batches, and it took 7 hours to bake all those biscuits, how many biscuits did Sugar make in each of her batches?

14. One day Sam Spade and Charlie Chan were sitting around the Famous Detectives Club comparing annual earnings.

"I made $60,000 last year," said Sam, peering down the barrel of his gun.

"Exactly how much I made" said Charlie, "yet I solved eight more cases than you did."

"I keep telling you that you should raise your rates, Charlie," said Sam, "I charge $2000 more per case than you. I get classier clients."

"Perhaps you are right, my friend," Charlie replied.

How much did Charlie Chan charge?

15. Dirk Gustav and his wife Cybil paint in a similar style using spraypaint on 20-foot-high steel cylinders. Cybil paints very quickly; she could finish a

cylinder by herself in 3 hours less time than Dirk would require. However, Dirk and Cybil always paint cylinders together, Dirk painting up from the bottom and Cybil painting down from the top. If it takes them 3 hours and 36 minutes to complete a cylinder, how far up the cylinder does Dirk paint?

16. In a proposed new Olympic Alpine event, each contestant must hike to the top of a 1-mile course laid out on the side of a mountain and then ski back down. Sven Olafson currently holds the record for this round trip: 1 hour and 2 minutes. On this run Sven averaged 29 mph faster coming down than he did going up. How fast did Sven go up the mountain?

17. After a Cabinet meeting concerning states' rights, the President decides to call all the governors to inform them of his decisions. He knows that if he does this by himself, it will take him a long time. He considers delegating the duty to his Secretary of State, who could complete the calls in 5 hours less time, but eventually decides to work with the Secretary in making the calls. Together they call all the governors in 200 minutes. How long would it take for the President to do the job by himself?

18. Burger Queen wants to market a "party-size" portion of french fries. The container will have a volume of 2250 cubic inches, and will be an open square box of depth 10 inches made out of a square piece of cardboard by cutting out the corners and folding up the edges (see Figure 15.2). Assuming that you are not allowed to pile french fries above the top of the container, find the length of the side of the original piece of cardboard.

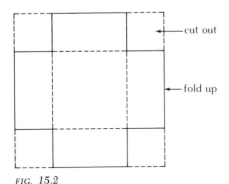

FIG. 15.2

19. Two years ago, Ivy College bought an old house and converted it into a library to house 30,000 books. At that time they bought just enough bookcases to hold all of the books. Unfortunately, the floors in the old house were not quite strong enough to support fully loaded bookcases, so the school purchased five more bookcases and put them in another room. This allowed them to have 300 fewer books in each of the cases. Assuming every bookcase holds the same number of books, how many bookcases did Ivy College buy originally?

20. The *Bronx Bugle* is printed on two presses, an old one and a new one that works at twice the rate of the old one. With both presses working it takes three hours to print the daily edition. One night the old press breaks down, and, owing to the commotion, printing does not start until 12:30 a.m. Does the new press work fast enough to get the day's edition out in time for the 5:15 a.m. commuter trains?

21. A man ekes out a meager living by carrying coals to Newcastle, a distance of 120 miles from his home. He finds he can make the trip in 50 minutes less time by night than by day because the traffic conditions allow him to travel 12 mph faster than he could by day. What is his average daytime speed for this futile trip?

22. Billy "Slugger" Robinson is the best home run hitter to play in the Kalamazoo Pee-Wee league since Jonny "Bongo" Smedlap back in the 1940s. In fact, if Billy had averaged one more home run every 20 games this year, and if he played in another 10 games, he would have tied "Bongo's" 1942 record of 40 home runs in one season. As it is he's 10 home runs behind the record. How many games did Billy play this season?

23. The power (W, in watts) used by an electrical appliance is equal to the voltage (V, in volts) across the appliance's circuit multiplied by the current (I, in amps) running in the circuit. A certain light fixture has a current regulator so that if the voltage across the circuit drops, the current will be increased to maintain the power used at 60 watts. If the voltage drops by 20 volts from its standard level, the current regulator will increase the current by 0.1 amp. What is the standard current used by the light?

24. The Kentucky Derby is $1\frac{1}{4}$ miles long. If Naval Jelly finishes second to Foolish Smile, averaging 1.5 mph less than the faster horse and crossing the finish line 5 seconds later, does Foolish Smile break the Derby record of 1 minute, 58 seconds?

25. Harriet Housewife, a harried suburban mother, served her infant son a 20-ounce bowl of Cream-of-Wheat, and then left the kitchen to wake her husband. The infant, who hated hot cereal, immediately began emptying his bowl by throwing $\frac{1}{2}$-ounce spoonfuls of the cereal at the wall. If he had thrown one more spoonful per minute, and if his mother had stayed out of the kitchen an extra 2 minutes, he would have just succeeded in disposing of his breakfast. As it turned out, however, he had 8 ounces of Cream-of-Wheat left when she returned. How long was Harriet out of the kitchen?

CHAPTER 15 REVIEW

What is the degree of the following equations?
1. $3x^2 - x^6 + 5x = 0$
2. $2t^3 - (5t^2)^2 = 10t$

3. What is the degree of

$$\frac{3}{p} + \frac{2}{p^3} + \frac{4}{p^4} = -1$$

after clearing of fractions?

4. What is the degree of $Bx^3 + A^2Bx - A^4B^3x^7 = 25$ as an equation for B regarding A and x as constants?

Find the roots of the following equations by inspection.

5. $0 = (2s - 3)(3s - 2)(3s - 1)(3s + 1)$

6. $(2t + 5)(\frac{1}{2}t + 5)(3t + 0.03)(4t - 10) = 0$

Find all values for the unknown that make the following expression 0.

7. $(R - 1)(3R - \frac{1}{4})(0.04 - 0.1R)R$

Find an equation with the following solutions.

8. $1, 2, -2, 3$

9. $2, 3, -1, \frac{1}{3}$

Find *all* the roots to the following equations.

10. $3Q^4 - 13Q^2 = 10$ 13. $x^4 - 16 = 0$

11. $2x^4 = 50$ 14. $(B^2 + 2)(B^2 - 1) + (5B^2 + 7) = 0$

12. $4y^3 - 9y^2 + 6y = 0$

Solve the following equations for real roots by any method.

15. $2y + \dfrac{7}{y} = 9$ 18. $\dfrac{2 - 4B - 5B^2}{B} = -13$

16. $\dfrac{3d + 1}{5d + 4} - \dfrac{d + 3}{4d + 5} = 0$ 19. $y + \dfrac{3y}{2} = \dfrac{-3}{2y} - 16y$

17. $q + 5 = \dfrac{12}{q - 3}$

Solve the following equations.

20. $\sqrt{3R} = \dfrac{5}{\sqrt{R - 2}}$ 21. $\sqrt{x^2 - 1} = \dfrac{3}{\sqrt{x^2 - 1}}$

22. The sum of the squares of two consecutive positive integers is 145. What are the two integers?

23. A PNU is a new atomic dishwasher. The demand for PNUs depends on their price. If they are free, the people of Krutesville will take 20 of them. For each $20 the price increases, demand falls by 1 PNU. Total revenue to M. A. Thacker & Company, the manufacturer, is the product of the price of an individual PNU and quantity sold.

    (a) For what price(s) is total revenue zero?

    (b) For what price(s) does total revenue equal $380?

24. The *Delta Queen* makes a trip 4 miles upstream against a 3 mph current, and returns to the jetty from which she started. If the round trip takes 1 hour, what speed does she go in still water?

25. Jerry Mander was the most crooked politician in the Senate. For a set bribe, his vote could be bought on any issue. Two years ago he made

$180,000 on at least 10 such bribes. Last year he raised his "price" by $5000 per bribe, took 3 *more* bribes, and made $315,000. How many bribes did he take last year? At what price?

26. A. Crylic is a contemporary artist who names his paintings according to their area. His two most recent works are "300 square feet," a giant super-realist rectangular hamburger, and "208 square feet," a predominately green abstract allegory of sleep. If "208" is 2 feet shorter and 4 feet narrower than "300," what are its dimensions?

27. Philadelphia and New York are 90 miles apart. A number of people commute from one city to the other each day. If one such commuter leaves New York traveling 8 mph faster than another commuter who starts in Philadelphia at the same time, and if these two drivers pass after $\frac{3}{4}$ hour, how fast is the commuter from New York driving?

28. Paul buys a balloon that is 6π feet around. He starts blowing more air into it so that every 4 seconds it is an extra π feet around. If the balloon contains only 100π square feet of rubber, when will it explode?

29. The San Francisco Giants have won 6 games out of 15; the Los Angeles Dodgers have won 8 out of 13. How many consecutive games must the Giants win from the Dodgers to make their averages the same?

30. A trained dog jumps through 60 hoops in a circus act. When in good shape, the dog can get through all of the hoops in 2 minutes less time than it takes him when he is out of practice. This is because he can average 8 hoops per minute more in condition than out of condition. What is his average speed when he is out of practice?

31. What is the degree of the following equation?

$$y^3[1 - (y^2 - 1) - (y^3)^2] = 1$$

32. What is the degree of the following equation after clearing the fractions?

$$\frac{4}{Q^2 - Q} + 3Q = Q - \frac{1}{Q^4 - Q^3}$$

33. Find an expression that becomes 0 when

$$x = 0 \quad \text{or} \quad x = -\pi \quad \text{or} \quad x = \frac{\sqrt{2}}{2}$$

is substituted into it.

34. Find an equation with two roots whose difference is 3.

35. Write a quadratic equation that has one rational and one irrational root.

Find all of the roots to the following equations.

36. $(s + 8)^4 - 3(s + 8)^2 + 2 = 0$

37. $0.6x^2(0.1 - x^2) = -0.03 + 0.03x^2$

38. $3\left(\dfrac{2x}{x-1}\right)^4 + 7\left(\dfrac{2x}{x-1}\right)^2 = -4$

39. $x^3 - 7x + 6 = 0$

Solve the following equations for all real roots.

40. $\dfrac{a+2}{a-1} - 1 - 2\left(\dfrac{a-1}{a+2}\right) = 0$

41. $\dfrac{y-3}{2-3y} + \dfrac{5y-1}{y+2} = \dfrac{88}{3y^2+4y-4}$

42. $\dfrac{3}{5(r+1)} = \dfrac{2 - \dfrac{r+3}{r+2}}{\dfrac{5}{r+2} + \dfrac{1}{1+\dfrac{2}{r}}}$

43. $\dfrac{5-x+2x^2}{-x^2+5} = 2$

Solve the following equations for the variable in parentheses.

44. $\dfrac{at+b}{t} = \dfrac{ts+b}{s^2}$ (t)

45. $y + a + 2 = \dfrac{4a}{y-a}$ (y)

46. $\dfrac{1}{p-R} - \dfrac{1}{p+R} = \dfrac{1}{Q+R} + \dfrac{1}{Q}$ (p)

47. $w^2x^2Q^2 = \dfrac{1-Q}{wx}$ (Q)

Solve each of the following equations. Be sure to check your answers.

48. $B - 5 + \sqrt{6-B} = 0$

49. $\sqrt{R-10} = 2(\sqrt{\tfrac{3}{4}R} - 1)$

Solve the following equation for m:

50. $\dfrac{m+n}{\sqrt{n-m}} = \sqrt{pm+n}$

51. The product of three consecutive integers exceeds twice their sum by 6. What are the numbers?

52. A group of children are playing a card game in which the entire deck is dealt out among the players. If one more child were playing, each child would receive 3 fewer cards, and there would be 2 cards left over. How many children are playing?

53. Show that the difference of the squares of two consecutive numbers is equal to the sum of those numbers. Use this fact to determine $(51)^2 - (50)^2$.

54. The current, (I, in amps) in an electrical circuit is equal to the voltage (V, in volts) across the circuit divided by resistance (R, in ohms) in the circuit. Standard house voltage is 110 volts. Ed Ison is frying an egg on his electric

range, and at the same time he makes toast in a toaster that has 12 more ohms of resistance than the range. If the current through the range and the current through the toaster add up to 16 amps, what is the resistance of the toaster?

55. Suppose you throw a rock straight up at time $T = 0$ with velocity V. Then its height H at any time T is $H = VT - \frac{1}{2}gT^2$, where g is the gravitational constant equal to 32 feet per second per second. If you throw the rock up at 64 feet per second, when will it hit the ground again?

56. I am a brazen lion; my spouts are my two eyes, my mouth, and the flat of my right foot. My right eye fills a jar in two days [1 day = 12 hours], my left eye in three, and my foot in four. My mouth is capable of filling it in six hours. Tell me how long all four together will take to fill it. [From *Greek Anthology*.]

57. The square root of half the number of bees in a swarm has flown out upon a jessamine bush, $\frac{8}{9}$ of the swarm has remained behind; one female bee flies about a male that is buzzing within a lotus flower into which he was allured in the night by its sweet odor, but is now imprisoned in it. Tell me, most enchanting lady, the number of bees. [From *Bhāskara*, Hindu, c. 1150]

58. A snake's hole is at the foot of a pillar which is 15 cubits high, and a peacock is perched on its summit. Seeing a snake, at a distance of thrice the pillar's height, gliding toward his hole, he pounces obliquely upon him. Say quickly at how many cubits from the snake's hole do they meet, both proceeding an equal distance? [From *Bhāskara*, Hindu, c. 1150] Modern-day hint: the distance to the snake is measured from the bottom of the pillar.

59. In an expedition to seize his enemy's elephants, a king marched 2 yojanas the first day. Say, intelligent calculator, with what increasing rate of daily march did he proceed, since he reached his foe's city, a distance of 80 yojanas, in a week? [From *Bhāskara*, Hindu, c. 1150] Modern-day hint: The king's speed is $2 + x$ on the second day, $2 + 2x$ on the third, etc.

60. Arjuna, exasperated in combat, shot a quiver of arrows to slay Carna. With half his arrows he parried those of his antagonist; with four times the square root of the quiverful he killed his horse; with six arrows he slew Salya (Carna's charioteer); with three he demolished the umbrella, standard, and bow; and with one he cut off the head of the foe. How many were the arrows which Arjuna let fly? [From *Bhāskara*, Hindu, c. 1150]

16 SIMULTANEOUS EQUATIONS

16.1 SIMULTANEOUS EQUATIONS: WHAT THEY ARE

A linear equation of the form $x + 5 = 3$ has exactly one variable and exactly one solution. Suppose now that we have a linear equation with two variables, such as $x + y = 3$. If we assume one variable is a constant, we can treat this as a literal equation and solve for one variable in terms of the other, giving:

$$y = 3 - x \quad \text{or} \quad x = 3 - y$$

If, however, we treat this as an equation with two unknowns, then there are infinitely many solutions. For example, $x = 1$, $y = 2$ is a solution, and so are $x = 2$, $y = 1$ and $x = 0$, $y = 3$, and $x = \frac{1}{2}$, $y = \frac{5}{2}$, and $x = -1$, $y = 4$.

So if we are given one equation with two unknowns in it, you cannot use it alone to find both of them. However, suppose besides knowing that

$$x + y = 3$$

we are told that

$$x - y = 1 \quad \text{for the same numbers } x \text{ and } y$$

The only pair of numbers that add to 3 and whose difference is 1 is $x = 2$, $y = 1$. In general, if we have two equations each with two unknowns, then there is (usually) a unique solution (meaning just one solution). The equations

$$\begin{cases} x + y = 3 \\ x - y = 1 \end{cases}$$

are called *simultaneous equations*, because they both must be satisfied at the same time by the values of x and y that we are looking for. The brace tells you that the equations have to be solved together; the solution is the pair of numbers $x = 2$, $y = 1$.

You won't be surprised that in order to find the values of three unknowns, x, y, z, we need a set of three simultaneous equations, such as

$$\begin{cases} x - 2y + 3z = 2 \\ x + y - z = 5 \\ 2x - y - z = 6 \end{cases}$$

A set of simultaneous equations is also sometimes called a *system of simultaneous equations*.

16.2   SOLVING SIMULTANEOUS EQUATIONS IN TWO VARIABLES

You can solve the equations

$$\begin{cases} x + y = 3 \\ x - y = 1 \end{cases}$$

by guesswork (getting $x = 2$, $y = 1$), but for more complicated simultaneous equations it is helpful to have a systematic method. There are in fact two such methods; use whichever you like best.

METHOD 1:   **Substitution Method**

Suppose we want to solve:

$$\begin{cases} 2x + y = 4 & \text{(equation 1)} \\ x - y = 5 & \text{(equation 2)} \end{cases}$$

The first equation allows us to solve for y in terms of x, giving

$$y = 4 - 2x \qquad \text{(equation 1)}$$

Substituting this value for y into the second equation gives an equation in x alone.

Substituting $y = 4 - 2x$ into

$$x - y = 5 \qquad \text{(equation 2)}$$

gives

$$x - (4 - 2x) = 5 \qquad \text{(equation 3) (this is a new equation, so it has to be given a new number)}$$

Since there are no longer any y's in this equation it can now be solved for x:

$$x - 4 + 2x = 5$$
$$3x = 9$$

so

$$x = 3$$

We now go back to either of the original equations (it doesn't matter which) and substitute the value we have just found for x (namely 3) to get an equation in y alone.

Substituting $x = 3$ into

$$2x + y = 4$$

gives

$$2 \cdot 3 + y = 4$$

Now solve for y:

$$y = -2$$

Therefore the solution is

$$x = 3 \qquad y = -2$$

CHECK: To check the solutions to simultaneous equations, you must show that the values you have found satisfy both equations.

Equation 1:   LHS $= 2x + y = 2(3) + (-2) = 6 - 2 = 4$
                RHS $= 4$

Equation 2:   LHS $= x - y = 3 - (-2) = 5$
                RHS $= 5$

Therefore both equations are satisfied by $x = 3$, $y = -2$.

METHOD 2:   **Elimination Method**

The elimination method depends on the "golden rule for equations," which says that you can do anything (except multiply or divide by zero) to an equation provided you do it to both sides. In particular, you can add the same thing to both sides of an equation.
Look at the equations:

$$\begin{cases} 2x + y = 4 & \text{(equation 1)} \\ x - y = 5 & \text{(equation 2)} \end{cases}$$

Equation 2 tells us that $(x - y)$ and 5 represent the same number. Therefore it is perfectly legal to add $(x - y)$ to one side of equation 1 and 5 to the other, giving:

$$2x + y + (x - y) = 4 + 5 \qquad \text{(equation 1 + equation 2)}$$

Simplifying:

$$2x + y + x - y = 9$$
$$3x = 9$$

Since the y's have cancelled out, we have ended up with an equation in x alone. This can be solved, giving

$$x = 3$$

As in method 1, we can substitute $x = 3$ into either of the original equations and solve for y, giving $y = -2$. Therefore the solution is $x = 3$, $y = -2$ as before.

CHECK: As in Method 1.

Let us do another, slightly harder, example:

$$\begin{cases} 3x + 2y = 17 & \text{(equation 1)} \\ 2x - 3y = 7 & \text{(equation 2)} \end{cases}$$

Adding these equations [meaning adding $(2x - 3y)$ to one side of equation 1 and 7 to the other] gives:

$$5x - y = 24 \qquad \text{(equation 1 + equation 2)}$$

Subtracting these equations [meaning subtracting $(2x - 3y)$ from one side of equation 1 and 7 from the other] gives:

$$x + 5y = 10 \qquad \text{(equation 1 - equation 2)}$$

Neither of these new equations is much help, since they contain both x and y and therefore can't be solved for either. Now $2x - 3y = 7$, so $2(2x - 3y) = 2 \cdot 7$, and therefore we can add twice the second equation to the first [meaning add $2(2x - 3y)$ to one side and $2 \cdot 7$ to the other] or subtract twice the second equation from the first. By the same argument, you can add or subtract any multiple of the second equation to any multiple of the first equation. In the case of

$$\begin{cases} 3x + 2y = 17 & \text{(equation 1)} \\ 2x - 3y = 7 & \text{(equation 2)} \end{cases}$$

if you multiply equation 1 by 3, and equation 2 by 2, the coefficients of y will be 6 and -6, respectively.

$$\begin{cases} 3(3x + 2y) = 3 \cdot 17 & \text{3(equation 1)} \\ 2(2x - 3y) = 2 \cdot 7 & \text{2(equation 2)} \end{cases}$$

or

$$\begin{cases} 9x + 6y = 51 & \text{3(equation 1)} \\ 4x - 6y = 14 & \text{2(equation 2)} \end{cases}$$

Adding the last two equations now *does* get rid of the y's:

$$9x + 4x + 6y - 6y = 51 + 14$$

So now we can find x:

$$13x = 65$$

$$x = 5$$

Substituting $x = 5$ into one of the original equations, say $3x + 2y = 17$, gives

$$3 \cdot 5 + 2y = 17$$

$$15 + 2y = 17$$

$$2y = 2$$

$$y = 1$$

Therefore the solution is $x = 5$, $y = 1$.

CHECK: When $x = 5$ and $y = 1$,

Equation 1:   LHS $= 3x + 2y = 3(5) + 2(1) = 15 + 2 = 17$
RHS $= 17$

Equation 2:   LHS $= 2x - 3y = 2(5) - 3(1) = 10 - 3 = 7$
RHS $= 7$

Therefore the solution is correct.

EXAMPLE: *Solve for u and v:*

$$\begin{cases} 3u + \dfrac{1}{3}v = \dfrac{25}{12} & \text{(equation 1)} \\ \dfrac{1}{2}u + 2v = \dfrac{5}{6} & \text{(equation 2)} \end{cases}$$

The best way to deal with an equation with fractional coefficients is to multiply through by its L.C.D. Equation 1 has an

L.C.D. of 12 and equation 2 has an L.C.D. of 6. Therefore the equations become

$$12\left(3u + \frac{1}{3}v\right) = 12 \cdot \frac{25}{12} \qquad \text{12(equation 1)}$$

$$6\left(\frac{1}{2}u + 2v\right) = 6 \cdot \frac{5}{6} \qquad \text{6(equation 2)}$$

or

$$36u + 4v = 25 \qquad \text{(still equation 1)}$$

$$3u + 12v = 5 \qquad \text{(still equation 2)}$$

This system can now be solved by either substitution or elimination.

METHOD 1:   SUBSTITUTION

Solve equation 1 for v:

$$v = \frac{25 - 36u}{4}$$

Substitute into equation 2:

$$3u + 12\left(\frac{25 - 36u}{4}\right) = 5$$

Solve:

$$3u + 75 - 108u = 5$$

$$70 = 105u$$

$$u = \frac{70}{105} = \frac{2}{3}$$

Substitute $u = \frac{2}{3}$ into equation 2 (because it has the smaller co-efficients) to find v:

$$3u + 12v = 5$$

$$3\left(\frac{2}{3}\right) + 12v = 5$$

$$2 + 12v = 5$$

$$12v = 3$$

$$v = \frac{1}{4}$$

Therefore the solution is $u = \frac{2}{3}$, $v = \frac{1}{4}$.

CHECK: When $u = \frac{2}{3}$ and $v = \frac{1}{4}$,

Equation 1: $\text{LHS} = 3u + \frac{1}{3}v = 3\left(\frac{2}{3}\right) + \frac{1}{3}\left(\frac{1}{4}\right) = 2 + \frac{1}{12} = \frac{25}{12}$

$\text{RHS} = \frac{25}{12}$

Equation 2: $\text{LHS} = \frac{1}{2}u + 2v = \frac{1}{2}\left(\frac{2}{3}\right) + 2\left(\frac{1}{4}\right) = \frac{1}{3} + \frac{1}{2} = \frac{5}{6}$

$\text{RHS} = \frac{5}{6}$

Therefore the solution is correct.

METHOD 2:   **ELIMINATION**

Looking at

$36u + 4v = 25$ (equation 1)

$3u + 12v = 5$ (equation 2)

You see that if you multiply the first equation by 3 and subtract the second one from it, you will eliminate the v's:

$3(36u + 4v) - (3u + 12v) = 3 \cdot 25 - 5$ 3(equation 1) − equation 2

$108u + 12v - 3u - 12v = 75 - 5$

$105u = 70$

$u = \frac{70}{105} = \frac{2}{3}$

To find v, as before, substitute $u = \frac{2}{3}$ into equation 2 and get

$v = \frac{1}{4}$

Therefore the solution is $u = \frac{2}{3}$ and $v = \frac{1}{4}$, which can be checked as before.

EXAMPLE: *Solve*

$\begin{cases} ax + by = 1 \\ bx - ay = 2 \end{cases}$ *for x and y in terms of a and b*

METHOD 1:   SUBSTITUTION

Solve the first equation for y:

$y = \frac{1 - ax}{b}$

Substitute into the second:

$$bx - a\frac{(1 - ax)}{b} = 2$$

Solve for x:

$$b^2x - a + a^2x = 2b \qquad \text{(multiplying through by } b\text{)}$$

$$(a^2 + b^2)x = a + 2b \qquad \text{(collecting terms)}$$

$$x = \frac{a + 2b}{(a^2 + b^2)}$$

Substituting for x into the first equation to find y:

$$a\frac{(a + 2b)}{(a^2 + b^2)} + by = 1$$

So

$$by = 1 - \frac{a(a + 2b)}{a^2 + b^2}$$

$$by = \frac{a^2 + b^2 - a(a + 2b)}{a^2 + b^2} \qquad \text{(subtracting fractions)}$$

$$by = \frac{b^2 - 2ab}{a^2 + b^2} = \frac{b(b - 2a)}{a^2 + b^2} \qquad \begin{array}{l}\text{(simplifying and}\\\text{factoring numerator)}\end{array}$$

$$y = \frac{b - 2a}{a^2 + b^2} \qquad \text{(dividing by } b\text{)}$$

Therefore $x = \dfrac{a + 2b}{a^2 + b^2}, \qquad y = \dfrac{b - 2a}{a^2 + b^2}$ is the solution.

This can (and should!) be checked.

METHOD 2:   ELIMINATION

Suppose we want to eliminate y between:

$$\begin{cases} ax + by = 1 & \text{(equation 1)} \\ bx - ay = 2 & \text{(equation 2)} \end{cases}$$

We multiply the first equation by a and the second by b (making the coefficients of y ab and $-ab$, respectively) and add:

$$a(ax + by) + b(bx - ay) = a + 2b \qquad a(\text{equation 1}) + b(\text{equation 2})$$

$$a^2x + aby + b^2x - aby = a + 2b$$

$$(a^2 + b^2)x = a + 2b$$

$$x = \frac{a + 2b}{(a^2 + b^2)}$$

Substituting for x in equation 1 and solving for y gives

$$a\left(\frac{a+2b}{a^2+b^2}\right) + by = 1$$

We have already solved this equation, and so we know that

$$y = \frac{b-2a}{a^2+b^2}$$

Therefore $x = \dfrac{a+2b}{a^2+b^2}$, $y = \dfrac{b-2a}{a^2+b^2}$ is the solution.

PROBLEM SET 16.2

Solve the following simultaneous equations using any method.

1. $\begin{cases} 3x + 2y = 2 \\ \quad x - y = 9 \end{cases}$

2. $\begin{cases} 3x - y = 5 \\ \quad x = 4 - 2y \end{cases}$

3. $\begin{cases} 5x + 7y - 3 = 0 \\ 6x - 5y - 17 = 0 \end{cases}$

4. $\begin{cases} 0.5s + 1.2p = 1.4 \\ -7.0p + 0.6s = 5.9 \end{cases}$

5. $\begin{cases} 6x - y = 5 \\ \quad y = 3x + 9 \end{cases}$

6. $\begin{cases} \quad y = 2x \\ 3x + 2y = 21 \end{cases}$

7. $\begin{cases} 3r + 4s + 2 = -4 \\ \quad r - 5s + 3 = 22 - r \end{cases}$

8. $\begin{cases} 3m + 5n = 1 \\ \quad 3n = m - 5 \end{cases}$

9. $\begin{cases} 2u + 3v = 13 \\ 3u + v = 7 - v \end{cases}$

10. $\begin{cases} \quad a = 2 - 2b \\ 3a = b - 8 \end{cases}$

11. $\begin{cases} 2m + n = 0 \\ 2m + n = 3n + 7 - m \end{cases}$

12. $\begin{cases} 3x - 2y = 1 \\ \quad y = 2 - x \end{cases}$

13. $\begin{cases} 3(2x + 3y - 2) = -x + y \\ \quad x + 5 = 2 - 5y \end{cases}$

14. $\begin{cases} 3r - y = 5 \\ 5r - y = 7 \end{cases}$

15. $\begin{cases} 0.5y - x = 1.7 \\ 0.3y + 0.4x = 1.3 \end{cases}$

16. $\begin{cases} y + 2 = 13 - 3x \\ 5x - 3y = 2 \end{cases}$

17. $\begin{cases} 2x + y = 15 \\ \dfrac{3x}{4y} + \dfrac{1}{y} = \dfrac{4}{7} \end{cases}$

18. $\begin{cases} 4(a + b) = 5 \\ \dfrac{1}{b-a} = 4 \end{cases}$

19. $\begin{cases} x - y = 1 \\ \dfrac{y}{x} = \dfrac{1}{2} \end{cases}$

20. $\begin{cases} \sqrt{2}x + y = 1 \\ \quad x - y = \sqrt{2} \end{cases}$

21. $\begin{cases} x - \dfrac{y+2}{7} = 5 \\ -4y = 3 + \dfrac{x+10}{3} \end{cases}$

22. $\begin{cases} \dfrac{1}{2}(3r - s) = r - \dfrac{1}{2} \\ \dfrac{1}{2r - s} = -1 \end{cases}$

23. $\begin{cases} \alpha - \beta = 5 \\ \dfrac{4}{\alpha + \beta} = \dfrac{10}{\alpha - \beta} \end{cases}$

24. $\begin{cases} \dfrac{3(a+b)}{5} = b \\ 7[a - 3(b - 2a)] = 17a + 1 \end{cases}$

25. $\begin{cases} \dfrac{x+y}{x-y} = \dfrac{1}{2} \\ 2x - 3y = 3 \end{cases}$

26. $\begin{cases} x + y = 28 \\ \dfrac{7}{x+y} = \dfrac{4}{x-y} \end{cases}$

27. $\begin{cases} \dfrac{15}{2} t = d - t \\ 9\left(t + \dfrac{1}{6}\right) = d + \dfrac{9}{4} \end{cases}$

Solve the following literal simultaneous equations for x and y.

28. $\begin{cases} cx + dy = c \\ c^2x + d^2y = c^2 \end{cases}$

29. $\begin{cases} x + ay = b \\ zx - by = a \end{cases}$

30. $\begin{cases} cx - dy = 0 \\ dx + cy - a = 0 \end{cases}$

31. $\begin{cases} rx + sy = t \\ 2rx - sy = 2t \end{cases}$

32. $\begin{cases} ax + by = a + b \\ abx + aby = a^2 + b^2 \end{cases}$

33. $\begin{cases} a_1x + b_1y = c_1 \\ a_2x + b_2y = c_2 \end{cases}$

34. $\begin{cases} \dfrac{x}{a + 2c} + \dfrac{y}{a - 2c} = 2 \\ x - y = 4c \end{cases}$

35. $\begin{cases} \dfrac{ax + y}{x - y} = \dfrac{1}{b} \\ bx + cy = -c \end{cases}$

36. If $2 = \dfrac{a}{b}$ find $\dfrac{a+b}{2b}$

37. If $\begin{cases} \dfrac{a}{4} = \dfrac{b}{3} - 3 \\ \dfrac{a}{3} - \dfrac{b}{2} = -5 \\ 9c(c + 2) = 2(3c - 2) \end{cases}$
find $a + b + c$.

38. If

$$R = x^2 - (a - 1)x - (b - 5) \quad \text{and}$$

$$Q = 2x^2 - (a + 2)x + (b + 1)$$

each have a factor of $(x - 2)$ (i.e., 2 is a root to the equations $R = 0$, $Q = 0$, find the values of a and b. Then factor R and Q completely.

39. If $a + 2b = 5$ and $3a - b = 1$, solve the following equation for x:

$$(a + b)x^2 - (a - 4b)x - (2b + 2a) = 0.$$

40. If $2a - b = -2$ and $\dfrac{b}{2} + a = 2$, does the following quadractic equation have real roots? Explain.

$$2abx^2 + (ab - a)x + 4a - b = 0$$

16.3   SOLVING SIMULTANEOUS EQUATIONS IN THREE VARIABLES

The methods for solving simultaneous equations in two variables can easily be extended to equations in three variables or to equations in any number of variables. The idea is to reduce a system of three equations in three variables to a system of two equations in two variables, which can then be solved as in the last section. The third variable is found by substituting back into one of the original equations. In the same way, if you start with a system of four equations in four unknowns, you first reduce it to three equations in three unknowns, then two equations in two unknowns, then solve and substitute back.

Suppose we want to find x, y, and z that satisfy each of the following three equations:

$$\begin{cases} x + y + z = 0 & \text{(equation 1)} \\ 2x - y - 2z = 6 & \text{(equation 2)} \\ x + 3y - z = 10 & \text{(equation 3)} \end{cases}$$

METHOD 1:

Substitution

Solving equation 1 for x will give an equation for x in terms of y and z that can be substituted into equations 2 and 3 to give equations in y and z only:

$$x = -y - z \qquad \text{(equation 1)}$$

Substituting into equation 2:

$$2(-y - z) - y - 2z = 6 \qquad \text{(call this equation 4)}$$

and into equation 3:

$$(-y - z) + 3y - z = 10 \qquad \text{(call this equation 5)}$$

Which simplify to:

$$-3y - 4z = 6 \qquad \text{(equation 4)}$$

$$2y - 2z = 10 \qquad \text{(equation 5)}$$

These can be solved by solving for y in equation 4 and substituting the result into equation 5:

$$y = \frac{-4z - 6}{3} \qquad \text{(equation 4)}$$

Substituting into equation 5:

$$2\frac{(-4z - 6)}{3} - 2z = 10 \qquad \text{(equation 5)}$$

Simplifying and solving for z:

$$\frac{-8z}{3} - 4 - 2z = 10$$

$$\frac{-14}{3}z = 14$$

$$z = \frac{-3}{\cancel{14}} \cdot \cancel{14}$$

$$z = -3$$

Substituting $z = -3$ into equation 4 to find y:

$$-3y - 4(-3) = 6$$

$$-3y + 12 = 6$$

$$6 = 3y$$

$$y = 2$$

Substituting $y = 2$, $z = -3$ into equation 1 to find x:

$$x + 2 + (-3) = 0$$

$$x = 1$$

Therefore the solution is $x = 1$, $y = 2$, $z = -3$.

CHECK: When $x = 1$, $y = 2$, and $z = -3$,

Equation 1:   LHS $= x + y + z = 1 + 2 + (-3) = 0$
RHS $= 0$

Equation 2:   LHS $= 2x - y - 2z = 2(1) - 2 - 2(-3) = 6$
RHS $= 6$

Equation 3:   LHS $= x + 3y \ - z = 1 + 3(2) - (-3) = 10$
RHS $= 10$

Therefore this solution is correct.

METHOD 2:   ## Elimination

These equations can equally well be solved by adding multiples of one equation to another to reduce the three-equation system to a set of two equations in two variables.

If you look at

$$\begin{cases} x + \ y + z = 0 & \text{(equation 1)} \\ 2x - y - 2z = 6 & \text{(equation 2)} \\ x + 3y - \ z = 10 & \text{(equation 3)} \end{cases}$$

you will see that subtracting the first equation from the third, and twice the first equation from the second will give you two equations that do not contain x:

$$2x - y - 2z - 2(x + y + z) = 6 - 2 \cdot 0 \qquad \text{[equation 2 − 2(equation 1)]}$$

$$x + 3y - z - (x + y + z) = 10 - 0 \qquad \text{(equation 3 − equation 1)}$$

or

$$-3y - 4z = 6 \qquad \text{(call this equation 5)}$$

$$2y - 2z = 10 \qquad \text{(call this equation 6)}$$

These two equations (5 and 6) can be solved by elimination. Subtract twice the second from the first to get rid of the z's

$$-3y - 4z - 2(2y - 2z) = 6 - 2 \cdot 10 \qquad \text{[equation 5 − 2(equation 6)]}$$

$$-7y = -14$$

$$y = 2$$

Substituting $y = 2$ into equation 5 to find z:

$$-3(2) - 4z = 6$$

$$-4z = 12$$

$$z = -3$$

Substituting $y = 2$, $z = -3$ into equation 1 to find x:

$$x + 2 + (-3) = 0$$

$$x = 1$$

Therefore the solution is $x = 1$, $y = 2$, $z = -3$, which can be checked as above.

PROBLEM SET   16.3

Solve the following systems of equations.

1. $\begin{cases} x + y + z = 3 \\ 2x + y - z = -6 \\ 3x - y + z = 11 \end{cases}$

2. $\begin{cases} 3r - 2s + t = 7 \\ 2r + s - 3t = 1 \\ r + 2s + 2t = 4 \end{cases}$

3. $\begin{cases} 2x - y + z - 7 = 0 \\ x - 3y + 2z - 4 = 0 \\ 3x + y - 2z - 6 = 0 \end{cases}$

4. $\begin{cases} 4x - 2y + z = -8 \\ x + 3y - 2z = 5 \\ 3x - y + 3z = -5 \end{cases}$

5. $\begin{cases} 4p - 6Q + 7r = -7 \\ 2p + 3Q - 5r = 7 \\ 8p - 2r = 6 \end{cases}$

6. $\begin{cases} 2x - 3y - 5z = 0 \\ x + 2y - 13z = 0 \\ 9x - 10y - 30z = 0 \end{cases}$

7. $\begin{cases} a + b + c = 2 \\ 4a + 5b - 3c = -15 \\ 5a - 3b + 4c = 23 \end{cases}$

8. $\begin{cases} 2l - 3m + n = -3 \\ -l + 4m + n = 5 \\ 4l - 6m + n = 5 \end{cases}$

9. $\begin{cases} 4x - y + z = 6 \\ 2x + y + 2z = 3 \\ 3x - 2y + z = 3 \end{cases}$

10. $\begin{cases} u + 3v - 2w = 12 \\ 3u - v + w = -1 \\ u + v + 4w = 0 \end{cases}$

11. $\begin{cases} 2a - b + c = 7 \\ a - 3b + 2c = 4 \\ 3a + b - 2c = 6 \end{cases}$

12. $\begin{cases} 3x - y + 2z = 4 \\ 2x + 3y - z = 14 \\ 7x - 4y + 3z = -4 \end{cases}$

13. $\begin{cases} x + y + z = 54 \\ \dfrac{x + y}{2} = \dfrac{y + z}{4} \\ \dfrac{z + x}{3} = \dfrac{x + y}{2} \end{cases}$

14. $\begin{cases} 2x - 3y + z = 5 \\ x + 2y - 3z = -15 \\ x - 4y + 2z = 12 \end{cases}$

15. $\begin{cases} 2(a + b) - 6c = 3 \\ a - b + 2c = 1 \\ \frac{7}{4}a + 2b + c = 13 \end{cases}$

16. $\begin{cases} 5p + 4q + r = 0 \\ 10p + 8q - r = 0 \\ p - q - r = 0 \end{cases}$

17. $\begin{cases} \dfrac{2}{u} - \dfrac{1}{v} - \dfrac{3}{w} = -1 \\ \dfrac{1}{u} + \dfrac{2}{v} - \dfrac{4}{w} = 17 \\ \dfrac{2}{u} - \dfrac{1}{v} + \dfrac{1}{w} = -9 \end{cases}$

(*Hint:* Solve for $\dfrac{1}{u}, \dfrac{1}{v}$, and $\dfrac{1}{w}$ first.)

18. $\begin{cases} \dfrac{x + y}{z} = -3 \\ 2x - y + 6z = -3 \\ -x + \frac{2}{3}y - z = -1 \end{cases}$

16.4   THINGS THAT CAN GO WRONG IN SOLVING SIMULTANEOUS EQUATIONS

Suppose we want to solve the equations:

$$\begin{cases} x + y = 1 & \text{(equation 1)} \\ x + y = 3 & \text{(equation 2)} \end{cases}$$

If we try to do it by substitution, we solve the first equation for x:

$$x = 1 - y$$

And substitute into the second:

$$(1 - y) + y = 3$$

which simplifies to:

$$1 - y + y = 3$$

or

$$1 = 3$$

But this, you say, is ridiculous. All the y's have dropped out, so we can't solve for y, and what is worse, we are left with the equation $1 = 3$, which isn't even true.

What has gone wrong? First let me assure you that I didn't make an algebraic mistake. Then, to convince you that something odd really is happening, we will try solving the same equations by elimination. If we want to eliminate the x's, we might decide to subtract the first equation from the second:

$$x + y - (x + y) = 3 - 1 \qquad \text{(equation 2 − equation 1)}$$

which simplifies to:

$$0 = 2$$

This is clearly just as bad as getting $1 = 3$!

As you may well have realized, the problem with solving these equations is that we are looking for values of x and y that satisfy both $x + y = 1$ and $x + y = 3$ at the same time. But whatever the values of x and y are separately, if $x + y$ is 1 then $x + y$ cannot be 3 also. Therefore there cannot possibly be any values for x and y that satisfy both $x + y = 1$ and $x + y = 3$. In other words, the system $x + y = 1$ and $x + y = 3$ has *no* solution because the equations are what is called *inconsistent*. The fact that we kept getting impossible equations like $1 = 3$ and $0 = 2$ was trying to tell us that the original equations were impossible.

Now consider the system:

$$\begin{cases} x + y = 1 & \text{(equation 1)} \\ 2x + 2y = 2 & \text{(equation 2)} \end{cases}$$

Suppose we want to solve these by substitution.

Solve the first equation for x:

$$x = 1 - y$$

and substitute into the second:

$$2(1 - y) + 2y = 2$$

which simplifies to:

$$2 - 2y + 2y = 2$$

or

$$2 = 2$$

Which is not much help, since the y's have cancelled out and so we can't solve for y. Unlike the last case, however, the equation we got is at least true—2 does equal 2.

What happens if we try elimination? To get rid of the x's, we should subtract twice the first equation from the second:

$$2x + 2y - 2(x + y) = 2 - 2 \cdot 1 \qquad [\text{equation } 2 - 2 \text{ (equation 1)}]$$

or

$$0 = 0$$

Again, this is true, but hardly any help! To find out what is wrong, look back at the original equations and you will see that the second is just the first one multiplied by 2. But this means that the second equation is just the first one in disguise, because *any* values of x and y that satisfy the first equation satisfy the second one. If $x + y = 1$, then $2x + 2y$, which is the same thing as $2(x + y)$, *must* be 2. So we really have only one equation, and one equation is not enough to determine x and y uniquely (see the beginning of this chapter).

Therefore the system

$$\begin{cases} x + y = 1 \\ 2x + 2y = 2 \end{cases}$$

has *infinitely* many solutions: for example, $x = 1$, $y = 0$; $x = \frac{1}{3}$, $y = \frac{2}{3}$; $x = \frac{1}{2}$, $y = \frac{1}{2}$; $x = \frac{3}{4}$, $y = \frac{1}{4}$; $x = 2$, $y = -1$; and so on. Such a system is called *degenerate*, because one of the equations has degenerated into the other.

To summarize: Two things can go wrong with solving simultaneous equations. The equations can be *inconsistent* and have *no* solutions, for example,

$$\begin{cases} x + y = 1 \\ x + y = 3 \end{cases}$$

or the equations can be *degenerate* and have infinitely many solutions, for example,

$$\begin{cases} x + y = 1 \\ 2x + 2y = 2 \end{cases}$$

Systems of equations with three (or more) variables can have exactly the same problems. For example,

$$\begin{cases} x + y - z = 1 \\ 2x + 3y + z = 4 \\ -x - y + z = 2 \end{cases}$$

is inconsistent. The reason that this system is inconsistent is that if we multiply the third equation through by -1, we get

$$x + y - z = -2$$

Therefore the first equation tells us that $x + y - z$ must be 1, and the third that $x + y - z$ must be -2, which is clearly impossible!

As another example,

$$\begin{cases} 3x - y + z = 9 \\ x + \dfrac{z}{3} = 3 + \dfrac{y}{3} \\ x + 3y = 3 + \dfrac{z}{3} \end{cases}$$

is degenerate. To see why this system is degenerate, rewrite the second equation as:

$$x - \frac{y}{3} + \frac{z}{3} = 3$$

Multiplying both sides by 3 gives

$$3x - y + z = 9$$

So you can see that the second equation is the first one in disguise. Thus, there are really only two different equations in this system, which is not enough to determine three variables uniquely.

In systems of equations with three or more unknowns, other kinds of difficulties can arise also. However, in a system with two unknowns these are the only two things that can go wrong.

In practice, you should not worry about whether or not you have an inconsistent or a degenerate set of equations. If you do, it will show up in the process of looking for the solution: An inconsistent set will give you a nonsense equation (like $1 = 3$), and a degenerate set will give you an obviously true equation (like $2 = 2$). So go about solving any set of equations in the usual way, and if you get an equation in which all the variables drop out, go back and check your original system for inconsistency or degeneracy.

PROBLEM SET   16.4

State whether the following simultaneous equations are inconsistent, degenerate or solvable. Explain.

1. $\begin{cases} 3x - 4y - 1 = 0 \\ 8y - 6x + 1 = 0 \end{cases}$

2. $\begin{cases} 3m - 4n - 1 = 0 \\ 6m - 8n - 2 = 0 \end{cases}$

3. $\begin{cases} 3x - y = 5 \\ x + 2y = 4 \end{cases}$

4. $\begin{cases} 10t + 2 = 2s \\ s - 5t = 1 \end{cases}$

5. $\begin{cases} 3x = 5 - 7y \\ \dfrac{15}{3} + \dfrac{-3\sqrt{3}}{\sqrt{3}}x = \dfrac{14}{2}y \end{cases}$

6. $\begin{cases} 6x - y = 5 \\ \quad\ \ y = 3x + 9 \end{cases}$

7. $\begin{cases} 3a - 2b = 4 \\ \quad\ \ 4b = 6a - 8 \end{cases}$

8. $\begin{cases} 3r - 4s = 8 \\ 6r - 8s = -4 \end{cases}$

9. $\begin{cases} a = 3b - 5 \\ b = 4a - 2 \end{cases}$

10. $\begin{cases} 2(x + y) + 3 = 2y - x + 4 \\ \qquad\ 5x + y = 3(1 - y) + 2x \end{cases}$

11. $\begin{cases} 4r + 12s + 16t = 4 \\ \ r + \ 8s + 11t = 1 \\ 3r + \ 4s + \ 5t = 3 \end{cases}$

12. $\begin{cases} \quad\ a + b - c = 2 \\ -a + 2b + 4c = 5 \\ \ 2a + 5b + c = 9 \end{cases}$

16.5   NONLINEAR SIMULTANEOUS EQUATIONS

So far we have solved only linear simultaneous equations, namely ones of the form

$$ax + by = e$$
$$cx + dy = f$$

where a, b, c, d, e and f are constants. However the methods we have used can also be used on some nonlinear ones. For example, suppose we want to solve

$$x + y = 10 \qquad \text{(equation 1)}$$
$$xy = 21 \qquad \text{(equation 2)}$$

Substitution works well here. Solve the second equation for y:

$$y = \frac{21}{x}$$

and substitute into the first:

$$x + \frac{21}{x} = 10$$

Now we have a quadratic equation:

$$x^2 - 10x + 21 = 0$$

which can be solved by factoring:

$$(x - 3)(x - 7) = 0$$

$$x = 3 \quad \text{or} \quad x = 7$$

Substituting into $x + y = 10$ to find y,

if $x = 3$, $y = 7$

and if $x = 7$, $y = 3$

Therefore there are two possible solutions to this system.
One is $x = 3$, and $y = 7$; the other is $x = 7$ and $y = 3$.

Note: It is *very important* which x goes with which y. For example, $x = 3$ and $y = 3$ is not a solution because these values do not satisfy the equations (try it!).

CHECK: When $x = 3$ and $y = 7$;

Equation 1:   LHS $= x + y = 3 + 7 = 10$
RHS $= 10$

Equation 2:   LHS $= xy = 3 \cdot 7 = 21$
RHS $= 21$

Therefore $x = 3$ and $y = 7$ is a solution. Similarly for $x = 7$ and $y = 3$.

Now let's consider the following set of equations:

$$\begin{cases} \dfrac{3}{s} + \dfrac{2}{t} = 12 & \text{(equation 1)} \\[2mm] \dfrac{2}{s} - \dfrac{1}{t} = 1 & \text{(equation 2)} \end{cases}$$

If we were to clear of fractions as we normally do, we would get two very messy equations containing terms in s, t, *and* st.

Suppose instead that we think of

$$\frac{1}{s} \quad \text{and} \quad \frac{1}{t}$$

as being the variables and solve by elimination. Adding twice the second equation to the first gives:

$$\frac{3}{s} + \frac{2}{t} + 2\left(\frac{2}{s} - \frac{1}{t}\right) = 12 + 2(1) \qquad \text{equation 1 + 2(equation 2)}$$

$$\frac{3}{s} + \frac{4}{s} = 14$$

$$\frac{7}{s} = 14$$

$$s = \frac{1}{2}$$

Substituting into equation 1 to find t:

$$\frac{3}{\frac{1}{2}} + \frac{2}{t} = 12$$

$$6 + \frac{2}{t} = 12$$

$$\frac{2}{t} = 6$$

$$t = \frac{1}{3}$$

Therefore the solution is $s = \frac{1}{2}$, $t = \frac{1}{3}$.

CHECK: When $s = \frac{1}{2}$, $t = \frac{1}{3}$,

Equation 1: $\text{LHS} = \dfrac{3}{\frac{1}{2}} + \dfrac{2}{\frac{1}{3}} = 6 + 6 = 12$
$\text{RHS} = 12$

Equation 2: $\text{LHS} = \dfrac{2}{s} - \dfrac{1}{t} = \dfrac{2}{\frac{1}{2}} - \dfrac{1}{\frac{1}{3}} = 4 - 3 = 1$
$\text{RHS} = 1$

So $s = \frac{1}{2}$, $t = \frac{1}{3}$ is a solution.

PROBLEM SET 16.5

Solve the following nonlinear simultaneous equations.

1. $\begin{cases} 2r + \dfrac{1}{s} = 12 \\ 2rs = 1 \end{cases}$

2. $\begin{cases} xy + y^2 = 5 \\ 2x + 3y = 7 \end{cases}$

3. $\begin{cases} a + b = 10 \\ ab = 21 \end{cases}$

4. $\begin{cases} \dfrac{1}{s} + \dfrac{1}{t} = 5 \\ \dfrac{2}{s} - \dfrac{1}{t} = 1 \end{cases}$

5. $\begin{cases} \dfrac{4}{3x} + \dfrac{1}{2y} = 1 \\ \dfrac{5}{3x} - \dfrac{3}{4y} = 4 \end{cases}$

6. $\begin{cases} \dfrac{3}{m} + \dfrac{4}{n} = 8 \\ \dfrac{6}{m} - \dfrac{3}{n} = 5 \end{cases}$

7. $\begin{cases} \dfrac{2}{u} + \dfrac{1}{t} = 7 \\ \dfrac{2}{u} - \dfrac{1}{t} = 1 \end{cases}$

8. $\begin{cases} \dfrac{3}{2a} - \dfrac{1}{3b} = 2 \\ \dfrac{3}{4a} + \dfrac{5}{6b} = -2 \end{cases}$

9. $\begin{cases} \dfrac{1}{3r} + p = 7 \\ \dfrac{1}{r} - 6p = 2 \end{cases}$

10. $\begin{cases} \dfrac{1}{m} + \dfrac{2}{n} = 2 \\ \dfrac{2}{m} - \dfrac{2}{n} = 1 \end{cases}$

11. $\begin{cases} \dfrac{2}{x} + \dfrac{3}{y} = 12 \\ \dfrac{3}{x} - \dfrac{2}{y} = 5 \end{cases}$

12. $\begin{cases} a+b-c=2 \\ a+10(b+c)=a-\frac{1}{2} \\ \dfrac{a}{b}=\dfrac{15}{4} \end{cases}$

13. $\begin{cases} y-\ 8=x^2-6x \\ 2y-18=x-y \end{cases}$

14. $\begin{cases} \dfrac{3p}{q}=2q+2 \\ \dfrac{1}{p}+\dfrac{1}{q}=\dfrac{11}{24} \end{cases}$

15. $\begin{cases} 4x^2+5y^2=21 \\ 2x+y=3 \end{cases}$

16. $\begin{cases} 2xy=1 \\ 3x^2+4y^2=4 \end{cases}$

17. $\begin{cases} 5a^2+4a-2=b-0.1 \\ 8a+4b=-2 \end{cases}$

18. Given $y=x^2+6x+4$ and $y=2x+6b$, where b is a constant real number, find the values of b that give one real solution. Find the values of b that give two real solutions; no real solutions.

16.6   SIMULTANEOUS WORD PROBLEMS

This section is about word problems that give rise to simultaneous equations. In solving a system of simultaneous equations, you are looking for the values of at least two variables, so a word problem that leads to simultaneous equations must contain at least two unknowns. Just about any kind of problem can contain two unknowns, so I will simply include a couple of typical examples.

EXAMPLE: *An energy-conscious person buys a solar-powered car, complete with a gasoline engine for night driving. She sets out to drive to the nearest city under solar power, but along the way, the sky clouds over, so she completes the journey under gasoline power. This entire trip takes 50 minutes. On the way home, the sky is clear and she uses solar power all the way. This trip takes 54 minutes. If she drives at 50 mph under solar power and 60 mph under gasoline power, find how far she drove before it became cloudy, and how far she drove while it was cloudy.*

(This is a distance problem and will be done by the method of section 13.2.)

STEP 2.   Let the distance she drove before it became cloudy be x miles and the remaining distance to the city be y miles. Before getting to the equations, let's put everything into the same units—which means saying 50 min $=\frac{5}{6}$ hour and 54 min $=\frac{9}{10}$ hour.

STEP 4.   Now, since we have two unknowns, we'd better have two equations. One comes from the outward journey:

$$\begin{pmatrix} \text{time (hours)} \\ \text{under solar} \\ \text{power} \end{pmatrix} + \begin{pmatrix} \text{time (hours)} \\ \text{under gas} \\ \text{power} \end{pmatrix} = \frac{5}{6}$$

The other comes from the return journey:

$$\begin{pmatrix} \text{time (hours) to} \\ \text{drive the whole} \\ \text{way under solar} \\ \text{power} \end{pmatrix} = \frac{9}{10}$$

STEP 5.   Clearly, we are going to need the formula

$$\text{time} = \frac{\text{distance}}{\text{speed}}$$

and we will use it for each segment of the journey.

STEP 6.   Outward journey: She goes x miles under solar power at a speed of 50 mph so

$$\text{time under solar power} = \frac{x}{50}$$

She goes y miles under gas power at 60 mph so

$$\text{time under gas power} = \frac{y}{60}$$

Return journey: She goes whole way, which is $(x + y)$ miles, under solar power at 50 mph so

$$\text{time under solar power} = \frac{x + y}{50}$$

STEP 7.   Therefore the equation for the outward journey is

$$\frac{x}{50} + \frac{y}{60} = \frac{5}{6}$$

and for the return journey is

$$\frac{x + y}{50} = \frac{9}{10}$$

STEP 8.   Multiplying the first equation by 300 and the second by 50 to clear of fractions:

$$\begin{cases} 6x + 5y = 250 \\ x + y = 45 \end{cases}$$

Subtracting five times the second equation from the first to eliminate y:

$$6x + 5y - 5(x + y) = 250 - 5(45)$$

$$x = 25$$

Substituting into $x + y = 45$ gives

$$y = 20$$

Therefore she drove 25 miles under solar power and 20 miles under gas power.

STEP 9.

Check: Outward trip:

$$\text{time to go 25 miles at 50 mph} = \frac{25}{50} = \frac{1}{2} \text{ hour}$$

$$\text{time to go 20 miles at 60 mph} = \frac{20}{60} = \frac{1}{3} \text{ hour}$$

$$\text{Therefore total time on outward trip} = \frac{1}{2} + \frac{1}{3} = \frac{5}{6} \text{ hour} \qquad \text{OK}$$

Return trip:

$$\text{time to go } 20 + 25 = 45 \text{ miles at 50 mph} = \frac{45}{50} = \frac{9}{10} \text{ hour} \qquad \text{OK}$$

EXAMPLE:

On three trips to an English tea shop, I first bought a cup of tea and two pieces of cake, and then a cup of tea, a piece of cake, and a penny bun (a kind of roll that doesn't cost a penny), and the last time two cups of tea, three penny buns, and two pieces of cake. I spent 36 pence, 31 pence, and 72 pence, respectively. How much did a cup of tea, a penny bun, and a piece of cake cost?

Besides the fact that it is reasonably straightforward, this problem is included for sentimental reasons. For some reason, all the simultaneous word problems in the book I was brought up on in England were either about trips to the tea shop or about the prices of different grades of coal. Back to solving the problem:

Let

t pence = price of a cup of tea

c pence = price of a piece of cake

b pence = price of a penny bun

Then, the first trip tells us that a cup of tea and two pieces of cake cost 36 pence:

$$t + 2c = 36 \qquad \text{(equation 1)}$$

The second trip tells us that:

$$t + c + b = 31 \qquad \text{(equation 2)}$$

and the last trip that:

$$2t + 2c + 3b = 72 \qquad \text{(equation 3)}$$

If twice equation 2 is subtracted from equation 3, we get:

$$2t + 2c + 3b - 2(t + c + b) = 72 - 2 \cdot 31$$

or

$$2t + 2c + 3b - 2t - 2c - 2b = 72 - 62$$

Therefore,

$$b = 10$$

Substituting $b = 10$ into equation 2 and equation 3:

$$\begin{cases} t + c + 10 = 31 \\ 2t + 2c + 3(10) = 72 \end{cases}$$

or

$$\begin{cases} t + c = 21 \\ 2t + 2c = 42 \end{cases}$$

Equation 2 and equation 3 have become the same at this point (equation 3 is just twice equation 2), so we will just solve equations 1 and 2 for t and c:

$$\begin{cases} t + 2c = 36 & \text{(equation 1)} \\ t + c = 21 & \text{(equation 2)} \end{cases}$$

Subtracting:

$$t + 2c - (t + c) = 36 - 21$$
$$c = 15$$

Substituting $c = 15$ into $t + c = 21$:

$$t + 15 = 21$$

so

$$t = 6$$

Therefore a cup of tea costs 6 pence, a piece of cake 15 pence, and a penny bun 10 pence.

CHECK: A cup of tea and two pieces of cake cost

$6 + 2(15) = 36$ pence

A cup of tea, a piece of cake, and a bun cost

$6 + 15 + 10 = 31$ pence

Two cups of tea, two pieces of cake, and three buns cost

$2(6) + 2(15) + 3(10) = 72$ pence

PROBLEM SET   16.6

1. You have eight coins consisting of nickels, dimes, and quarters and totaling $1.10. If the nickels and dimes alone total 35¢, how many quarters do you have? How many dimes? And nickels?

2. If a war breaks out somewhere in the world, U.S. military industries often prosper. Recently, one side in a conflict bought 6 U.S. planes and 8 tanks, while the other side bought 2 planes and 56 tanks. Although exact prices for weapons are classified, it is rumored that each side spent $8 million. What does a plane cost?

3. The length of the Serumore Hotel's lobby exceeds its width by 4 feet. A rug covers the floor of the lobby except for a border 2 feet wide all around it. If the area of this border of exposed floor is 68 square feet, what is the area of the rug?

4. What are the two positive numbers whose ratio is 5 and whose difference is 12?

5. Two bottles of a certain wine and a six-pack of beer cost $7.42. A bottle of the same wine with two six-packs of beer cost $5.87. How much does a six-pack of beer cost?

6. In a computing center, the records indicate that during the daytime 5 large programs and 80 small ones are executed during a typical hour. During the night, 11 large programs and 32 small ones are processed in an hour. How long does it take to run a large program? A small one?

7. Bruce "Fingers" Gerlinger is a pickpocket who has spent $\frac{2}{3}$ of his life in jail. He is there currently, with 2 more years to serve. If at the end of those 2 years he stays out of jail entirely for as many years as he will have spent in jail, he will be 44 years old. How old is he now?

8. A rectangular field is enclosed by 1320 feet of fencing. A man walking at a speed of 3 miles per hour takes 4 minutes to walk around three of the four sides of the field. What are the lengths of each side of the field? (1 mile per hour = 88 feet per minute.)

9. (a) Two guys go into a bar. One orders a scotch and the other a martini, which together cost $2.50. Instead of asking the bartender how much each drink costs, they decide to figure it out. To this end the first guy orders three more scotches and his friend orders four more martinis. The tab for the second round is $9.00. What is the price of a martini?

(b) Some time later in the same bar the same two guys decide to try something new. This time one orders a screwdriver and the other a bloody mary, and they pay $3.00. Then, in order to figure out the price of each drink, each guy orders three more of his respective drink, and they pay $9.00. Can they find the price of a bloody mary?

10. Reed and his brother Ryan can put 9 miles between them in 20 minutes when starting from the same point and sprinting at top speed in opposite directions. It takes the same time for Reed to draw a mile ahead of Ryan when they're running in the same direction (again starting from the same point). What is each one's top speed?

11. Make a crown of gold, copper, tin, and iron weighing 60 minae: Gold and copper shall be two-thirds of it; gold and tin three-fourths of it; and gold and iron three-fifths of it: Find the weights of gold, copper, tin, and iron required. [From *Greek Anthology*]

12. One night at the Hunt Club, Col. Smith-Smythe-Smith and Lord Abercrombie Fitch were exchanging safari stories.

"Last time I was in the Congo," boasted the Colonel, "I bagged 12 leopards and 4 zebras. It took 24 natives just to carry them out of the jungle."

"That's nothing," replied Lord Abercrombie. "Two years ago I got 7 leopards and 8 zebras. That took 31 natives to carry out."

If a native can carry 50 pounds, how much does a zebra weigh?

13. An overworked student is told by a doctor that four tablets each of aspirin and phenacetin per day is a good remedy for his throbbing headaches. He calculates that $3.60 will suffice for a 30-day supply. The student develops an annoying tingling in his ears and is told to halve his aspirin dosage while keeping his phenacetin intake the same. A 30-day supply now costs $3.00. How much does a single tablet of aspirin and phenacetin cost?

14. While it is common knowledge that Martians are green and have seven heads, few people realize that a Martian has eleven eyes. At a recent party, while I was conversing with a Martian, I innocently remarked that if one-half of the Martians at the party suddenly became Earthlings (while the other half remained Martians), there would be 295 eyes at the party. Somewhat of-

fended, the Martian noted that if one-half of the Earthlings at the party became Martians (while the other half stayed Earthlings), there would be 410 heads at the party. How many Martians and how many Earthlings were at the party?

15. A woman who lives in the suburbs normally gets to her job downtown by bicycling to the nearest subway stop and then taking the train from there. When she times it exactly right and doesn't have to wait for a train, the journey takes her 30 minutes. One day the subways are on strike and she has to bicycle the whole way, which takes her 50 minutes. If her average speed on her bicycle is 12 mph and the average speed of the subway is 24 mph, find how far it is from her home to the subway stop and how far she usually travels by subway.

16. An ice cream parlor sells two kinds of milk shakes, regular and superthick. A quart of the regular shake costs \$1.00, while a quart of the super-thick costs \$1.60. The manager decides to offer a medium-thick shake, made by blending the regular and superthick. How many quarts of each should be mixed to produce 9 quarts of a blend that costs \$1.40 per quart?

17. Bruce McGinnis and Bruce McGiver are the fastest sheep shearers in Australia. Working together for 3 hours, they can shear a standard herd of 105 sheep. Although the two men never competed to see who was faster, there was one occasion when the two men worked for different periods of time. On that occasion McGinnis worked for 4 hours and McGiver finished the herd in $2\frac{1}{4}$ hours. How long does it take McGinnis to shear a sheep?

18. A thinking young student develops a bad cold and buys some Excedrin and Sudafed, a decongestant. He takes 6 Excedrin and 3 Sudafed tablets a day until he is better, and being economically minded, notes that he has been spending 30¢ per day. The next time he gets a cold he buys a bottle of less expensive aspirin, which costs only two-thirds of what Excedrin costs per pill. Again he takes 6 aspirins and 3 Sudafed tablets per day, and sees that his costs are now 24¢ per day. What are the costs of the Excedrin, the Sudafed, and the aspirin per tablet?

19. A company has decided to break into the market by manufacturing three new types of tires. The cheapest has two nylon belts, and costs \$12.80 to manufacture. The medium-priced model, which they build for \$18.95, has three nylon and two steel belts. The expensive tire has six steel belts and costs \$23.00 to make. If the cost differences are due entirely to the costs of the belts, what does a steel belt cost? A nylon belt? The basic tire (no belts)?

20. (a) International Blodgett Corp., the world's largest producer of blodgetts, has decided to open a third manufacturing plant. In the first of its existing plants, IBC has three machines of type 1 for manufacturing blodgetts and two machines of type 2, and produces 40 blodgetts weekly. The second

and larger plant has five type 1 machines and six of type 2, and produces 96 blodgetts weekly. If both types of machines cost the same amount, which type should be installed in IBC's new plant? If the new plant installs ten machines of this type, what will be its weekly output?

(b) Suppose you are the vice president in charge of new construction, and IBC's president has informed you that you will be fired if any mistakes are made in designing the new plant. After the new machines have been installed, you discover that the second plant in fact produces only 88 blodgetts weekly. Should you be looking for a new job?

21. A certain number of small trucks and large truck carriers are driving in convoy from a G.M. truck plant in Cleveland to a truck dealer in Chicago. At the end of the Ohio Turnpike, each truck is charged a toll of $1.50 per axle. The total bill for the convoy is $120.00. Angered by this expense, the drivers decide to save money on the Indiana Turnpike by putting one small truck on each of the large truck carriers. Some of the small trucks still have to be driven, but the amount saved on the carried trucks is substantial. At the Indiana toll booths, where they are charged $1.20 per axle, the convoy pays $72.00. If a large truck has three axles on the road, and a small truck has two, how many small trucks were driven piggyback through the Indiana toll booths?

22. One hundred years ago, a master silversmith could make a set of tankards in 8 days less time than his apprentice. However, the smith would charge $15 per day, while the apprentice would charge only $2. Working together, the two could complete such a set of tankards in 3 days. Would it cost less to have the master do the job alone, to have the apprentice do it alone, or to have both work together? State the cost in each case.

23. An oil field has two wells, well 1 and well 2, which are operated intermittently. Both wells feed into the same pipe, whose flow is carefully metered. The data from two test days is recorded below:

	Time Operated		Total Oil Flow
	Well 1	Well 2	
First test day	3 hours	7 hours	18 barrels
Second test day	18 hours	14 hours	66 barrels

The field superintendent wants to know how much oil could be produced daily if both wells were operated around the clock. How much is it?

24. Oil wells often produce water mixed with the oil. A chemist receives samples S_1 and S_2 of the fluids from two wells, but unfortunately the field hands have accidentally mixed them such that S_1 is 80% from well 1 and 20%

from well 2, while S_2 is 20% from well 1 and 80% from well 2. The chemist has analyzed the two samples and found that S_1 is 45% water while S_2 is 30% water. What is the percentage of water that each well produces?

CHAPTER 16 REVIEW

Solve if possible. If impossible, say why.

1. $\begin{cases} x + 3y = 11 \\ x - \ y = 19 \end{cases}$

2. $\begin{cases} 0.07n - 0.04m = 0.01 \\ 0.2n - 0.05m = 0.035 \end{cases}$

3. $\begin{cases} \frac{2}{7}x + \frac{1}{8}y = 0 \\ \frac{3}{4}x - \frac{1}{3}y = 0 \end{cases}$

4. $\begin{cases} 2s + t + 14 = 7(s - 4t + 2) \\ 3(2s - 3t + 4) = 5s + 12 \end{cases}$

5. $\begin{cases} a + 2b - 1 = \dfrac{-9a + b}{5} \\[2mm] 4a + 3b = \dfrac{5 - 2a}{3} \end{cases}$

6. $\begin{cases} \dfrac{x + y}{4} + \dfrac{x - y}{2} = 1 \\[2mm] \dfrac{3x - y}{4} + \dfrac{4x + 2y}{11} = 3 \end{cases}$

7. $\begin{cases} -2x + 2y - 3 = 5 \\ x - y - 8 = 0 \end{cases}$

8. $\begin{cases} 2m - 3n - 1.25 = 0 \\ 1.5n + 0.625m - 1 = 0 \end{cases}$

9. $\begin{cases} j = 2k \\ (k - 1) + (j - 2) = 3 \end{cases}$

10. $\begin{cases} x + 2y = 14 \\ 4x - 7y = 26 \end{cases}$

Solve for x and y.

11. $\begin{cases} a_1x + b_1y = c_1 \\ a_2x + b_2y = c_2 \end{cases}$

12. $\begin{cases} c^2x + 2y = 1 - x \\ cx + y = 1 \end{cases}$

13. $\begin{cases} bay - ay + 2x = x \\ -cx - 4 = -y \end{cases} \quad (b \neq 1)$

14. $\begin{cases} tx + (1 - t)y = 1 \\ sx + (1 - s)y = 1 \end{cases}$

Solve for x, y, z, if possible.

15. $\begin{cases} 3x - 4y + 10z = -7 \\ 2x - 3y - z = -21 \\ x + y + z = 0 \end{cases}$

16. $\begin{cases} 0.3x + 0.2y = 1 \\ 0.2x + 0.3y + 0.1z = 0.5 \\ x + y + z = 1.8 \end{cases}$

17. $\dfrac{x + y}{4} = \dfrac{y + z}{3} = \dfrac{2x + 5z}{3} = 14$

18. $\begin{cases} 3(x - 4) - 2(y + \frac{3}{2}) + 4(z - 1) = 0 \\ 3(y + 1) + z = 0 \\ z - 3 = 0 \end{cases}$

19. $\begin{cases} x + y = z \\ -\frac{1}{2}x - y + 11 = z \\ -6x + 3y + 6 = z \end{cases}$

20. Both of the shapes in Figure 16.1 have areas of 20. What is b?

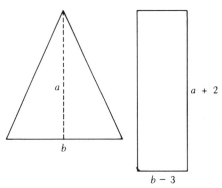

FIG. 16.1

21. I have $1.65 in my pocket, consisting entirely of quarters and nickels. If I have twice as many quarters as nickels, how many subway rides can I take? (A subway ride costs 25¢, and the transit authority will only accept quarters.)

22. John is 8 years older than Mike. In 3 years, he will be twice as old as Mike. How old are Mike and John now?

23. A small, spluttering motorboat takes an hour to go six miles downstream and takes two hours to return. Find her speed in still water and the speed of the river's current.

24. To celebrate the end of the Depression, Ty Coon decided to throw a party for his friends. Unfortunately, Ty's fortune was still somewhat depleted, so he couldn't afford to serve straight champagne. He decided to serve a punch that was one-half root beer, and one-half a mixture of champagne and ginger ale. He makes 36 quarts of the stuff. Champagne costs $2.25 a quart, ginger ale and root beer both cost $0.75 a quart, and Ty wants the final mixture to cost $30.

    (a) How many quarts of champagne and ginger ale together are there in the whole mixture?

    (b) How much did the champagne and ginger ale cost together?

    (c) Exactly how many quarts of champagne are there in the final mixture?

25. The owner of a shooting gallery charges 5 cents for each miss and pays out 25 cents for each hit. After 30 shots a poor marksman has to pay 90 cents. How many hits and misses did he make?

26. For years Todd and Alice Liudahl supplemented their modest income by blackmailing their neighbors. Neither of them made very much money at this until two years ago, when Alice found out about Linda Larsen's sordid past. That year, Todd and Alice together extorted $108,000. The following year the Internal Revenue Service found out about Alice, so although she took

in the same amount of money, she had to pay half of it in taxes. Meanwhile, Todd's activities went unnoticed, and he took in $3000 more than he had the year before. Together they netted $65,000. If Linda was paying Alice $75,000, how much was Alice getting from other neighbors?

27. Two robbers hold up the same two banks. The first robber takes half of bank A's assets and one-third of bank B's assets. The second robber then takes all the remaining assets from both banks. If the second robber ends up with half again as much money as the first, what was the ratio of bank A's assets to bank B's assets?

28. A drug company decides to market a new headache remedy consisting of aspirin, phenacetin, and caffeine. They calculate that a 1-gram tablet made from 600 milligrams (mg) of aspirin, 300 mg of phenacetin, and 100 mg of caffeine would cost 1.5¢ per tablet to produce. Since caffeine is three times as expensive as aspirin, they finally decide to market a tablet consisting instead of 650 mg of aspirin, 300 mg of phenacetin, and 50 mg caffeine costing 1.4¢ a tablet. Assuming that the cost of the chemicals is the entire cost of producing a tablet, how much do phenacetin, aspirin, and caffeine cost per gram? (1 gram = 1000 milligrams)?

29. A large contingent of soccer fans left Munich in Volkswagen vans to attend a match in Düsseldorf. Halfway there, 10 vans broke down, so it was necessary for each remaining van to carry one more person. After the game, they discovered that 15 more vans were out of commission, so on the return trip there were three persons more in each van than when they left Munich. How many fans went to the game?

30. Yuri walks into the Siberian Post Office to buy some stamps. He buys both airmail and regular stamps, with the airmail stamps costing 3 rubles more than the regular ones, per stamp. If Yuri spends 60 rubles on the regular stamps and 120 rubles on the airmail stamps, and takes home 27 stamps altogether, how many airmail stamps did he buy?

31. If a rectangular field is enlarged by making it 10 yards longer and 5 yards wider, its area is increased by 1050 square yards. If its length is decreased by 5 yards and its width by 10 yards, its area is decreased by 1050 square yards. Find the original dimensions of the field.

32. Dr. Susan Sefcik is a private physician who teaches part-time at the State Medical School. Two years ago she spent one-fourth of her time teaching, and grossed (from both jobs together) $75,000. Last year she taught half-time, practiced half-time, and grossed $65,000. Obviously, private practice pays better than teaching. But half of the money Sue grosses from her private practice must be spent to support her office, something the teaching job supplies for free. In the long run, would she make more money practicing full time or teaching full time?

17 INEQUALITIES

17.1 INEQUALITIES AND THE NUMBER LINE

Having spent the last four chapters solving equations, or equalities, we will spend this chapter solving inequalities. In other words, instead of looking for values of a variable that make two expressions equal, we will be interested in values that make expressions less than or greater than one another.

The Inequality Symbols $<, >, \leq, \geq$

$2 < 3$ means 2 is less than 3

$x > 1$ means x is greater than 1

$a \leq 10$ means a is less than or equal to 10

$q \geq r$ means q is greater than or equal to r

Any of these statements is called an *inequality*. Remembering these symbols should not be too hard since the larger number always goes with the larger end of the sign. Also, $\leq$ is just a contraction of "$<$" and "$=$."

Some examples of inequalities are

$$3 > 0 \quad \text{and} \quad -3 < 7 \quad \text{and} \quad -2 > -5$$

If the last example seems peculiar, the following analogy may be helpful. Think of the numbers as temperatures, with a larger number corresponding to a higher temperature. For example, 10° is warmer than 6°, so $10 > 6$; now ask yourself which is warmer, −5° or −2°? Clearly −2° is warmer and so −2 is larger than −5; that is, $-2 > -5$.

Graphs of Inequalities

I find that a mathematical idea is easier to understand if I can somehow make a picture of it, and "see" it. Certainly a picture is much easier to remember than a string of words. It turns out to be easy to draw a picture, or *graph*, of an inequality on the number line. Such graphs give you a quick, clear way of seeing what is going on, and can actually be used to avoid some tedious calculations (this is especially true when we get to absolute values). So here's how we represent, or graph, an inequality on the number line.

Consider the inequality $5 < 12$. Looking at 5 and 12 on the number line (Figure 17.1) shows us that 5 (the smaller number) is to the left of 12 (the larger). Taking another example, say $7 > 1$ (see Figure 17.1), the smaller number, 1, is again to the left of the larger, 7.

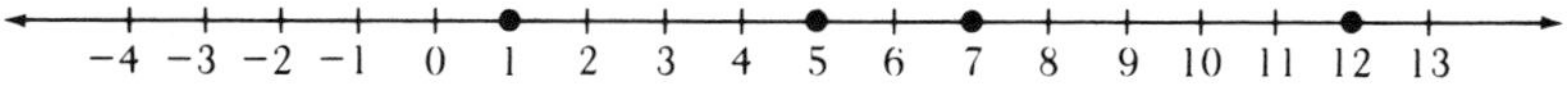

FIG. *17.1*

In any pair of numbers you pick, the smaller will always be to the left of the larger (or the larger to the right of the smaller). In other words,

$$\left. \begin{array}{c} a < b \\ \text{or} \\ b > a \end{array} \right\} \quad \text{means} \quad \left\{ \begin{array}{l} a \text{ is to the left of } b \\ \text{or} \\ b \text{ is to the right of } a \end{array} \right.$$

Similarly

$$\left. \begin{array}{c} a \le b \\ \text{or} \\ b \ge a \end{array} \right\} \quad \text{means} \quad \left\{ \begin{array}{l} a \text{ is to the left of } b, \text{ or at the same point} \\ \text{or} \\ b \text{ is to the right of } a, \text{ or at the same point} \end{array} \right.$$

EXAMPLE: $-5\frac{1}{2} < -5$ and $-\frac{1}{2} < 0$.

Since $-5\frac{1}{2}$ is to the left of -5, this shows you that $-5\frac{1}{2} < -5$. The number $-\frac{1}{2}$ is also to the left of 0, so $-\frac{1}{2} < 0$.

Graphs of Inequalities Containing a Variable

If x is a variable, an inequality such as

$$x < 3$$

means that x can be any number that is less than 3, such as $1, 2, \frac{3}{2}, 2.9, -1$, or -100. Therefore x can be any one of the numbers marked on the number line in Figure 17.2. The hollow circle at 3 means that x cannot be 3 itself. x can be anything to the left of 3, and such x's are said to *satisfy the inequality $x < 3$.*

FIG. 17.2

Notice that this inequality, like many others, has infinitely many solutions, covering a whole stretch of the number line.

The inequality

$$x \le 3$$

means that x can be any number that is less than or equal to 3, such as $1, 2, \frac{3}{2}$, or 3 itself.. These x's are marked on the number line in Figure 17.3. Here the solid circle at 3 means that 3 is included among the possible x's.

FIG. 17.3

We may also come across an inequality such as

$$1 < x < 3$$

which means that x must be both greater than 1 and less than 3—that is, something like $\frac{3}{2}$, 2, or 2.7. This inequality can be represented on the number line as in Figure 17.4, where the open circles mean that 1 and 3 are not included in the allowable x's. So x must be somewhere between 1 and 3, but not actually at 1 or 3. The numbers between 1 and 3 are said to *satisfy the inequality*

$$1 < x < 3$$

FIG. 17.4

You can guess that

$$1 \le x \le 3$$

means x must be both greater than or equal to 1 and less than or equal to 3. This is shown on the number line in Figure 17.5, where the solid circles mean that 1 and 3 are possible values for x. Again, an x that lies in the region shown on the number line is said to *satisfy* $1 \le x \le 3$.

FIG. 17.5

By now it will be no surprise that

$$1 < x \le 3$$

means that x must be both greater than 1 and less than or equal to 3. In other words, x must be found in the region shown in Figure 17.6.

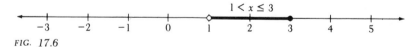

FIG. 17.6

"And's" and "Or's"

Suppose x is to satisfy both

$$1 < x \quad and \quad x < 3$$

The "and" means that the *same* x must satisfy *both* inequalities; x must therefore be both greater than 1 and less than 3—exactly what is meant by

$$1 < x < 3$$

Therefore

$$1 < x \quad and \quad x < 3$$

and

$$1 < x < 3$$

mean the same thing, and in fact the second is just a condensed form of the first.

By the same reasoning, if

$$x \ge 1 \quad and \quad x \le 3$$

(meaning the *same* x must be both greater than or equal to 1 *and* less than or equal to 3), then x satisfies

$$1 \le x \le 3$$

EXAMPLE: *Show on the number line the x's that satisfy*

$$x > -1 \quad and \quad x < 2$$

This means that x must be both greater than (to the right of) -1 and less than (to the left of) 2. So the x's we are looking for must be between -1 and 2, as in Figure 17.7.

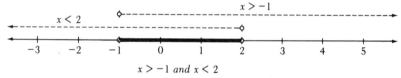

FIG. 17.7

You should notice that the graph of $x > -1$ *and* $x < 2$ is exactly that part of the number line where the graphs of $x > -1$ and of $x < 2$ overlap. This will always happen when there is an "and" in the problem.

Of course, another way of expressing

$x > -1$ and $x < 2$

is to write

$-1 < x < 2$

EXAMPLE: *Show on the number line the x's that satisfy*

$x < 1$ *and* $x > 3$

This is asking for the x's that are both less than 1 and greater than 3. Since there clearly aren't any numbers that are both less than 1 and greater than 3, there are no x's that satisfy this inequality. So there's really nothing to draw on the number line. You might predict this from the fact that the graphs of $x < 1$ and $x > 3$ do not overlap (see Figure 17.8).

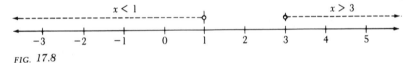

FIG. 17.8

EXAMPLE: *Graph (meaning draw) on the number line the x's that satisfy*

$x > 2$ *and* $x > 4$.

We want the x's that are both greater than 2 and at the same time greater than 4. This can only be achieved by taking x greater than 4. See Figure 17.9.

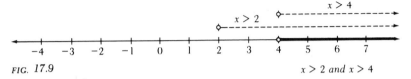

FIG. 17.9 $x > 2$ *and* $x > 4$

Notice that the graph of $x > 2$ *and* $x > 4$, being the overlap of the graphs of $x > 2$ and $x > 4$, is just the graph of $x > 4$.

EXAMPLE: *Graph x < 1 or x > 3*

This means x is *either* less than 1 *or* greater than 3 *or both.* Here x doesn't have to satisfy both conditions at once, and so the allowable x's are as shown in Figure 17.10. Here the graph of $x < 1$ or $x > 3$, instead of being the overlap of the graphs of $x < 1$ and of $x > 3$, is all the points in either one of these two graphs.

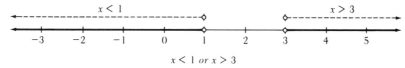

FIG. *17.10*

Please note that $x < 1$ or $x > 3$ *cannot* be condensed into the form $3 < x < 1$, because $3 < x < 1$ would imply that the *same* x had to be both greater than 3 and less than 1.

EXAMPLE: *Graph x < −4   or   x ≥ −1*

Here we want the x's that are either less than -4 or greater than or equal to -1. The graph consists of the graph of $x \geq -1$ together with the graph of $x < -4$, as shown in Figure 17.11.

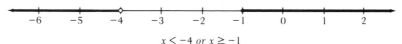

FIG. *17.11*

EXAMPLE: *Graph x > 2 or x > 4.*

Now we want the x's that are either greater than 2 or greater than 4 (or both). So we need all the numbers greater than 2. See Figure 17.12.

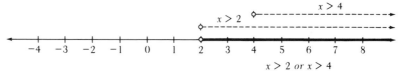

FIG. *17.12*

The point of these examples is to make it clear that "and" and "or" have very specific meanings when used with inequalities, and it really matters which you write. In summary:

$x > 2$ *and* $x < 9$ ⎫ $2 < x < 9$ ⎬	means the *same* x must be *both* greater than 2 *and* less than 9
$x < 1$ *or* $x > 7$	means x is *either* less than 1 *or* greater than 7 *or both*

PROBLEM SET   17.1

Fill in the proper inequality or equality ($>$, $<$, or $=$).

1. $0 \underline{} 3$

2. $-4 \underline{} -6$

3. $\frac{4}{5} \underline{} \frac{2}{3}$

4. $\frac{6}{17} \underline{} \frac{2}{5}$

5. $-\frac{32}{?} \underline{} -\frac{49}{11}$

6. $8\frac{1}{2} \underline{} \frac{51}{6}$

7. $\dfrac{7}{5\sqrt{10}} \underline{} \dfrac{6\sqrt{5}}{25\sqrt{2}}$

8. $\dfrac{2}{\sqrt{3}+\sqrt{12}} \underline{} \dfrac{\sqrt{2}\sqrt{3}}{\sqrt{18}}$

9. $\dfrac{(2)\frac{4}{3}}{\sqrt[3]{2}} \underline{} 2$

10. $\dfrac{0.01}{\sqrt{1.44}-\sqrt{1.21}} \underline{} 0.98$

Express the inequality or equality represented by the following graphs.

11.

12.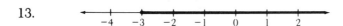

13.

14.

15.

Graph the following inequalities on the number line.

16. $x > 3$

17. $x \leq -2$

18. $-\frac{3}{4} - x < 1$

19. $x < 2$ and $x > -2$

20. $x > -2$ or $x < -3$

Fill in the proper inequality ($>, \geq, <, \leq, =$).

21. (a) $a^2 \underline{\quad} b^2$ where $a > b > 1$
    (b) $a^2 \underline{\quad} b^2$ where $a < b < -1$
22. $(a + b)^2 \underline{\quad} a^2 + b^2$ where $a \leq -1$ and $b \geq 1$
23. $a^{-b} \underline{\quad} a^{(1/b)}$ where $a \geq 1$ and $b > 1$
24. $x^5 \underline{\quad} x^{-5}$ where $0 < x \leq 1$
25. $2a^{-1} \underline{\quad} (2a)^{-1}$ where $-1 \leq a \leq 1, a \neq 0$

Express the inequality represented by the following graphs.

26.

27.

28.

29.

30.

31.

32.

Graph the following inequalities on the number line.

33. ($x < 3$ and $x > 0$) or $x < -5$ 36. $x > -\frac{1}{2}$ and $x < \frac{3}{2}$
34. $x = 2$ or $x > 3$ 37. $x > 2$ or $x < 5$
35. $x < 2$ and ($x > -1$ or $x < -3$)

17.2   SOLVING LINEAR INEQUALITIES

An equation consists of two expressions that are equal to one another for certain values of the variable. Solving the equation means finding these values. An inequality consists of two expressions that may be less (or greater) than one another for certain values of the variable. Solving the inequality, then, means finding these values—there are usually infinitely many of them. The point of this section is to show you how to do this.

First, remember that if x satisfies the inequality

$$x < 7$$

then we know that x is any number less than, but not equal to, 7 (see Figure 17.13).

FIG. 17.13

What if we are now told that x satisfies

$$x + 2 < 7$$

To find the possible values for x, look at the values of $(x + 2)$ at a variety of points on the number line in Figure 17.14.

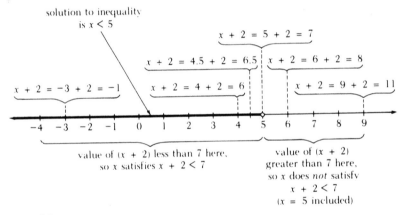

FIG. 17.14

From the figure you can see that if x is to the left of 5, then x satisfies $x + 2 < 7$. Therefore,

$$x < 5 \quad \text{is the solution to} \quad x + 2 < 7.$$

Now consider the inequality

$$x - 5 \geq 1$$

Look at values of $(x - 5)$ in Figure 17.15. From the figure we see that the solution to $x - 5 \geq 1$ is all the numbers at or to the right of 6. That is,

$$x \geq 6 \quad \text{is the solution to} \quad x - 5 \geq 1.$$

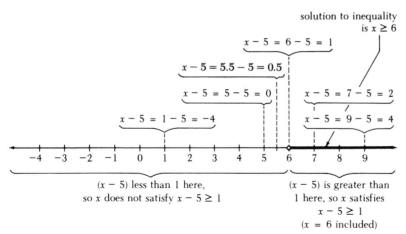

FIG. *17.15*

You will probably have noticed that both the above two examples could be solved as though they were equations. In the case of $x + 2 < 7$, subtracting 2 from each side gives

$$x + 2 - 2 < 7 - 2$$

or

$$x < 5$$

which is the solution. In the case of $x - 5 > 1$, adding 5 to both sides gives

$$x - 5 + 5 > 1 + 5$$

or

$$x > 6$$

which is the solution.

It seems to be legal to add or subtract in an equality, and you therefore shouldn't be surprised by the following:

Rule 1 for Solving Inequalities

You can add anything to or subtract anything from an inequality, provided you do it to both sides.

So far so good. Inequalities look as though they may be going to behave just like equations! The next thing to investigate is whether you can multiply or divide an inequality on both sides by a number, as you can an equation.

For example, suppose $2x < 6$. Is the solution to this inequality $x < 3$? Look at the number line again. From Figure 17.16, you can see that if x is to the left of 3, then x satisfies $2x < 6$ and therefore $x < 3$ is the solution to $2x < 6$.

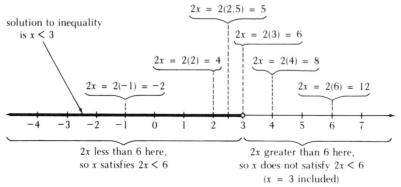

FIG. *17.16*

Again, this looks exactly like what happens with equations, because we appear to have multiplied through by $\frac{1}{2}$ (or divided through by 2). But, needless to say, life is not quite as simple as it might be—there is one thing that goes wrong in the equation/inequality analogy.

Look at the following example:

$$-x < -4$$

Searching on the number line for the numbers that satisfy this, we see from Figure 17.17 that if x is to the right of 4, then x satisfies $-x < -4$ and therefore $x > 4$ is the solution of $-x < -4$. This solution should be a surprise because it is *not* what we get by treating $-x < -4$ like an equation. To solve the equation $-x = -4$ you multiply both sides by -1. Trying that on $-x < -4$ gives $x < 4$, which is *not* the right answer; $x > 4$ is. What has happened?

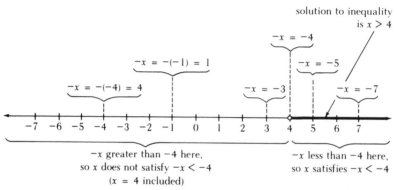

FIG. *17.17*

The answer is that it turns out to be fine to multiply both sides of an inequality by a positive number (such as $\frac{1}{2}$), but if you multiply by a negative number (such as -1) then you must reverse the inequality sign. This gives us the following rules:

Rule 2 for Solving Inequalities

You can multiply or divide both sides of an inequality by any *positive* (and nonzero) number.

Rule 3 for Solving Inequalities

You can multiply or divide both sides of an inequality by any *negative* (and nonzero) number, *provided* you reverse the inequality sign.

It's worth looking back to the example to see why you have to reverse the inequality when multiplying by a negative number. If $x > 4$, then x is to the right of 4, and farther away from the origin than 4. When you multiply by (-1), both x and 4 go over to the other side of the origin, and they each remain the same distance away (see Figure 17.18). So $-x$ is now to the *left* of -4, making $-x < -4$. In other words,

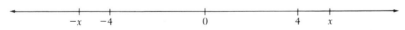

FIG. 17.18

if you start with $x > 4$ then you end up with $-x < -4$.

To summarize:

You can solve linear inequalities just like linear equations
Except
when you multiply or divide by a negative number, you must remember to reverse the inequality sign.

EXAMPLE: *Solve $2(x + 3) \geq (x + 18) - 2x$ and graph the solution on the number line.*

$$2(x + 3) \geq (x + 18) - 2x$$

Therefore

$$2x + 6 \geq 18 - x \qquad \text{(using the distributive law)}$$

$$3x + 6 \geq 18 \qquad \text{(adding } x \text{ to both sides by rule 1)}$$

$$3x \geq 12 \qquad \text{(subtracting 6 from both sides by rule 1)}$$

$$x \geq 4 \qquad \text{(multiplying both sides by } \tfrac{1}{3} \text{ by rule 2)}$$

So the solution is $x \geq 4$, which is graphed in Figure 17.19.

FIG. 17.19

Alternatively, let us see what would have happened if (for some unclear reason) you had decided to put the x's and numbers on opposite sides of the inequality:

$$2(x + 3) \geq (x + 18) - 2x$$

$$2x + 6 \geq 18 - x$$

$$6 \geq 18 - 3x$$

$$-12 \geq -3x$$

Now, to get rid of the coefficient of x, we must divide both sides of the inequality by -3, which is a *negative* number. Negative numbers should be ringing bells in your head by now, and so you remember to reverse the inequality sign when you divide by -3:

$$\frac{-12}{-3} \leq x$$

$$4 \leq x$$

Since $4 \leq x$ means the same as $x \geq 4$, this is the same answer as we got before.

EXAMPLE: *Solve $2x - 1 \leq 8 - x$ and $2 - 2x < 1 - x$.*

The "and" means that we are looking for the x's that satisfy *both* inequalities at once. The way we'll get them is to find the x's that satisfy just the first inequality and find those that satisfy just the second, and then look to see which x's come into both categories and so satisfy both inequalities. Solving

$2x - 1 \leq 8 - x$	Also $2 - 2x < 1 - x$
gives	gives
$3x \leq 9$	$1 < x$
so	so
$x \leq 3$	$x > 1$

Therefore, to satisfy both inequalities we must have

$x \leq 3$ *and* $x > 1$

So x must be less than or equal to 3 and greater than 1, which means between 1 and 3. In other words,

$1 < x \leq 3.$

The graph is in Figure 17.20.

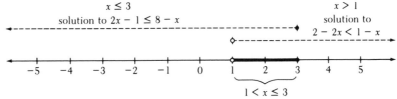

FIG. *17.20*

EXAMPLE: *Solve*

$$2x - 1 < 8 - x \quad or \quad \frac{x}{2} - \frac{5}{6} > \frac{x}{3}$$

Here we want to find the x's that satisfy either of the inequalities. As before, we will solve each separately, and then see which x's satisfy either one or the other (or both). Solving

$2x - 1 < 8 - x$

gives

$x < 3$

To solve

$\dfrac{x}{2} - \dfrac{5}{6} > \dfrac{x}{3}$

we should first clear of fractions so that we can see what's going on. That means multiplying by the L. C. D. which is 6:

$3x - 5 > 2x$

Solving gives

$x > 5$

Therefore, to satisfy either inequality we must have

$x < 3$ or $x > 5$

which means that x can be either greater than 5 or less than 3.

Graphically, $x < 3$ is all the points to the left of 3, and $x > 5$ all the points to the right of 5. The solution to this problem consists of both these sets of points put together; it is shown in Figure 17.21.

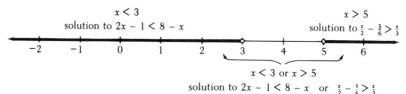

FIG. *17.21*

EXAMPLE: *Solve $ax + b < c$ for x in terms of a, b, and c.*

We set about this in the usual way by subtracting b from both sides:

$$ax < c - b$$

Now we want to divide through by the coefficient of x, which is a. Unfortunately, however, we don't know whether a is positive or negative, and so we don't know whether to reverse the inequality sign or not. Since there's no way of finding out whether a is positive or negative (because a could stand for any number), we will have to do the problem twice, once for each case.

If a is positive, $ax < c - b$ gives

$$x < \frac{c - b}{a}$$

Therefore this is the solution if a is positive. If a is negative, $ax < c - b$ gives

$$x > \frac{c - b}{a}$$

Therefore this is the solution if a is negative.

PROBLEM SET   17.2

Solve and plot on a number line.

1. $x - 3\tfrac{1}{2} > 0$
2. $-x > 2 + x$
3. $2x + 5 > 4x - 9$

4. $-3x - 5 < 2$
5. $\tfrac{5}{3}x - \tfrac{3}{2} < \tfrac{9}{5}x$
6. $1.1(y + 1.1) > 1.1(1.1 - y)$

7. $-\frac{3}{2}x \geq -2$

8. $2(1 - 2x) < 8$

9. $2(x + 5) > (x - 5)$

10. $3x - 7 > -4(x + 2)$

11. $7x + x(1 + 3) > 2(1 - x)$

12. $x > -3x + 2$

13. $3x - 5 > 16$

14. $\frac{5}{3} - 2x < -\frac{3}{2}$

15. $2x + 5 > 7x - 3 - 5x$

16. $12x + 5 < -3x + 7$

17. $x - 4 \leq 3x + 1$

18. $2x + 5 > 4(x + 3) - 2x + 8$

19. $\frac{7}{3}(3x - 5) < \frac{2}{5}(x + 5)$

20. $3(x + 2) - 3x + 2(x - 1) < 5 - 3x + \frac{2}{3}$

Solve and plot on a number line.

21. $2x - 1 < 8 - x$ or $3x - 5 > 2x$

22. $x < 3$ and $x + 2 \geq -1$

23. $2x + 3 \leq 9$ and $x \geq 0$

24. $2 + x < -1(1 + x) + 2$ or $x - 2 > 3x - 4$

25. $x + 2 \leq 4$ and $2(x - 1) < -7 - 11(1 - x) - 6x$

26. $2x + 8 < 6$ and $3 < 5 - x$

27. $2x + 9 \leq 3 - 2x$ or $2 - x < \frac{4}{3}x$

28. $\dfrac{3x + 1}{2} - 3 + 2x < -3(2 - 3x) + 1$ or $2x - 3 > 5x + 1$

Solve the following literal inequalities for x.

29. $\dfrac{x}{a} < b$

30. $ax + b > cx + d$

31. $m(x + b) \leq nx + c$

32. $x + 2b \geq 3x - ax$

33. $\dfrac{x + a}{b} < \dfrac{x - c}{d}$

34. Prove: If $a \neq b$ and both a and b are real numbers, then $a^2 + b^2 > 2ab$. [*Hint:* Examine $(a - b)^2$.]

35. Prove that if

$$\frac{a}{b} < \frac{c}{d} \quad \text{and} \quad b, d > 0 \quad then \quad \frac{a}{b} < \frac{a + c}{b + d} < \frac{c}{d}$$

36. Show that $\sqrt{3} > \sqrt{2}$.

17.3 SOLVING INEQUALITIES INVOLVING POWERS AND FRACTIONS

The linear inequalities of the last section can all be solved by pretty much the same methods as linear equations. Inequalities that are not linear are considerably more awkward, and we have to resort to a number of different methods. In this section I will give examples of various types of arguments that are used.

EXAMPLE: *Solve $x^2 < 4$.*

If $x^2 < 4$, the first thing that jumps into my mind is that x must be less than 2, because $2^2 = 4$. Now $x < 2$ includes all the numbers between 2 and the origin—all of which do have squares less than 4. However $x < 2$ also includes all the negative numbers, some of which *don't* have squares less than 4 [for example, the square of -3 is 9; the square of -5 is 25 and so on].

In Figure 17.22, as you move to the left from the origin, the numbers get more negative and their squares get larger and more positive. If we are looking for the numbers whose squares are less than 4, we should include only those negative numbers down to -2 (and not including -2 since we want the square to be strictly less than 4).

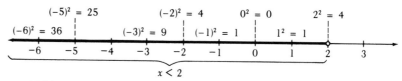

FIG. 17.22

Therefore the numbers whose squares are less than 4 are exactly those between 2 and -2, as in Figure 17.23. So the solution to

$$x^2 < 4$$

is

$$-2 < x < 2$$

that is $-2 < x$ and $x < 2$.

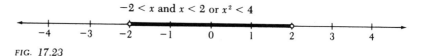

FIG. 17.23

EXAMPLE: *Solve $x^2 \geq 25$.*

$x^2 \geq 25$ suggests that $x \geq 5$, and certainly all numbers greater than or equal to 5 do have squares greater than or equal to 25. Are there any other numbers whose squares are above 25? The numbers between -5 and 5 have squares below 25, and are the solution to the opposite inequality $x^2 < 25$. However, the numbers to the left of and including -5 all have squares greater than or equal to 25. Therefore the numbers we are looking for are those that are to the right of 5 together with those that are to the left of -5. This means x must satisfy either $x \geq 5$ or $x \leq -5$, as in Figure 17.24.

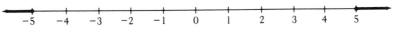

FIG. *17.24*

Therefore the solution to

$x^2 \geq 25$

is

$x \leq -5$ or $x \geq 5$.

EXAMPLE: *Solve $3 - 2u^2 < 6(2 - 3u^2)$.*

First treat this as a linear equation in u^2, and get the u's alone on one side.

$3 - 2u^2 < 6(2 - 3u^2)$

$3 - 2u^2 < 12 - 18u^2$

$\quad 16u^2 < 9$

$\quad\quad u^2 < \dfrac{9}{16}$

This is exactly like the $x^2 < 4$ example, and the solution is

$-\dfrac{3}{4} < u < \dfrac{3}{4}$

that is,

$-\dfrac{3}{4} < u \quad \text{and} \quad u < \dfrac{3}{4}$

See Figure 17.25.

FIG. *17.25*

EXAMPLE: *Solve $(x - 1)(x - 2) < 0$.*

This example looks different from those above and needs a new idea. We might multiply out, giving $x^2 - 3x + 2 < 0$, but that doesn't seem to help much. Then you might remember that when you had to solve a quadratic equation like $x^2 - 3x + 2 = 0$, you factored it, giving $(x - 1)(x - 2) = 0$, so that you could say that either $(x - 1) = 0$ or $(x - 2) = 0$ [this depends on the Useful Fact (Section 16.2) that if $ab = 0$ then $a = 0$ or $b = 0$ (or both)]. Since the factored form was useful for equations, perhaps it is useful here.

In this problem, instead of having $ab = 0$, we have something of the form $ab < 0$ [here $a = (x - 1)$, $b = (x - 2)$]. What does this tell us about a and b?

First notice that:

If y is positive, then $y > 0$.

If y is negative, then $y < 0$.

So, if $ab < 0$, then ab is negative. This means that a and b must have opposite signs, because if they are both positive or both negative, their product is positive. Therefore, if $ab < 0$, then either a is positive and b is negative, or a is negative and b is positive. In other words,

if $ab < 0$, then

either $a > 0$ and $b < 0$

   or $a < 0$ and $b > 0$

Therefore, if $(x - 1)(x - 2) < 0$, then

either $(x - 1) > 0$ and $(x - 2) < 0$

   or $(x - 1) < 0$ and $(x - 2) > 0$

This looks reasonably promising, since we have managed to replace a quadratic inequality by a collection of linear inequalities, and these we know how to solve.

Let us see what the first condition

$(x - 1) > 0$ and $(x - 2) < 0$

gives us, by solving each inequality separately and remembering to keep the "and" in the middle. We have

$x - 1 > 0$ and $x - 2 < 0$

so

$x > 1$ and $x < 2$

This means we want the x's that are both greater than 1 and less than 2; in other words, $1 < x < 2$. See Figure 17.26.

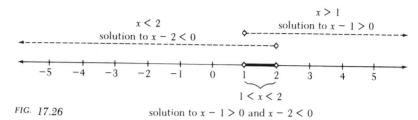

FIG. 17.26   solution to $x - 1 > 0$ and $x - 2 < 0$

Now let us see what the second condition gives us by solving:

$(x - 1) < 0$ and $(x - 2) > 0$

to get

$x < 1$ and $x > 2$

This means we want the x's that are both less than 1 and greater than 2. Such numbers, unfortunately, are no more real than a unicorn, and therefore the second condition gives us nothing. Looking at the graphs of $x < 1$ and of $x > 2$ in Figure 17.27, you see that they do not overlap at all, which is another way of seeing that there are no x's that satisfy these two inequalities at once.

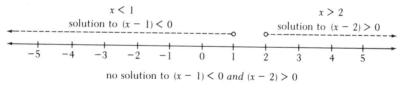

no solution to $(x - 1) < 0$ *and* $(x - 2) > 0$

FIG. 17.27

So let us see what we have just done: If $(x - 1)(x - 2) < 0$, then we showed that

either $(x - 1) > 0$ and $(x - 2) < 0$

  or $(x - 1) < 0$ and $(x - 2) > 0$

The first condition led to $1 < x < 2$, and the second led to nothing.

Therefore all this maneuvering adds up to:

If $(x - 1)(x - 2) < 0$, then $1 < x < 2$.

EXAMPLE: *Solve $x^2 + 4x > -3$.*

This example can be done in the same way as $(x - 1)(x - 2) < 0$ if it is rewritten with the product of two factors on one side and a zero on the other.

$$x^2 + 4x > -3$$

$$x^2 + 4x + 3 > 0$$

$$(x + 1)(x + 3) > 0$$

Now realize that if the product of two factors [here $(x + 1)$ and $(x + 3)$] is positive, then either both factors must be positive or both must be negative. So if

$$(x + 1)(x + 3) > 0$$

then

either $(x + 1) > 0$ and $(x + 3) > 0$

or $(x + 1) < 0$ and $(x + 3) < 0$

In the first case, $x > -1$ and $x > -3$, which amounts to $x > -1$. In the second case, $x < -1$ and $x < -3$, which amounts to $x < -3$. Therefore the solution is

$$x > -1 \quad \text{or} \quad x < -3$$

EXAMPLE: *Solve*

$$\frac{x}{x + 1} > 0$$

The last example relied on the fact that if $ab > 0$, then either $a > 0$ and $b > 0$ or $a < 0$ and $b < 0$. Here we have a fraction of the form $\frac{a}{b}$, and the inequality tells us that $\frac{a}{b} > 0$. But a fraction can be positive only if the numerator and denominator are both positive or both negative, that is,

if $\frac{a}{b} > 0$, then

either $a > 0$ and $b > 0$

or $a < 0$ and $b < 0$

Therefore,

if $\dfrac{x}{x + 1} > 0$ then

either $x > 0$ and $x + 1 > 0$

or $x < 0$ and $x + 1 < 0$

So this problem, too, has come down to solving a collection of linear inequalities. Since that's exactly the way the last two ended up, I'm leaving it up to you to show that the solution is

$x < -1$ or $x > 0$

PROBLEM SET   17.3

Solve and plot on the number line.

1. $x^2 > 9$

2. $2x^2 - 8 < 0$

3. $\dfrac{x - 2}{2x + 5} > 0$

4. $(x + 3)(x - 4) \geq 0$

5. $x^2 \leq 25$ and $x > 1$

6. $7 < \dfrac{3}{x}$

7. $(x - 1)(x + 2)(x - 3) < 0$

8. $x^2 > 4x + 5$

9. $(x + 1)(x - 3) < -4x$

10. $(x + 2)(1 - x) > 0$

11. $\frac{1}{2}x^2 \leq 9$

12. $\frac{9}{4} \leq x^2$

13. $x^2 < 0.04$

14. $\dfrac{1}{x} < 3$

15. $\dfrac{(x - 1)(x + 2)}{x - 4} < 0$

16. $\dfrac{-1}{x + 3} > \dfrac{3}{x - 1}$

17. $\dfrac{1}{x + 1} + \dfrac{2}{x} < 0$

18. $\dfrac{5x - 3}{2x + 1} > 1$

19. $x + 1 > \dfrac{x + 1}{x + 3}$

20. $\dfrac{12}{5} < \dfrac{3}{2x + 1}$

21. $(x - 1)^2 > 4$

22. $x^2 - \frac{1}{2}x + \frac{1}{16} \leq \frac{1}{4}$

23. $3x^2 - 2x - 6 < 2x^2 - 6x - 1$

24. $3x^2 + 2x + 2 < 2x^2 + x + 4$

25. True or false:

$$\dfrac{x - 2}{2 + x} > 0 \quad \text{when } x < -2 \text{ or } x > 2$$

CHAPTER 17 REVIEW

Fill in the proper inequality or equality ($>$, $<$, $=$).

1. $\dfrac{6}{17} \;\underline{\quad}\; \dfrac{17}{30}$

2. $\dfrac{7}{\sqrt{3}} \;\underline{\quad}\; \dfrac{8\sqrt{3}}{3}$

3. $0.035 \;\underline{\quad}\; \dfrac{25}{1000}$

Express the inequality or equality represented by the following graphs.

4.

FIG. 17R-4

5.

FIG. 17R-5

6.

FIG. 17R-6

Graph the following inequalitites.

7. $x \geq -1$
8. $2\frac{1}{2} < x \leq 5\frac{1}{2}$
9. $x \leq 2$ or $x \geq 5$
10. $x = 5$ and $\frac{x}{3} < 1$

Solve and plot on a number line:

11. $2 - 3x \geq 5x$
12. $\frac{4}{3}x - 1 \leq 3$
13. $3(x - 2) \leq 5(3 - x)$
14. $-2x \geq 4 - x$
15. $\dfrac{1 - x}{2} \leq \dfrac{3 - 5x}{7}$

16. $x + 2 \geq 4$ and $x - 7 \leq -3$
17. $\dfrac{x}{3} < -1$ or $x = \dfrac{3}{4}$
18. $0.5x > 0.25$ and $-2x + 10 > 5x$

Solve the following literal inequalities for x.

19. $cx \geq d$ where $c \neq 0$
20. $\dfrac{x}{q} < \dfrac{1}{r}$
21. $n(x + m) > p(r - x)$
22. Show that if a and b are ≥ 0, then $(a - b)^2 \leq a^2 + b^2$.

Solve and plot on the number line.

23. $81 \geq x^2$
24. $3x^2 - 1 > 26$
25. $x^2 < \frac{1}{4}$
26. $(x + 1)(x - 2) > 0$
27. $x^2 - 5x < -6$

28. $\dfrac{1}{x + 1} > 2$
29. $\dfrac{x}{x - 2} > 0$
30. $x^2 - \dfrac{2x}{3} + \dfrac{1}{9} > 0$

Fill in the proper inequality or equality ($>$, $<$, $=$).

31. $\dfrac{0.018}{5\sqrt{20} - \sqrt{5}} \underline{} \dfrac{\sqrt{5}}{100}$
32. $y^z \underline{} y^{1/z}$ where $y \geq 1$, $z < -1$

Write the inequality or equality represented by the following graphs.

33.

FIG. 17R-33

34.

FIG. *17R-34*

Graph the following inequalitites.
  35. $-1 \leq x \leq \frac{1}{2}$ and $2 \leq x \leq 5$
  36. $-x \geq 3$

Solve and plot on a number line:
  37. $-5(x - 2) \leq -3(x + 2)$
  38. $0.01(0.1x - 0.1) \geq -0.02(x - 0.02)$
  39. $13 - 3x + 4(x - 2) \geq 2x + 1$
  40. $2x + \dfrac{3(1 - 2x)}{4} < \dfrac{3}{2}(4 - 3x) + 1$
  41. $\dfrac{3}{5}(2x - 1) \geq \dfrac{x - 6}{3}$ and $3x < 5(2 + x) - 1$
  42. $\dfrac{2x - 3}{-5} + 1 \leq \dfrac{7 - x}{4}$ or $\dfrac{0.1x - 0.03}{0.01} \geq 2 - 0.02x$

Solve the following literal inequalities for x.
  43. $\dfrac{a - x}{b} \geq \dfrac{c - dx}{f}$ 44. $b - ax < \dfrac{b}{c}x + a$
  45. If $\dfrac{a}{b} < \dfrac{a}{c}$, when is $c^2 - b^2 < 0$?
  46. Show that

$$\frac{1}{\sqrt{2}} > \frac{1}{\sqrt{5}}$$

Solve the following inequalities.
  47. $0.03x^2 \geq \frac{1}{3}$
  48. $-6x^2 \geq -0.0216$
  49. $3x^2 - 5 \geq 2(x^2 + 10)$
  50. $(x - 3)(x - 4) > -15x$
  51. $0.1x(0.6 - 0.2x) \leq 0.3(0.1x - 0.3)$
  52. $\dfrac{10x - 3}{x + 2} > -1$
  53. $\dfrac{-1}{x - 7} < \dfrac{6}{x + 2}$
  54. $(x - 1)(x - 2)(x - 5) \leq 0$
  55. $(x - 2)(x - 1)(x + 6) > 0$

18   ABSOLUTE VALUE

18.1   DEFINITION OF ABSOLUTE VALUE

There are many occasions when we are interested in the size of some quantity but do not really care about its sign. When we are looking at the difference between two quantities we often care only about the magnitude of the difference, and not about which quantity is larger. For example, when we are estimating the error in an experiment, we can usually determine only how much we might be off by, and not whether we are over or under the true result. In other words, we are interested only in how far apart the experimental and true results are.

This leads us to define the absolute value of $(a - b)$, written $|a - b|$, as the distance between a and b on the number line.

$$|a - b| = \text{distance between } a \text{ and } b$$

For example,

$$|8 - 5| = \text{distance between 8 and 5} = 3$$

$$|5 - 8| = \text{distance between 5 and 8} = 3$$

Since distance is always positive, $|a - b|$ is always positive and represents the *magnitude*, or size, of $(a - b)$ without regard to sign. Since any number is unchanged by subtracting zero, $|a|$ can be thought of as $|a - 0|$, and so

$$|a| = \text{distance between } a \text{ and the origin}$$

For example,

$$|6| = \text{distance between 6 and the origin} = 6$$

$$|-6| = \text{distance between } -6 \text{ and the origin} = 6$$

$$\left|\tfrac{7}{2}\right| = \tfrac{7}{2}$$

$$\left|-\tfrac{3}{5}\right| = \tfrac{3}{5}$$

$$|\pi| = \pi$$

Again, $|a|$ is always positive and represents the magnitude of a without regard to sign. We now have two ways, one using the definition of $|a|$ and one using the definition of $|a - b|$, to work out things like $|10 - 6|$:

First,

$$|10 - 6| = \text{distance between 10 and 6} = 4$$

Second,

$$|10 - 6| = |4| = \text{distance between 4 and the origin} = 4$$

Another example:

$$|6 - 10| = \text{distance between 6 and 10} = 4$$

and the other way:

$$|6 - 10| = |-4| = \text{distance between } -4 \text{ and the origin} = 4$$

In general, you can see that

$$|a| = |-a| \quad \text{and} \quad |a - b| = |b - a|$$

But what about $|a + b|$? This can also be interpreted as a distance, if you use the trick of writing $b = -(-b)$. Then

$$|a + b| = |a - (-b)| = \text{distance between } a \text{ and } -b$$

or

$$|a + b| = |b - (-a)| = \text{distance between } -a \text{ and } b$$

Summarizing, we have:

> $|a|$ = distance between a and the origin
>
> $|a - b|$ = distance between a and b
>
> $|a + b| \begin{cases} = \text{distance between } a \text{ and } -b \\ = \text{distance between } -a \text{ and } b \end{cases}$

Alternative Definition of $|a|$

The definition of $|a|$ that I have given is a geometric one, because it involves distances along a line. There is also an algebraic definition, which goes like this:

> If a is positive or zero, $|a| = a$
>
> If a is negative, $|a| = -a$

The point of this definition is that no matter what the sign of a, $|a|$ will always come out positive (or zero). If a is negative, then $-a$ is positive, so $|a|$ is positive. For example,

$$3 \text{ is positive, so } |3| = 3$$

$$-3 \text{ is negative, so } |-3| = -(-3) = 3$$

There is one other place where this always happens. Look at $\sqrt{a^2}$:
For $a = 2$,

$$\sqrt{a^2} = \sqrt{2^2} = \sqrt{4} = 2 \qquad (\sqrt{4} \text{ means the positive root})$$

For $a = -2$,

$$\sqrt{a^2} = \sqrt{(-2)^2} = \sqrt{4} = 2$$

Therefore, if you square a number and then take its square root, since "$\sqrt{}$" is defined to be positive, you'll always get a positive number. Also, the number you get has the same magnitude as the one you started with, and therefore is the absolute value of the one you started with. Hence

> $$\sqrt{a^2} = |a|$$

It is possible to solve absolute value problems both geometrically and algebraically. By and large, I think the geometric viewpoint is more illuminating, so I suggest that you look at all the problems in terms of distance. However, for some of the problems I will give an algebraic method as well as the geometric, and you can compare them.

PROBLEM SET   18.1

Evaluate the following expressions.

1. $|3(-\frac{1}{2}) + \frac{1}{2}|$
2. $|4|$
3. $|-3|$
4. $-|-3|$
5. $-(-|3|)$
6. $|6 - 2|$
7. $|2 - 6|$
8. $|3| - |7|$
9. $-3 + |4| - |5|$
10. $-|-(-2)|$

11. $|5| - |3 - 4|$
12. $|7 - |3 - 5||$
13. $|2| - 3|-3|$
14. $4|2 - 3| + |-5|$
15. $|2 + |3|(-2) + |4 - 7||$
16. $-|7 + 3(-5)| + 2$
17. $|4 - (3)^2|$
18. $-|(-4)^3|$
19. $[-|(-2)^2| - |-3|^3]^2$
20. $-|2^5| - 3(|-2^3| - 4^2)$

Simplify.

21. $|a - b|$ if $a > b$ and if $a < b$
22. $|-mn|$ if $mn > 0$ and if $mn < 0$

Write an expression that represents the distance between:

23. x and 3
24. -3 and y
25. a and z
26. 0.02 and b

27. -1.3 and $-x$
28. $-y$ and 2
29. 0.0042 and -0.0042

Write an expression that represents the distance between:

30. $\dfrac{x + apq}{2}$ and $\dfrac{apx}{4}$
31. $xy + ab$ and the origin
32. y and $-y$
33. $\dfrac{x^2 + 2x}{3}$ and $\dfrac{x}{2}$

34. b^2 and $(-b)^2$
35. $x^2 + y^2$ and $x^2 - y^2$
36. $\dfrac{x - y}{x + y}$ and $\dfrac{y - x}{y + x}$
37. $\dfrac{xy - 2zx}{4x + y}$ and $\dfrac{2zx - xy}{y + 4x}$

$|x - 3|$ represents "the distance between x and 3." In the same way, write in words the distance represented by the following expressions.

38. $|x + 2|$
39. $|3 - x|$
40. $\left|\dfrac{0.2 + x}{3}\right|$
41. $3|x|$

42. $\dfrac{1}{5}\left|x - \dfrac{2}{7}\right|$
43. $\left|\dfrac{0.092 - x}{0.13}\right| \cdot 2$

44. $\left|x - \dfrac{b}{c}\right|$
45. $\left|\dfrac{x + ab}{d}\right|$

18.2   **THINGS YOU CAN AND CAN'T DO WITH ABSOLUTE VALUES**

1. You can't add them in general:	$\|a\| + \|b\| \neq \|a + b\|$
2. You can't subtract them in general:	$\|a\| - \|b\| \neq \|a - b\|$
3. You can multiply them:	$\|a\| \cdot \|b\| = \|ab\|$
4. You can divide them:	$\dfrac{\|a\|}{\|b\|} = \left\|\dfrac{a}{b}\right\|$

To justify statements like (1) or (2), all you need to do is to produce *one* pair of numbers a and b for which the two expressions are not equal.

For example, let $a = 7$, $b = -2$:

1.
$$\|a\| + \|b\| = \|7\| + \|-2\| = 7 + 2 = 9$$
$$\|a + b\| = \|7 + -2\| = \|5\| = 5$$

So $\|a\| + \|b\|$ and $\|a + b\|$ are not equal for $a = 7$, $b = -2$. However $\|a\| + \|b\|$ and $\|a + b\|$ are equal if a and b are both positive or both negative, or one is zero. They are not equal otherwise, and so are not equal *in general*.

2.
$$\|a\| - \|b\| = \|7\| - \|-2\| = 7 - 2 = 5$$
$$\|a - b\| = \|7 - (-2)\| = \|9\| = 9$$

So $\|a\| - \|b\|$ and $\|a - b\|$ are not equal for $a = 7$, $b = -2$. However $\|a\| - \|b\|$ and $\|a - b\|$ are equal if a and b are both positive and a is greater than or equal to b, or if a and b are both negative and a is less than or equal to b. They are not equal otherwise, and so are not equal *in general*.

To justify statements like (3) and (4), you must show that they are true for *any* a and b. We cannot just give an example for which the statement is true because there always might (somewhere) be another pair of numbers, a and b, for which the statement is not true, and we are trying to show that the statement is true for *every* a and b.

3. To see why $\|a\| \cdot \|b\| = \|ab\|$, remember how you multiply numbers. If you have one or more negative numbers, you ignore the signs at first, multiply, and then attach the correct sign. For example, to find $(-4) \cdot 6$, you just find $4 \cdot 6 = 24$, and then attach a minus, giving -24. To find $(-2) \cdot (-11)$, you just find $2 \cdot 11 = 22$ and since there are two minuses, this is the answer. Therefore the magnitude of a product is the product of magnitudes. And, since the absolute value of a number is just its magnitude, the absolute value of a product is the product of the absolute values. That is,

$$|a| \cdot |b| = |ab|$$

4. The same reasoning works for division. When you divide two numbers you first divide their magnitudes and then attach the correct sign, and therefore the magnitude of the quotient is just the quotient of the magnitudes. Hence:

$$\frac{|a|}{|b|} = \left|\frac{a}{b}\right|$$

PROBLEM SET 18.2

For each of the following equations, give a pair of numbers a and b that will make it true. (There will be many possible answers.)

1. $|a| + |b| = |a + b|$
2. $|a| - |b| = |a - b|$
3. $a|b| = |ab|$
4. $b = |b|$
5. $b^3 = |b^3|$
6. $b^2 = |b^2|$
7. $\dfrac{|a|}{b} = \dfrac{a}{b}$
8. $\left|\dfrac{1}{b}\right| = \dfrac{1}{b}$
9. $|a| + |b| + |-2| = |a| + |b - 2|$
10. $|a| \, |b^2| = |ab^2|$

11–20. For each of the equations in Problems 1 through 10, give a pair of numbers a and b that will make the equation false. (There may be many or no answers.)

21. Show that $|3t - 6| = 3|t - 2|$.
22. Show that $\left|\dfrac{2t + 4}{-2}\right| = |t + 2|$.
23. Explain why $|a + b| = |b + a|$.
24. Explain why $|a - b| = |b - a|$.
25. Explain why $|ab| = |b| \, |a|$.

Simplify the following.

26. $\dfrac{|4t + 8y + 4|}{4}$
27. $2|x| - (|x| - |-x|)$
28. $\dfrac{|u + v| + |u + v|}{2}$
29. $a - b + |a - b| - |b - a|$
30. $|7r - 7s| - |r - s| - 2\,|3(r - s)|$
31. $\dfrac{|m|}{m}$
32. $\sqrt{b^2} + |a + b| - a - |b|$
33. $|2| - \big|p + |\, p - |-p| - p\,| + |-2|\big|$

For what signs $>, <, \geq, \leq, =$ are the following always true?

34. $|a - b| \; \underline{\quad} \; |a| - |b|$
35. $|a + b| \; \underline{\quad} \; |a| + |b|$
36. $\dfrac{|a|}{|b|} \; \underline{\quad} \; \dfrac{a}{b}$
37. $|a| \; \underline{\quad} \; b$ if $a < b$ and $-a < b$
38. $|a| \; \underline{\quad} \; b$ if $-a > b$ and $a < b$
39. $\left|\dfrac{1}{x}\right| \; \underline{\quad} \; 1$ if $-1 < x < 0$

18.3   ABSOLUTE VALUES
IN EQUATIONS

Geometric Method

Suppose you are trying to find some quantity z, and suppose that instead of being told anything about z itself, you are told something about the absolute value of z. This might enable you to write an equation involving the absolute value of z, from which you would want to find z itself. In such a situation it becomes important to know how to solve equations involving absolute values.

For example, suppose you know $|z| = 2$. What is z? The easiest way to do this is to think of absolute value on the number line. $|z| = 2$ means that the distance between z and the origin is 2. So where can z be if it is exactly 2 units from 0? Clearly, it can be either at 2 or at -2, as in Figure 18.1.

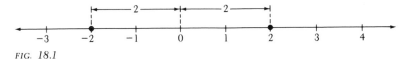

FIG. 18.1

Therefore, the solution to $|z| = 2$ is $z = 2$ or $z = -2$.

EXAMPLE: *Solve $3(|x| - 1) + 2 = 2 - (1 - |x|)$ for x.*

This is a linear equation in $|x|$, and we can use the usual methods to find $|x|$.

$$3(|x| - 1) + 2 = 2 - (1 - |x|)$$
$$3|x| - 3 + 2 = 2 - 1 + |x|$$
$$3|x| - 1 = 1 + |x|$$
$$2|x| = 2$$
$$|x| = 1$$

But $|x| = 1$ means that x is a distance of 1 away from the origin, so it must be either at 1 or at -1. Therefore the solution is

$$x = 1 \quad \text{or} \quad x = -1$$

CHECK: If $x = 1$,

$$\text{LHS} = 3(|x| - 1) + 2 = 3(|1| - 1) + 2 = 3(1 - 1) + 2 = 2$$
$$\text{RHS} = 2 - (1 - |x|) = 2 - (1 - |1|) = 2 - (1 - 1) = 2 - 0 = 2$$

and therefore $x = 1$ is a solution.

If $x = -1$,

LHS $= 3(|x| - 1) + 2 = 3(|-1| - 1) + 2 = 3(1 - 1) + 2 = 2$

RHS $= 2 - (1 - |x|) = 2 - (1 - |-1|) = 2 - (1 - 1) = 2 - 0 = 2$

and therefore $x = -1$ is a solution.

EXAMPLE: *Solve $|x| = -3$.*

This question asks us to find the numbers whose distance from the origin is -3. But there is no such thing as a negative distance, and so there are no x's that are -3 away from 0. Therefore, this equation has no solution.

Note on the number of solutions in an absolute value equation: The equations involving absolute values that we have solved so far have either had two or no solutions; there are also equations, for example, $|x| = 0$, that have one solution. But in general, look for two solutions in a linear equation that contains absolute values.

EXAMPLE: *Solve $|x - 5| = 2$.*

Since $|x - 5|$ is the distance between 5 and x, this equation tells us that the distance between 5 and x is 2. Therefore x is a number that is exactly 2 units away from 5. Now x can be either 2 units to the right of 5, which puts it at 7, or 2 units to the left of 5, which puts it at 3. See Figure 18.2.

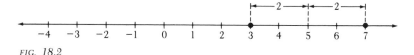

FIG. *18.2*

Therefore, the solutions are

$x = 3$ or $x = 7$

CHECK: When $x = 3$,

LHS $= |x - 5| = |3 - 5| = |-2| = 2$

RHS $= 2$

and therefore $x = 3$ is a solution.

When $x = 7$,

LHS $= |x - 5| = |7 - 5| = |2| = 2$

RHS $= 2$

and therefore $x = 7$ is a solution.

Alternative Method: Algebraic

The way of solving the equation that we used above mainly involves the number line and lengths. It is also possible to do the problem entirely algebraically by using the definition:

$$|a| = a \text{ if } a \geq 0$$

$$|a| = -a \text{ if } a < 0$$

Since the definition involves two different cases, solving $|x - 5| = 2$ will also have to be done by cases.

Now

$$|x - 5| = x - 5 \quad \text{if } x - 5 \geq 0, \text{ that is, if } x \geq 5$$

$$|x - 5| = -(x - 5) \quad \text{if } x - 5 < 0, \text{ that is, if } x < 5$$

So, *first we will assume that* $x \geq 5$. Then

$$|x - 5| = 2$$

gives

$$x - 5 = 2$$

so

$$x = 7 \qquad \text{one possible solution}$$

At this stage we have to look back and check that the solution we have does in fact satisfy the assumptions that we made. Fortunately, we assumed that $x \geq 5$, which is certainly satisfied by $x = 7$, so $x = 7$ is indeed a solution.

Now assume $x < 5$. Then

$$|x - 5| = 2$$

gives

$$-(x - 5) = 2$$

$$-x + 5 = 2$$

so

$$x = 3 \qquad \text{another possible solution}$$

$x = 3$ does satisfy the assumption $x < 5$, so $x = 3$ really is a solution.

Therefore, the solutions, as before, are

$$x = 3 \quad \text{or} \quad x = 7$$

EXAMPLE: *Solve* $|2x + 4| = 3$.

Remember that $|2x + 4|$ can be written as $|2x + 4| = |2x - (-4)| =$ the distance between -4 and $2x$, so this equation tells us that the distance between -4 and $2x$ is 3. Therefore $2x$ can either be 3 to the right of -4, that is, at -1, or 3 to the left of -4, that is, at -7. See Figure 18.3.

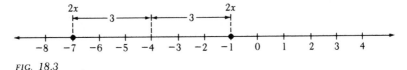

FIG. 18.3

Therefore, either $2x = -1$ or $2x = -7$, which means that

$$x = -\frac{1}{2} \quad \text{or} \quad x = -\frac{7}{2}$$

Therefore, the solutions are $x = -\frac{1}{2}$ or $x = -\frac{7}{2}$.

A Minor Variation on This Method of Solution   From Section 18.3, you know you can multiply absolute values, so:

$$|2x + 4| = |2(x + 2)| = |2| \, |x + 2| = 2|x + 2|$$

So if

$$|2x + 4| = 3$$

then

$$2|x + 2| = 3$$

Therefore,

$$|x + 2| = \frac{3}{2}$$

This tells us that the distance between -2 and x is $\frac{3}{2}$, which puts x at $-2 + \frac{3}{2} = -\frac{1}{2}$, or at $-2 - \frac{3}{2} = -\frac{7}{2}$. See Figure 18.4.

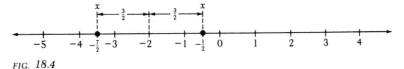

FIG. 18.4

Therefore, the solutions are $x = -\frac{1}{2}$ or $x = -\frac{7}{2}$.

EXAMPLE: *Solve $|a^2 - 7a + 12| = 0$.*

This looks totally peculiar and quite unlike anything we've seen before until we realize that the only way that the absolute value of something can be zero is for that thing to be zero. Therefore

$$|a^2 - 7a + 12| = 0$$

means

$$a^2 - 7a + 12 = 0$$

This is an ordinary quadratic, which factors to $(a - 3)(a - 4) = 0$.

Therefore, the solutions are

$$a = 3 \quad \text{and} \quad a = 4$$

EXAMPLE: *Solve $|x - 1| = |x - 5|$.*

This equation tells us to look for an x that is the same distance from 1 as it is from 5. The only number that is equidistant from 1 and 5 is the one halfway in between, namely 3. See Figure 18.5.

FIG. 18.5

Therefore, the solution is $x = 3$.

Oddly enough, there is no second solution in this case.

CHECK: When $x = 3$,

$$\text{LHS} = |3 - 1| = |2| = 2$$
$$\text{RHS} = |3 - 5| = |-2| = 2$$

Therefore $x = 3$ is the correct solution.

Alternative Method: Algebraic   This problem can also be done using the algebraic definition of $|x - 1|$ and $|x - 5|$.

$$|x - 1| = x - 1 \quad \text{if } x - 1 \geq 0 \quad \text{i.e. if } x \geq 1$$
$$= -(x - 1) \quad \text{if } x - 1 < 0 \quad \text{i.e. if } x < 1$$

and

$$|x - 5| = x - 5 \quad \text{if } x - 5 \geq 0 \quad \text{i.e. if } x \geq 5$$

$$= -(x - 5) \quad \text{if } x - 5 < 0 \quad \text{i.e. if } x < 5$$

Since the expression used to represent $|x - 1|$ changes at $x = 1$, and the one used to represent $|x - 5|$ changes at $x = 5$, the number line divides naturally into three regions (see Figure 18.6).

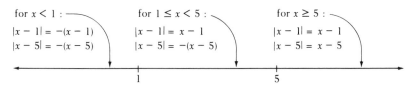

FIG. 18.6

We have to solve the equation separately for each region.

For $x \geq 5$, the equation $|x - 1| = |x - 5|$ becomes

$$x - 1 = x - 5$$

which is no help at all, since the x's drop out and leave you with

$$-1 = -5$$

So we will try $1 \leq x < 5$ and hope for better luck.   For this region the equation becomes

$$x - 1 = -(x - 5)$$

or

$$2x = 6$$

So

$$x = 3$$

Fortunately, 3 is between 1 and 5, so $x = 3$ does not contradict the initial assumption that $1 \leq x < 5$.   Therefore, $x = 3$ really is a solution.

Lastly, we look at $x < 1$.   In this case, $|x - 1| = |x - 5|$ becomes

$$-(x - 1) = -(x - 5)$$

which reduces to $1 = 5$, another useless equation.

Therefore, the only solution is $x = 3$.

From this example you can see that the geometric method seems to be easier than the algebraic, because you don't have to break things up into so many separate cases and assumptions.   Being able to "see" what is going on makes your common sense more useful and makes you much less likely to make a mistake. In most cases you are better off — and you'll certainly be quicker — using the geometric method rather than the algebraic.

PROBLEM SET   18.3

Solve the following equations.

1. $|x| = 3$
2. $|-x| = 0$
3. $|2x| = -1$
4. $|x - 2| = 6$
5. $\frac{1}{2}|x| + 1 = 3$
6. $|3x + 1| = 4$
7. $|x| - 1 = \frac{|x|}{2}$
8. $|x + 3| = 2$
9. $\left|\dfrac{x-3}{-3}\right| = 1$
10. $\left|\dfrac{x}{3}\right| = 2$
11. $4 - |x| = \frac{1}{2}|x| + 1$
12. $\dfrac{|x|}{2} - 1 = |x|$

13. $\dfrac{|2 - x|}{3} = 4$
14. $|x| = |-3|$
15. $|x| = 3$
16. $-x = |3|$
17. $\dfrac{|2x + 4|}{5} = 10$
18. $2|x| - 3 = |x|$
19. $3|x|^2 + 5|x| - 2 = 0$
20. $\dfrac{|4 - x|}{-3} = 2|x - 4|$
21. $|x - 6| = |x - 4|$
22. $2|x - 5| = |x + 1|$
23. $4|x - 1| = 2|x + 7|$
24. $|x^2 - 9| = 7$
25. $\dfrac{|x - 1|}{-3} = |1 - x| + 2$

26. Show that the larger of a and b is

$$\frac{a + b + |a - b|}{2}$$

27. Show that the smaller of a and b is

$$\frac{a + b - |a - b|}{2}$$

18.4   ABSOLUTE VALUES IN INEQUALITIES

Geometric Method

In practice, absolute values turn up most often in inequalities. For example, we are often interested in finding the x's that satisfy:

$$|x| < 3 \quad \text{or} \quad |x + 1| < |x + 7|$$

Let us look at the first example, namely, $|x| < 3$. This is asking us for the x's whose distance from the origin is less than 3. If you can't see which x's fall into that category, imagine a goat tied onto a rope whose other end is fixed at the origin. If the rope is 3 units long, the goat can be no more than 3 units away from the origin, and so the numbers we are looking for are the points to which the goat can go.

Starting from the stake at 0, the goat can walk in the positive direction up to 3, and in the negative direction down to -3. See Figure 18.7. Therefore, the solution to this inequality is the numbers between -3 and 3. Since the original inequality is $|x| < 3$, rather than $|x| \leq 3$, the end points 3 and -3 are not included. So the solution to $|x| < 3$ is $-3 < x < 3$, that is $x > -3$ and $x < 3$.

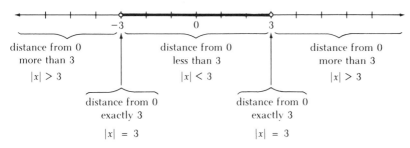

FIG. 18.7

Another standard problem is to find the x's that satisfy $|x| > 5$. This inequality asks for the numbers whose distance from 0 is greater than 5. Imagine a goat on a rope of length 5; this time you want the points to which the goat *can't* go. It can go up to 5 in the positive direction, and down to -5 in the negative, but it can't go either to the x's greater than 5 or to those less than -5. See Figure 18.8.

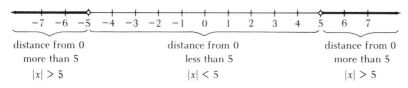

FIG. 18.8

Therefore, the solution to $|x| > 5$ is $x > 5$ or $x < -5$.

Alternative Method: Algebraic

The two inequalities, $|x| < 3$ and $|x| > 5$, can also be solved using the definition

$$|x| = x \text{ if } x \geq 0$$
$$= -x \text{ if } x < 0$$

EXAMPLE: *Solve* $|x| < 3$.

The definition shows that if x is a positive solution of $|x| < 3$, then $x \geq 0$ (because x is positive) and $x < 3$ (because x satisfies the inequality). So

$$0 \leq x < 3$$

If x is a negative solution of $|x| < 3$, then x satisfies

$$x < 0 \quad \text{and} \quad -x < 3$$

that is,

$$x < 0 \quad \text{and} \quad x > -3 \qquad \text{(multiplying through by } -1)$$

so

$$-3 < x < 0$$

Therefore, if x is any solution of $|x| < 3$, x satisfies

$$0 \leq x < 3 \quad \text{or} \quad -3 < x < 0$$

which amounts to the same thing as

$$-3 < x < 3.$$

See Figure 18.9.

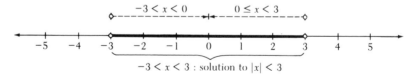

FIG. 18.9

Therefore, the solution to $|x| < 3$ is $-3 < x < 3$.

The algebraic solution to $|x| > 5$ goes along exactly the same lines.

EXAMPLE: *Solve* $|z - 6| < 2$.

This tells us that the distance between 6 and z is less than 2. So if you tie up your goat on a rope of length 2 fixed at a stake at 6, it will only be able to move at most 2 units away from 6. Since it can move 2 units to the right, it can go up to 8, and since it can move 2 units to the left it can go down to 4. See Figure 18.10.

FIG. 18.10

The inequality reads $|z - 6| < 2$, rather than $|z - 6| \leq 2$, and so we are interested just in those points that are actually less than 2 units away from 6. That means 8 and 4 are not included as possible points for z, because they are exactly 2 units from 6. So the solution to $|z - 6| < 2$ is

$4 < z < 8$, or in other words, $z > 4$ and $z < 8$

Alternative Method: Algebraic   This example can also be done by using the fact that

$$|z - 6| = z - 6 \quad \text{if } z - 6 \geq 0 \quad \text{i.e. } z \geq 6$$
$$= -(z - 6) \quad \text{if } z - 6 < 0 \quad \text{i.e. } z < 6$$

Therefore, if $z \geq 6$, the inequality becomes

$$|z - 6| = z - 6 < 2$$

So we have to solve

$$z \geq 6 \quad \text{and} \quad z - 6 < 2$$

giving

$$z \geq 6 \text{ and } z < 8 \quad \text{that is} \quad 6 \leq z < 8$$

If $z < 6$, the inequality becomes

$$|z - 6| = -(z - 6) < 2$$

So we have to solve

$$z < 6 \quad \text{and} \quad -(z - 6) < 2$$

giving

$$z < 6 \quad \text{and} \quad -z + 6 < 2$$

or

$$z < 6 \text{ and } 4 < z \quad \text{that is} \quad 4 < z < 6$$

If z is any solution of the inequality, then

$$6 \leq z < 8 \quad \text{or} \quad 4 < z < 6$$

which is the same thing as

$$4 < z < 8$$

See Figure 18.11 on page 408.

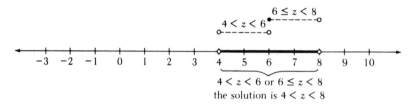

FIG. *18.11*

EXAMPLE: *Solve* $|2t + 8| > 2.$

Since we can write

$$|2t + 8| = |2t - (-8)|$$

this inequality tells us that the distance between -8 and $2t$ is more than 2. Imagine the goat tied on at -8 with a rope of length 2 and you will see that it can move between -10 and -6. See Figure 18.12.

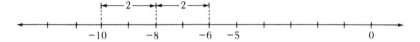

FIG. *18.12*

Therefore, if $2t$ is to be farther from -8 than 2, then $2t$ must be either to the right of -6 or to the left of -10. So either

$$2t > -6 \quad \text{or} \quad 2t < -10$$

Dividing both of these inequalities by 2, we see that either

$$t > -3 \quad \text{or} \quad t < -5$$

See Figure 18.13.

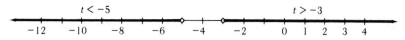

FIG. *18.13*

Therefore,

$$t < -5 \quad \text{or} \quad t > -3$$

is the solution to $|2t + 8| > 2.$

EXAMPLE: *Solve* $|10 - u^2| < 6.$

The presence of a square makes this inequality look a bit different, but all it really means is that the inequality tells us something about the position of u^2 rather than u. We can therefore use the inequality to find the range of possible values for u^2, just as in the previous examples we found the range of possible values for x, z, and $2t$. Then we'll have to figure out the range of possible values for u, which is something we covered in the last chapter (Section 17.3), and which can equally well be done by common sense.

Now $|10 - u^2| < 6$ tells us that the distance between 10 and u^2 is less than 6, which means that u^2 must be between 4 and 16. See Figure 18.14. Therefore

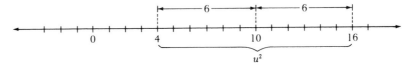

FIG. 18.14

$4 < u^2 < 16$

But if $u^2 > 4$, then u must be greater than 2 or less than -2. That is,

$u > 2$ or $u < -2$

And if $u^2 < 16$, then u must be between -4 and 4. That is,

$-4 < u < 4$

We are interested in the u's that satisfy both $u^2 > 4$ *and* $u^2 < 16$. Therefore we want just the u's that satisfy both $\{u > 2$ or $u < -2\}$ *and* $\{-4 < u < 4\}$. This means we want exactly those values that are covered by two shaded regions in Figure 18.15.

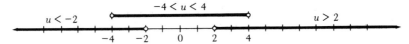

FIG. 18.15

That is,

$-4 < u < -2$ or $2 < u < 4$

See Figure 18.16.

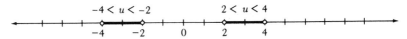

FIG. 18.16

EXAMPLE: *Solve $|x + 1| < |x - 7|$.*

On the number line, this means that the distance between -1 and x is less than the distance between 7 and x. You find the point midway between -1 and 7 (which is 3) and realize that any number to the left of 3 is closer to -1, and any number to the right of 3 is closer to 7. Therefore, x must be to the left of 3. See Figure 18.17. So the solution is $x < 3$.

FIG. *18.17*

PROBLEM SET   18.4

Write an equality or inequality with absolute values describing the following sentences, then solve it and graph the solution on the number line.

1. x is less than 4 units away from $\frac{1}{2}$.
2. x is less than 2 units away from 3.
3. x is equidistant from 5 and 11.
4. x is more than 3 units away from 5.
5. $2x$ is less than 8 units from -2.
6. x is more than twice as far from 1 as from 5.
7. $2x$ is less than $\frac{1}{3}$ unit away from 0.01.
8. x is less than twice as far from 0.02 as from 0.34.

Solve the following inequalities and plot the solution on the number line.

9. $|x| > \frac{3}{4}$
10. $|x| > 2$
11. $|x| > 0$
12. $|x - 8| < 2$
13. $|1 - x| < 4$
14. $2|x| < 9$
15. $|x + 5| > 2$
16. $|6 + x| > 20$
17. $|x + 1| > 1$
18. $|3 - x| > 2$
19. $|x| + 2 < \frac{1}{5}$
20. $2|x| + 1 < 3|x| - 2$
21. $-|3 - x| < -5$
22. $|2 - 2x| < 5$
23. $|x| < -2$
24. $|x^2 - 9| \geq 0$

25. $|3x + 1| > \frac{5}{3}$
26. $|2x - 3| \leq 3$
27. $|2x - 8| < 8$
28. $|3x + 9| > 3$
29. $|0.01 - 0.02x| \geq 1.02$
30. $0 \leq |y + 2|$
31. $|5 - 2x| \leq \frac{13}{2}$
32. $\dfrac{1}{|3 - 4x|} \geq 1$
33. $\left|\dfrac{x - 1}{3}\right| < 2$
34. $|x| \leq |x - 4|$
35. $\left|\dfrac{5}{3x}\right| > \dfrac{4}{3}$
36. $|3 - 4x| \geq 1$
37. $0 \leq |4 - x| \leq 2$

38. $x^2 > 16$ and $|x| < 20$
39. $|x| < 1$ or $3x - 4 > 1$
40. $|2x + 6| < 4$ and $|x| > 3$
41. $|2x + 6| < 4$ or $|x| > 3$
42. $|x + 2|\,|x| \geq 0$
43. $|-4x| > 6$ or $|x - 3| \leq 4$

Which of the following are true for $x < -1$?

44. $|x| + 1 > x + 1$
45. $|x| + 1 < |x + 1|$
46. $\dfrac{1}{|x| + 1} < \dfrac{1}{|x + 1|}$

Explain why there are no solutions to the following.

47. $|x - 2| < 2$ and $|x + 5| < 1$
48. $|x - 3| < 3$ and $|x - 1| > 5$
49. $|x + 1| > 8$ and $|x - 4| < 2$
50. A goat is tied by a 5-unit length of rope to a stake pounded into the number line at point "2." The poor goat can only wander along the line within the bounds the rope establishes. Which one of the following accurately describes the interval the goat can travel? (G is the goat's position.)

    (a) $|G + 2| \leq 5$ (d) $|2G - 4| \leq 10$
    (b) $|G - 5| \leq 2$ (e) $|3G - 6| \leq 5$
    (c) $|2 - G| \geq 5$

51. Prove that

$$|a + b| \leq |a| + |b|.$$

Note that $-|x| \leq x \leq |x|$ and also that $-a < x < a$ is equivalent to $|x| < a$.

CHAPTER 18 REVIEW

Evaluate the following expressions.

1. $|-2|$
2. $|-4| + |3|$
3. $-|-2|^3$
4. $|6| - |2 - 3| \cdot |-4|$

Write an expression that represents the distance between

5. x and $\frac{1}{2}$
6. Q and 0

$|x - 3|$ represents "the distance between x and 3." In the same way describe the following absolute value expressions.

7. $|6\frac{1}{2} - x|$
8. $|x + 4|$

For each of the following equations, give a pair of numbers a and b that will make it true. (There will be many possible answers.)

9. $|a| + b = |a + b|$
10. $\dfrac{|a|^3}{b} = \dfrac{a^3}{|b|}$

11. Explain why $\dfrac{|b|}{|a|} = \left|\dfrac{b}{a}\right|$

12. Show that $|4 - 5Q| = \tfrac{1}{2}|8 - 10Q|$.

Simplify the following.

13. $\dfrac{|3Q - 6R + 12|}{3}$

14. $2|B| - |-B|$

15. $\dfrac{|P + Q| + |3P + 3Q|}{4}$

For what signs $>, <, \geq, \leq, =$ are the following always true?

16. $\left|\dfrac{1}{a}\right| \underline{\quad} 1$ if $a < -1$

17. $\left|\dfrac{a}{b}\right| \underline{\quad} \dfrac{|a|}{b}$

Solve the following absolute value equations.

18. $|2 - x| = 5$

19. $6 - 2|x - 3| = 0$

20. $-2|x| = 6$

21. $\dfrac{|3x - 15|}{2} = 1$

22. $5(1 - |x - 1|) = 3 - 2(|x - 1|)$

Write an equality or inequality with absolute values describing the following sentences, then solve it and graph it on the number line.

23. x is less than 7 units away from $\tfrac{25}{4}$.

24. $3x$ is more than $\tfrac{8}{8}$ unit away from $\tfrac{7}{2}$.

Solve the following inequalities and plot the answers on the number line.

25. $|x| > \tfrac{7}{6}$

26. $|x + 7| < 3$

27. $|14 - x| > 2$

28. $|4x + 20| < 12$

29. $\left|\dfrac{x + 1}{-3}\right| < \dfrac{4}{5}$

30. $2|x + 10| \geq \dfrac{1}{3}$

Evaluate the following expressions.

31. $-2|-3(2)| + 7$

32. $|\tfrac{1}{3} - \tfrac{1}{5} + |\tfrac{1}{2}(-\tfrac{3}{4})| - \tfrac{2}{5}|$

Write an expression that represents the distance between

33. a^2 and -2^2

34. $\dfrac{p - qt}{r - s}$ and $\dfrac{qt - p}{s - r}$

$|x - 3|$ represents "the distance between x and 3." In the same way, describe the following absolute value expressions.

35. $\left|\dfrac{x - 1}{7}\right|$

36. $3\left|\dfrac{x - q}{p}\right|$

37. For the following equation give a pair of numbers a and b that will make it true. (There will be many possible answers.)

$$|a + b| + |a - b| + |-1| = |2a - 1|$$

38. Explain why

$$\left|\frac{2Q - 7}{-4}\right| = |2Q - 7|\left(\frac{1}{4}\right)$$

Simplify the following.

39. $4C - 4D - |D - C| - 5|-3(C - D)|$

40. $\dfrac{|m^3|}{-m}$

41. $-\left|-s - \left|s + |3s - |-s||\right| - 2s\right| + 3s$

42. For what signs $>, <, \geq, \leq, =$ is the following always true?

$$\frac{|a|}{b} \quad \frac{a}{|b|} \text{ if } b < a < 0$$

Solve the following equations.

43. $\left|\dfrac{x - 1}{3}\right| = 1$

44. $3|x| = |x - 9|$

45. $\dfrac{2|x - 6|}{-5} = \dfrac{|1 - x|}{2}$

46. Write an equality with absolute values describing the following sentence, then solve it and graph it on the number line.

x is 2 units away from $\frac{20}{3}$ and 4 units away from $\frac{2}{3}$

Solve the following inequalities and plot the answers on the number line.

47. $0 \leq |x - 3| \leq 5$

48. $|x - 1| > 2$ or $2x - 5 > -3$

49. $|x| < 2$ and $5x - 5 < 0$

50. $|2 - x| < 5$ and $3x + 1 < 2x - 4$

51. $|a^2 - 8a + 15| = 0$

52. $|8 - 2B^2| < 10$

53. $|x + 1| \cdot |x - 4| < 0$

Explain why there are no solutions to the following:

54. $|x + 3| < 1$ and $|3x - 18| < 12$

55. $\left|\dfrac{x + 2}{7}\right| < \dfrac{1}{14}$ and $|x + 1| = |x - 4|$

56. The spy in Figure 18.8 is trying to escape from his enemies in the guard tower at A and in the house at C by running along a number line. In order not to be caught, he must be in a place that is either more than twice as far from the guard tower as from the fence at B *and* be south of the house, *or* be no

nearer to the tower than 360 yards. Express algebraically the information statements given and then combine to find all of the places where the spy can safely be. Graph your results on a number line.

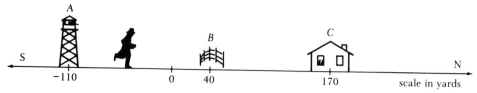

FIG. *18.18*

19 THE CARTESIAN PLANE

19.1 COORDINATES AND AXES: WHAT'S WHERE IN THE PLANE

The previous chapters have been concerned mainly with doing things with one variable. We have solved for them, substituted for them, graphed them on the number line, and simplified, reduced, and cancelled all kinds of expressions containing them. But in the "real world" (or at any rate the world in which math is used) we are often interested in the relationship between two variables—how the value of one affects or depends on the value of the other. That means that we concern ourselves with two variables and the equations that define the relation between them.

Without a doubt the best way of "seeing" the relationship between two variables is to draw a graph of them, called *plotting a graph*. The chapter on inequalities and absolute values was full of graphs of a single variable on the number line; now we have to work on graphing pairs of numbers. This is done using the *Cartesian plane*, named after René Descartes, who invented it.

Imagine two number lines at right angles to one another, crossing at the zero on both lines, and with the positive numbers going up on the vertical line and to the right on the horizontal line, as in Figure 19.1. The point at which the lines cross is called the *origin*; the horizontal number line is the *x axis*, the vertical one is the *y axis*, and together they are called the *axes*.

The points on the plane are now labeled as follows. Suppose you go from the origin to some point, P, by traveling 4 units to the right along the x axis and then 3 units up. P is then denoted by (4, 3), as shown in Figure 19.2.

Of course you get to the same point P if you first go 3 units up, and then 4 units in a direction parallel to the x axis; as in Figure 19.3.

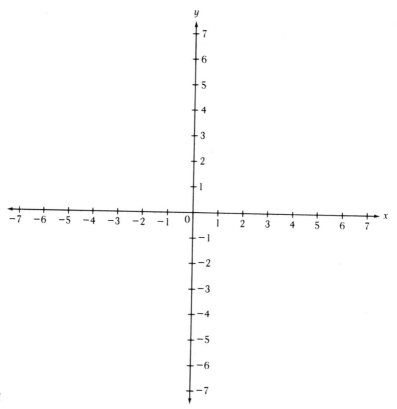

FIG. *19.1*

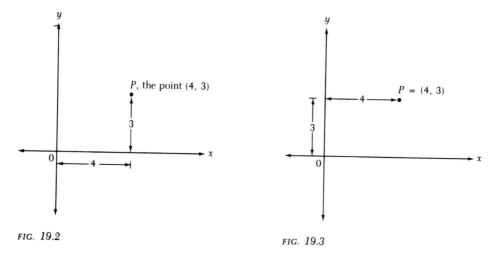

FIG. *19.2*

FIG. *19.3*

However, if you go 3 units horizontally and 4 vertically, you definitely do *not* end up at the same point, but at a point Q, which is denoted by (3, 4). See Figure 19.4.

Therefore, the order in which the numbers are written is of the greatest importance: (3, 4) and (4, 3) represent different points. For this reason symbols like (3, 4) are called *ordered pairs* of numbers.

If the first number in an ordered pair is negative, this means go so many units to the left instead of the right. If the second number is negative, then go downward instead of upward. Thus (−2, 5), (3, −5), and (−3, −6) are as shown in Figure 19.5.

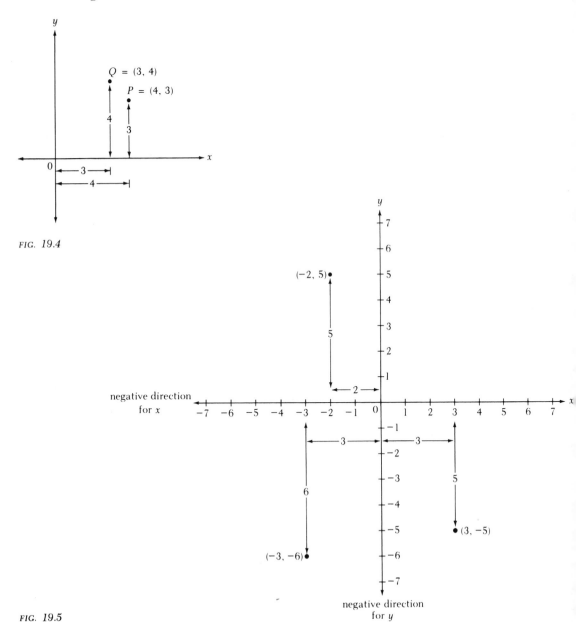

FIG. 19.4

FIG. 19.5

In general:

> (x, y) represents the point reached by going x units horizontally, and y units vertically.

Marking on the plane the point represented by a given ordered pair is called *plotting the point*. x is sometimes called the *abscissa* of the point and y the *ordinate*. It is more usual (and more reasonable) to call x the x *coordinate*, and y the y *coordinate*. Using this method, coordinates can be assigned to any point on the plane, which is then called a Cartesian plane.

The axes divide the plane into four quarters, or *quadrants*, which for some reason are always labeled by the Roman numerals shown in Figure 19.6.

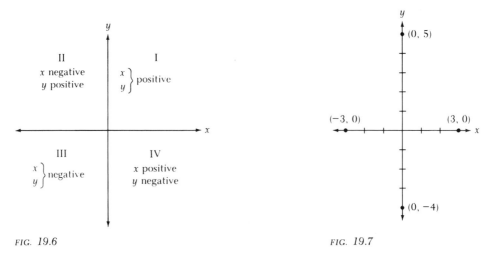

FIG. 19.6   FIG. 19.7

Therefore $(1, 7)$ is in quadrant I; $(2, -6)$ is in quadrant IV; $(-\frac{1}{2}, -\frac{1}{2})$ is in quadrant III; and $(-5, \frac{1}{2})$ is in quadrant II.

A point on one of the axes has 0 as one of its coordinates, and the origin has coordinates $(0, 0)$. To get to the point $(3, 0)$ from the origin, you move 3 units in the x direction and none vertically, and so end up on the x axis; for similar reasons, $(0, 5)$ represents a point on the y axis. See Figure 19.7 above.

In general,

$(x, 0)$ represents a point lying on the x axis

$(0, y)$ represents a point lying on the y axis

Where Things Are in the Plane

Having learned how to find individual points in the plane, we now move on to finding a whole group of points.

For example: *Where are all the points whose x coordinate is 3?* The points we want are ones like $(3, 1)$, $(3, 5)$, $(3, 0)$, $(3, -1)$, $(3, 3)$. If we plot them on a graph, a pattern emerges, as shown in Figure 19.8

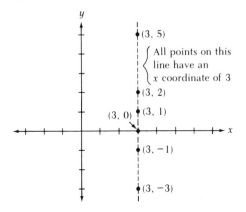

FIG. *19.8*

Clearly, all the points with an x coordinate of 3 lie on a vertical line through the point $(3, 0)$. If we draw in the whole line, rather than just mark off the particular points we chose, we have a graph that represents *all* the points with an x coordinate of 3. Since all the points on this line have $x = 3$, we say that *the line is the graph of the equation $x = 3$.*

EXAMPLE: *Graph all the points with $y = -5$.*

It will be no huge surprise that this comes out to be a horizontal line through $(0, -5)$, as shown in Figure 19.9.

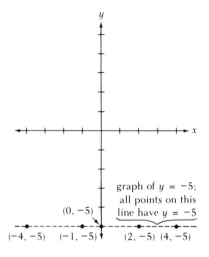

FIG. *19.9*

EXAMPLE: *Graph all the points with x ≥ 3.*

This time *x* doesn't have to be exactly equal to 3—greater than 3 will do as well. Consequently, possible points are (3, 1), (4, 1), (7, 2), (5, −3), and (8, −5). In fact, anything to the right of the line *x* = 3 will do, and so the graph of *x* ≥ 3 is a whole region; see Figure 19.10.

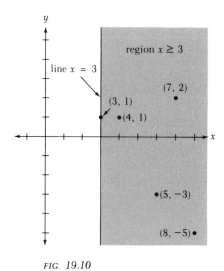

FIG. 19.10

PROBLEM SET   19.1

1. Plot the following points on a Cartesian plane and determine the quadrant that each point lies in.

(a) (2, 3) (b) (4, −1) (c) (0, 2)   (d) (−3, 4) (e) (−1, −6)
(f) (4,0)   (g) (−8, 1) (h) (0, −3) (i) (5, 5)   (j) (−5, −5)
(k) (1, 11) (l) (9, 2)   (m) (−8, $\frac{3}{2}$) (n) ($\frac{5}{3}$, −$\frac{2}{3}$) (o) (−$\frac{11}{4}$, −$\frac{9}{2}$)

Plot and label each of the following sets of points on a Cartesian plane.
2. All points having an *x* coordinate of 2
3. All points having a *y* coordinate of −3
4. All points having a *y* coordinate of 6
5. All points having an *x* coordinate of −$\frac{4}{3}$

6. All points having an x coordinate of 0
7. All points having a y coordinate of $\frac{7}{2}$

For problems 8–15, plot each region on a different Cartesian plane.

8. The region where x is greater than or equal to 0
9. The region where $y \leq 0$
10. The region where $x \leq 2$
11. The region where $y \geq -3$
12. The region where y is greater than 1 and less than 8
13. The region where $x \geq 2$ and y is between 3 and 4
14. The region where $-3 \leq x < \frac{1}{2}$ and y is greater than -1
15. The region where $-4 \leq x < -1$ and $-\frac{3}{2} < y < 8$

Draw the following figures on different Cartesian planes, and label the coordinates of all the vertices (corners):

16. A rectangle with sides 6 and 7, lying entirely in the first quadrant, with one vertex at the origin

17. A rectangle with sides 3 and 4, lying in the first and fourth quadrants, with two vertices at (2, 1) and (5, 1).

18. A square that is centered at the origin and has one vertex at $(-4, 4)$.

For Problems 19–26, use the graph in Figure 19.11.

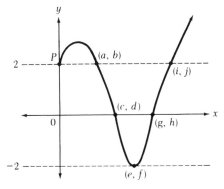

FIG. *19.11*

19. What are the coordinates of P?
20. If $x = a$, what is y?
21. What is f?
22. $h = ?$
23. What value(s) of x make $y = 2$?
24. For what value(s) of x is $y < 0$?
25. For what value(s) of x is y between -2 and 2?
26. For what value(s) of x is $|y| < 2$?

For Problems 27–34 use the graph in Figure 19.12.

27. For what values of x is y less than 2?
28. For what values of x is $y < 0$?
29. What is the value(s) of x when $y = 0$?
30. What is the value(s) of y when $x = 0$?
31. Where does the line $y = 2$ cut the graph?
32. For what values of x is $|y| < 2$?
33. What is y when x is 3?
34. What is the least value y attains on the portion of the graph shown? Where does it attain this value (that is, at what point)?

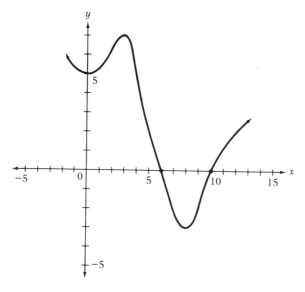

FIG. *19.12*

19.2   GRAPHING EQUATIONS: BASIC SHAPES

As you move around in the Cartesian plane, the coordinates of the point representing your position vary. The end of the last section showed how restricting one of the coordinates of your position gives a graph that is a line or region (for example, restricting the x coordinate to 3 gave a line). The real power of graphing, however, is in showing relationships between two variables; in this section we will see how to get a graph by restricting the coordinates to those satisfying a given relationship. For example, the relationship might be that each y coordinate should be equal to its corresponding x coordinate, which is usually expressed by the equation $y = x$. To graph this relationship, we pick out and mark exactly those points on the plane whose x and y coordinates are equal, and these make up the graph. In practice, we first list a number of typical points whose coordinates do satisfy the given relationship (in this case, that they be equal), and then plot them on a graph and see what pattern emerges. Some typical points with $y = x$ are

$$(0, 0), (1, 1), (2, 2), (5, 5), (-1, -1), (-2.5, -2.5)$$

These points are plotted in Figure 19.13.

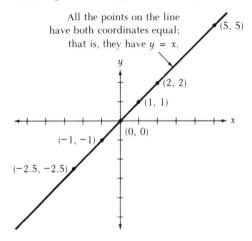

FIG. 19.13

All the points with x and y coordinates equal clearly lie on a diagonal line through the origin. This line is called the graph of the equation $y = x$, and the coordinates of the points lying on the line are said to satisfy the equation $y = x$.

The Equation $y = x + 1$

If the relationship between the coordinates had been more complicated than in the above example, it might not have been so easy to pick out the "typical points." For example, if the relationship were $y = 2x^3 - 3$, or worse, $4y - (x - 1)^2 = 0$, it would be pretty impossible to list off the top of your head a bunch of points whose coordinates satisfied the equation.

What we do then is choose some values for one of the coordinates (usually *x*) and calculate values for the other one from the relationship. It is usual to list these pairs of numbers in a *table of values* with corresponding values alongside one another, so that you can see what points to plot.

For example, to graph the equation $y = x + 1$:

$$\text{If } x = 0, \quad y = 0 + 1 = 1$$
$$x = 1, \quad y = 1 + 1 = 2$$
$$x = 2, \quad y = 2 + 1 = 3$$
$$x = 3, \quad y = 3 + 1 = 4$$
$$x = -1 \quad y = -1 + 1 = 0$$
$$x = -2, \quad y = -2 + 1 = -1$$

So Table 19.1 is the table of values.

Table 19.1

x	$y = x + 1$
0	1
1	2
2	3
3	4
−1	0
−2	−1

Now plot the points, as in Figure 19.14. Since they are obviously on a straight line, and since any other points that you decided to fill in would ob-

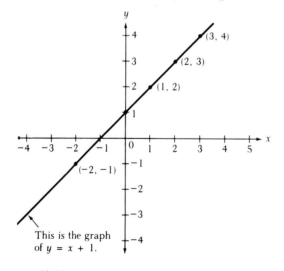

This is the graph of $y = x + 1$.

FIG. 19.14

viously be on the same straight line, you can just draw in the whole line. This will then go through every point whose y coordinate is one more than its x coordinate.

Now, what is worth noticing about the graph? First, that it is a straight line, which indicates (in this case) that x and y increases steadily together. Second, the graph cuts the x axis at the point $(-1, 0)$, which means that we say that its *x intercept is −1*. Analogously, we say that its *y intercept is 1* because the graph cuts the y axis at $(0, 1)$.

But the most important thing to understand is how to "read" the graph. A graph can tell you many of the same things as the equation, and sometimes much more quickly. So it is important to know how to get information directly off the graph.

For example, you can use the graph to find out what x and y values correspond to one another. From the equation you already know that the value corresponding to $x = 4$ is $y = 4 + 1 = 5$, but here's how you get it from the graph.

You are looking for the y value of the point on the graph that has $x = 4$. To find that you start at 4 on the x axis and go vertically up until you hit the graph at P (see Figure 19.15). You then find the y coordinate of P by traveling horizontally to the y axis. Since you end up at $y = 5$, this is the value you need.

The graph can equally well be used to get x values from y. Starting with $y = 4$, travel horizontally over to the graph, meeting it at Q, and then go downward to the x axis. You find yourself at $x = 3$, which is therefore the value corresponding to $y = 4$. This is just what we expect from the equation, since $y = 4$ means $4 = x + 1$ or $x = 3$.

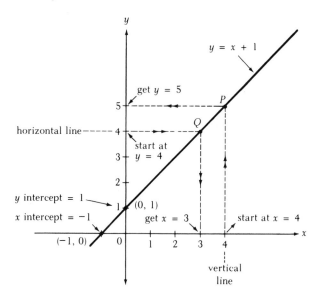

FIG. 19.15

The Equation $y = x^2$

To plot a graph of $y = x^2$ we need to mark the points whose y coordinates are the square of their x coordinates—for example, $(1, 1)$ or $(2, 4)$. We will again make a table of values, mark the points on a graph, and see what pattern emerges. The table of values is shown in Table 19.2

Table 19.2

x	$y = x^2$
-3	9
-2	4
-1	1
0	0
1	1
2	4
3	9

Now plot the points, as shown in Figure 19.16. Since they all lie on a smooth curve, you can believe that any other points that you might mark would lie on this curve. Therefore if you draw in the whole curve, you will include all the points whose y coordinates are the square of their x coordinates.

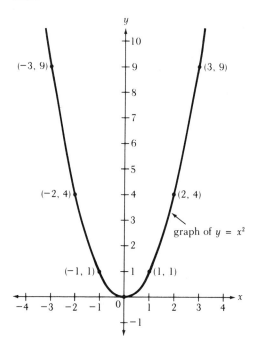

FIG. *19.16*

The graph of $y = x^2$ is *symmetric about the y axis*, which means that if you imagine a mirror along the y axis, the left-hand side of the graph is the reflection of the right in the mirror. Both the x and y intercepts are zero, because the graph cuts the axes only at the origin.

You will also notice that the graph never goes below the x axis. This means that nowhere on the graph is y negative, because only the points below the x axis are those with negative y's. This should come as no surprise when you realize that y is defined as the square of x, and so cannot be negative.

Reading off the graph of $y = x^2$, you see that if $x = 1$, then $y = 1$ (which is reasonable since $x = 1$, gives $y = (1)^2 = 1$). Using the graph the other way round, and starting with $y = 4$, we discover that there are two ways we can travel over to the graph, one putting us at P and one at Q (see Figure 19.17). P clearly has an x coordinate of 2, and Q of -2. So we conclude that there are two x values (namely, 2 and -2) corresponding to $y = 4$.

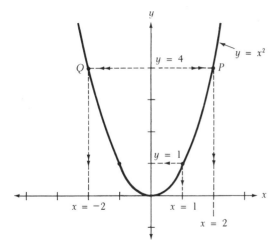

FIG. *19.17*

The algebra leads to exactly the same result. If we substitute $y = 4$ into the relationship, we get the equation $4 = x^2$ to solve for x, and, being quadratic, this equation has two roots, namely, 2 and -2. You can also see that if we started with $y = -4$ and tried to find the corresponding x's from the graph, we'd be in trouble. Moving horizontally from -4 on the y axis does not bring you to the graph at all, so it seems to be impossible to find an x corresponding to $y = -4$. But again this is just what you would expect from the algebra. Substituting $y = -4$ gives $-4 = x^2$, an equation that has no real solutions and so no x values.

The Equation $y = x^2 - 1$

This can be graphed by making a table of values (Table 19.3) and plotting points as usual.

Table 19.3

x	$y = x^2 - 1$
-3	8
-2	3
-1	0
0	-1
1	0
2	3
3	8

The points can all be connected by a smooth curve; as in Figure 19.18.

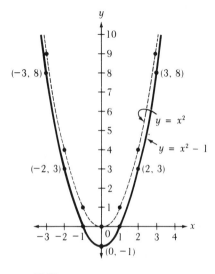

FIG. 19.18

The point of this example is to show that the graph of $y = x^2 - 1$ is exactly the same shape as $y = x^2$, only moved down by 1. This means that all the way along the graphs corresponding points (that is, ones with the same x values) are exactly 1 unit apart.

The Equation $y = x^3$

First make a table of values; as given in Table 19.4.

Table 19.4

x	$y = x^3$
-2	-8
-1	-1
0	0
1	1
2	8

Then plot points and join them to get the graph shown in Figure 19.19.

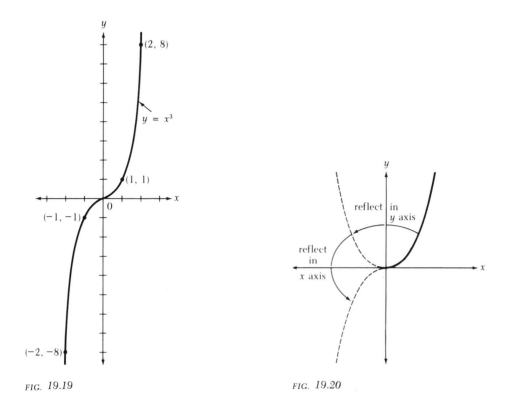

FIG. *19.19*   FIG. *19.20*

This graph clearly climbs much faster than that of $y = x^2$. At $x = 3$, $y = x^3$ has reached 27, while $y = x^2$ is only 9. The x and y intercepts are both 0, since the only place the graph cuts either axis is at the origin.

The graph of $y = x^3$ is not symmetric about the y axis in the same way as the graph of $y = x^2$. However, $y = x^3$ clearly does have some kind of symmetry, and is said to be *symmetric about the origin*. Geometrically, this means that the part of the graph in the third quadrant can be obtained by reflecting the part in the first quadrant twice, first in the y axis and then in the x axis (see Figure 19.20 above).

The Equation $y = \dfrac{1}{x}$

If we make a table of values for this equation, things go fine until we decide to put in $x = 0$. Then

$$y = \frac{1}{0}$$

which is undefined, so clearly we can't let x be 0. Otherwise the table of values looks like Table 19.5.

Table 19.5

x	$y = \frac{1}{x}$
-4	$-\frac{1}{4}$
-3	$-\frac{1}{3}$
-2	$-\frac{1}{2}$
-1	-1
0	?
1	1
2	$\frac{1}{2}$
3	$\frac{1}{3}$
4	$\frac{1}{4}$

Look at Figure 19.21, where we have plotted the points so far.

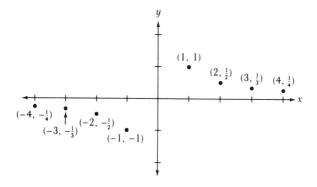

FIG. *19.21*

The problem now is how to join them up. Plotting more points to the right of $(4, \frac{1}{4})$, and to the left of $(-4, -\frac{1}{4})$ will convince you that the graph creeps closer and closer to the x axis as x gets larger and larger (either positively or negatively). The x axis is said to be an *asymptote*, meaning that the graph gets ever closer to it, but never actually hits it. So we now have Figure 19.22.

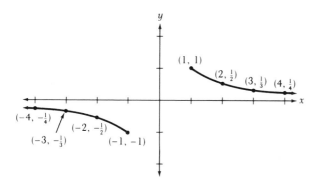

FIG. *19.22*

But what happens when x is between -1 and 1? The first thing we have to realize is that there is *no y* value when $x = 0$, and therefore there can be no point on the graph whose x coordinate is 0. Since all the points on the y axis have x coordinates 0, this means that there is no point on the y axis on the graph—in other words, the graph does not cut the y axis. Now this poses a problem, because how can we join up the part of the graph in the first quadrant with that in the third quadrant without crossing the y axis? The answer is that we can't, and so the graph must be in two parts.

However, there is certainly more to the graph than we have drawn so far. In order to find out what happens between $x = 0$ and $x = 1$, let us find y when $x = \frac{1}{2}, \frac{1}{3}$, and so on. Some values are listed in Table 19.6.

Table 19.6

x	$y = \frac{1}{x}$
$\frac{1}{2}$	2
$\frac{1}{3}$	3
$\frac{1}{4}$	4

As we move from $x = 1$ back toward $x = 0$, y is clearly growing. The closer we get to 0, the larger y gets, and so the graph must go up indefinitely as x decreases toward 0. Much the same thing happens when x is between -1 and 0; as shown in Table 19.7.

Table 19.7

x	$y = \frac{1}{x}$
$-\frac{1}{2}$	-2
$-\frac{1}{3}$	-3
$-\frac{1}{4}$	-4

Therefore, as x increases from -1 to 0, y gets more and more negative, and the graph goes down and down, as shown in Figure 19.23.

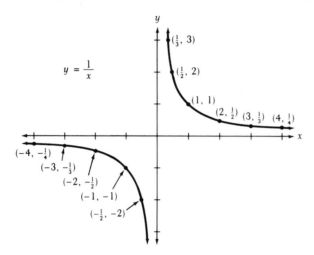

FIG. 19.23

The y axis is also an asymptote, since the graph gets closer and closer to it, but never reaches it. Notice that the graph has come out symmetric about the origin.

Another way of saying that as x gets larger, y gets closer to 0 is to write

$$\text{As } x \to \infty, \quad y \to 0.$$

The arrow, $\to$, means "tends to" or "goes toward," and ∞ means infinity, which is to make you think of some number way out at the right-hand end of the number line. We can also write that as x gets more and more negative, y gets closer to 0 as

$$\text{As } x \to -\infty, \quad y \to 0.$$

Here $-\infty$ means negative infinity, which is to make you think of some number way out to the left-hand end of the number line. We can also write

$$\text{As } x \to 0^+, \quad y \to \infty$$

meaning that when x is greater than 0 but getting closer to it, y gets very large and positive. The symbol $\to 0$ means getting closer to zero, and the $+$ means that x is always to the right of zero. Therefore

$$\text{As } x \to 0^-, \quad y \to -\infty$$

means that when x is smaller than 0 but getting closer to it, then y is getting very negative.

The Equation $y = |x|$

A table of values is shown in Table 19.8.

Table 19.8

x	$y = |x|$
-3	3
-2	3
-1	1
0	0
1	1
2	2
3	3

Plotting them gives Figure 19.24.

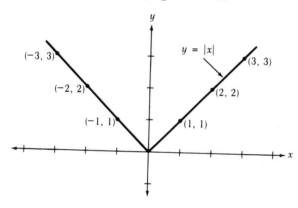

FIG. *19.24*

Notice that 1 and -1 give the same value of y, as do 2 and -2, and 3 and -3, which makes this graph symmetric about the y axis. The fact that $|x|$ is always positive or 0 means that the graph is entirely above the x axis, where y is positive.

These examples show that graphs of different equations can have all kinds of different shapes. We now pick up a couple useful tools—the distance and midpoint formulas (Chapter 20)—before investigating lines in great detail in Chapter 21.

PROBLEM SET 19.2

Graph each of the following by plotting points.

1. $y = x + \frac{3}{2}$
2. $y = x - 3$
3. $y - 1 = x^2$
4. $y = (x - 2)^2$
5. $y = |x| + 2$
6. $y = |x - 3|$

7. $y^2 = x - 3$

8. $(y + 1)^2 = x$

9. $y = \dfrac{1}{x} + 3$

10. $y = \dfrac{1}{x - 2}$

11. $x^2 + y^2 = 1$

12. $x^2 + y^2 = 4$

13. $y = \sqrt{9 - x^2}$

14. $y = (x + 3)^3$

15. $y = x^3 + 1$

CHAPTER 19 REVIEW

For Problems 1 to 5 use Figure 19.25.

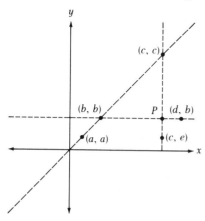

FIG. 19.25

1. Which is larger, a or b?
2. Which is larger, d or c?
3. Which of the five labeled points has the smallest x coordinate?
4. Which of the five labeled points has the largest y coordinate?
5. What are the coordinates of P?

For Problems 6–8 use Figure 19.26, and assume d and r are positive.

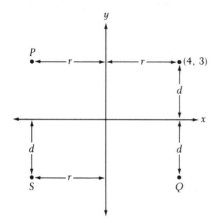

FIG. 19.26

6. What are the coordinates of P?
7. What are the coordinates of Q?
8. What are the coordinates of S?

For Problems 9 and 10 use Figure 19.27.

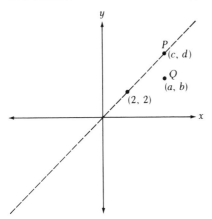

FIG. 19.27

9. Which is smallest: a, b, c, or d?
10. If $d = 5$ and $b = a - 1$, what are the coordinates of P? Of Q?

For Problems 11 and 12 use Figure 19.28.

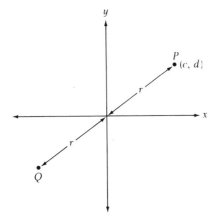

FIG. 19.28

11. What are the coordinates of Q?
12. What are the dimensions of the rectangle centered at the origin and with vertices at P and Q?

For Problems 13–22, plot each region on a different Cartesian plane.

13. The region where $x \geq 2$ or y is less than 5.
14. The region where x is between -3 and 3, but y is not between -1 and 1

15. The region where $-8 < x < -6$ or $-2 < y < 7$
16. The region where $|x| < 2$ and $|y| < 1$
17. The region where $|x| > 4$ and $|y| > 2$
18. The region where $|x| > -1$ and $|y| < 3$
19. The region where $y > x$
20. The region where $x > y$
21. The region where $|x| < 2$ and $x < y$
22. The region where $|x| < y$ and $y \geq 2$

Draw the following figures on different Cartesian planes, and label all vertices.

23. A square of area 25 which is centered at $(2, 3)$
24. A rectangle with sides 4 and 5 which is centered at $(-1, -5)$

For Problems 25–31 use the graph in Figure 19.29

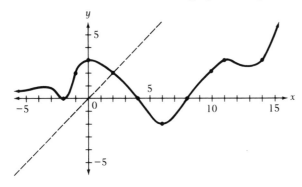

FIG. 19.29

25. Find y when $x = 0$.
26. Find x when $y = 3$.
27. Find y when $x = 11$.

28. Find the x's such that $y = 0$.
29. Find the x's such that $y > x$.
30. Find the x's such that $y > 2$.

31. How many times does the line $y = -1$ cut the graph?

Graph each of the following by plotting points.

32. $y = 2x + 1$
33. $y = -x^2 + 1$
34. $y = -x^3 - 2$
35. $y = \dfrac{2}{x}$
36. $x = 2y^2 + 1$
37. $(3x)^2 + (3y)^2 = 81$
38. $3x^2 + 108y^2 = 27$

39. $x^2 + \dfrac{y^2}{4} = 1$
40. $x = |y| + 3$
41. $x = \dfrac{-1}{y} + 1$
42. $(1.1x)^2 + (1.1y)^2 = 4.84$
43. $|y| = |x|$
44. $y = \sqrt{4 - x^2}$
45. $x = \sqrt{16 - y^2}$

20 DISTANCES AND MIDPOINTS IN THE PLANE

20.1   DISTANCE

One of the tools that we needed for working on the number line was a way of finding the distance between two points, and this was the reason that we bothered with absolute values. It turns out to be just as necessary to know about distance in the plane, and that's why this section exists.

If you have two points, a and b, on the number line, you know how to find the distance between them: It is $|a - b|$. The problem now is to find the distance between two points in the plane.

Suppose we look at the points (2, 3) and (7, 3) in Figure 20.1. Since they both lie on the same horizontal line (which goes through 3 on the y axis), we can pretend they lie on a number line. The distance between the points is therefore the difference in the x coordinates, or 5.

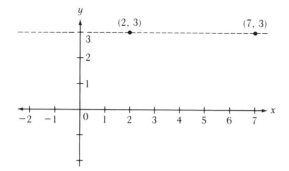

FIG. 20.1

To generalize, the distance between $(a, 3)$ and $(c, 3)$, which lie on the same horizontal line, is $|c - a|$. The absolute value comes in because we are measuring distance.

Now suppose we look at the points (2, 2) and (2, 6) in Figure 20.2. These two points can again be thought of as lying on a number line, only this time a vertical one. The distance between them is the difference between their y coordinates, or 4.

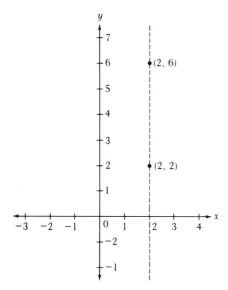

FIG. *20.2*

In general, the distance between (2, b) and (2, d) is $|d - b|$.

Now, what about the distance between two points that do not lie on either a vertical or a horizontal line? Consider, for example, the points $P = (1, 2)$ and $Q = (5, 5)$.

Since we want to find a distance along a diagonal line, and only know how to find distances along lines parallel to the axes, let us draw in a triangle with two sides parallel to the axes, and the third side joining (1, 2) to (5, 5), as in Figure 20.3. The third vertex of the triangle, R, is on the same horizontal line as P, and therefore has the same y coordinate as P—namely, 2. R is on the same vertical line as Q, and so has an x coordinate of 5. Therefore R's coordinates are (5, 2), as shown in the figure.

Triangle PQR is a right triangle, because in a Cartesian plane the axes are at right angles to one another, so angle PRQ is a right angle. Remember that in a right triangle with sides a, b, c, we can apply Pythagoras' theorem, which says:

$$c^2 = a^2 + b^2$$

where c, a, and b are as shown in Figure 20.4.

Therefore in our triangle:

$$(PQ)^2 = (PR)^2 + (RQ)^2 \qquad \text{(where } PQ = \text{distance from } P \text{ to } Q\text{)}$$

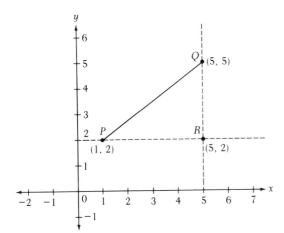

FIG. 20.3

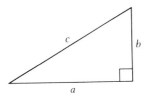

FIG. 20.4

which is very useful because it relates the distance that we want to find, PQ, to two distances, PR and RQ, that we can find because they are measured parallel to one of the axes. Since P and R are on the same horizontal line, $PR = |5 - 1| = 4$; since R and Q are on the same vertical line, $RQ = |5 - 2| = 3$. Therefore the triangle PQR looks like Figure 20.5.

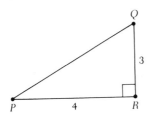

FIG. 20.5

Pythagoras' theorem tells us that

$$(PQ)^2 = 4^2 + 3^2$$
$$(PQ)^2 = 16 + 9$$
$$(PQ)^2 = 25$$
$$PQ = 5 \quad \text{or} \quad -5$$

Since a distance can't be negative,

$$PQ = 5$$

The Distance Formula

It would be most useful to have a formula for the distance between an arbitrary pair of points. We will therefore use the same method as in the example above to find the distance between a point P with coordinates (x_1, y_1) and a point Q with coordinates (x_2, y_2).

Let us draw a diagram and put in the third point R. Since R is on the same horizontal line as P, it has the same y coordinate as P, namely, y_1; and since R is on the same vertical line as Q, it has x coordinate x_2. Therefore R is the point (x_2, y_1), as shown in Figure 20.6.

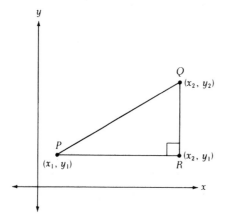

FIG. 20.6

The distance from P to R is being measured on a horizontal line along which only x is changing, so

$$PR = |x_2 - x_1|$$

Similarly, R and Q are on the same vertical line, and therefore

$$RQ = |y_2 - y_1|$$

We now have the triangle in Figure 20.7.

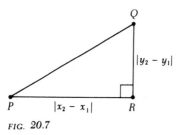

FIG. 20.7

Pythagoras tells us that

$$(PQ)^2 = (|x_2 - x_1|)^2 + (|y_2 - y_1|)^2$$

But since the square of a number and the square of its negative are the same, the absolute value signs are now unnecessary, and so

$$(PQ)^2 = (x_2 - x_1)^2 + (y_2 - y_1)^2$$

Since PQ is a distance, we must take the positive square root:

$$PQ = \sqrt{(x_2 - x_1)^2 + (y_2 - y_1)^2}$$

This is called the *distance formula*.

EXAMPLE: *Find the distance between $(-1, 3)$ and $(4, -9)$.*

Let $(-1, 3)$ be $P = (x_1, y_1)$ and $(4, -9)$ be $Q = (x_2, y_2)$. Then

$$PQ = \sqrt{(x_2 - x_1)^2 + (y_2 - y_1)^2}$$
$$= \sqrt{[4 - (-1)]^2 + (-9 - 3)^2}$$
$$= \sqrt{(4 + 1)^2 + (-12)^2}$$
$$= \sqrt{25 + 144}$$
$$= \sqrt{169}$$
$$= 13$$

You should notice that it doesn't matter in the least which point you call (x_1, y_1) and which you call (x_2, y_2). All that matters is that you are consistent—that x_1 and y_1 are coordinates of the same point.

Just so that you're convinced that it doesn't matter which point is which, try the last example the other way round. Let $(4, -9)$ be $P = (x_1, y_1)$ and $(-1, 3)$ be $Q = (x_2, y_2)$. Then

$$PQ = \sqrt{(x_2 - x_1)^2 + (y_2 - y_1)^2}$$
$$= \sqrt{(-1 - 4)^2 + [3 - (-9)]^2}$$
$$= \sqrt{(-5)^2 + (3 + 9)^2}$$
$$= \sqrt{25 + 144}$$
$$= \sqrt{169}$$
$$= 13$$

EXAMPLE: *Find the distance between (a, 3) and (c, 3).*

We already know the answer to this one—it's $|c - a|$—because $(a, 3)$ and $(c, 3)$ lie on the same horizontal line. We'll do it again to show that the distance formula gives the same answer.

Let $(a, 3) = (x_1, y_1)$ and $(c, 3) = (x_2, y_2)$. Then

$$\text{distance} = \sqrt{(x_2 - x_1)^2 + (y_2 - y_1)^2}$$

$$= \sqrt{(c - a)^2 + (3 - 3)^2}$$

$$= \sqrt{(c - a)^2}$$

Now, squaring a number and then taking the square root gives the same result as taking the absolute value. Therefore,

$$\sqrt{(c - a)^2} = |c - a|$$

so

$$\text{distance} = |c - a|$$

PROBLEM SET   20.1

Find the distance between:

1. $(0, 5)$ and $(0, -1)$
2. $(3, 0)$ and $(-2, 0)$
3. $(4, 2)$ and $(-\frac{3}{2}, 2)$
4. $(3, 6)$ and $(3, -\frac{7}{2})$
5. (a, b) and (a, c)
6. (d, e) and (f, e)
7. $(2, -1)$ and $(5, 3)$
8. $(5, -1)$ and $(2, 3)$
9. $(6, 2)$ and $(1, -10)$
10. $(3, 2)$ and $(6, 7)$
11. $(4, -4)$ and $(-4, 4)$
12. $(\frac{5}{2}, -\frac{3}{4})$ and $(\frac{7}{4}, -\frac{3}{2})$
13. $(\frac{19}{3}, \frac{1}{2})$ and $(-\frac{4}{3}, -\frac{5}{2})$
14. $(1.4, 2.3)$ and $(3.1, -1.0)$
15. $(-2.7, -3.61)$ and $(\frac{3}{2}, -\frac{7}{4})$
16. $(\frac{16}{3}, -\frac{1}{3})$ and $(-\frac{8}{3}, 2)$
17. $(1.1, 1.1)$ and $(-1.3, 1.8)$
18. $(\pi, -\frac{\pi}{2})$ and $(-\frac{3}{2}\pi, 2\pi)$
19. $(\frac{1}{2}, \frac{2}{3})$ and $(\frac{3}{4}, \frac{4}{5})$

20. Find the length of the diagonal of a rectangle whose sides are 5 and 12.
21. Find the length of the diagonal of a square of area 16.
22. Find the length of the hypotenuse of an isosceles right triangle whose other sides are $4\sqrt{2}$.

23. Show that the three points $(2, 3)$, $(-4, -3)$, and $(6, -1)$ are the vertices of a right triangle. *Hint:* Use the distance formula three times and show that the Pythagorean theorem holds.

24. What is the length of the longest straight line that can be drawn on a sheet of $8\frac{1}{2} \times 11$ typing paper?

20.2   MIDPOINT

Midpoints on the Number Line

If we are given two points on the number line, we find their *midpoint* (the point half way between them) by averaging. For example, the average of 3 and 9 is

$$\frac{3+9}{2} = 6$$

and 6 is certainly halfway between 3 and 9 (see Figure 20.8).

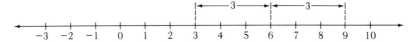

FIG. *20.8*

This even works when one of the numbers is negative. For example, the average of -8 and 4 is

$$\frac{-8+4}{2} = \frac{-4}{2} = -2,$$

and you can see from Figure 20.9 that -2 is indeed halfway between -8 and 4.

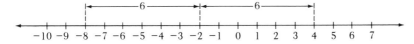

FIG. *20.9*

In general:

> The midpoint of a and b is
> $$\frac{a+b}{2}$$

Justification of the Midpoint Formula   Let us suppose that a is less than b. Then a is to the left of b, and the distance between them is $b - a$. The midpoint is then

$$\left(\frac{b-a}{2}\right)$$

to the right of a, as shown in Figure 20.10.

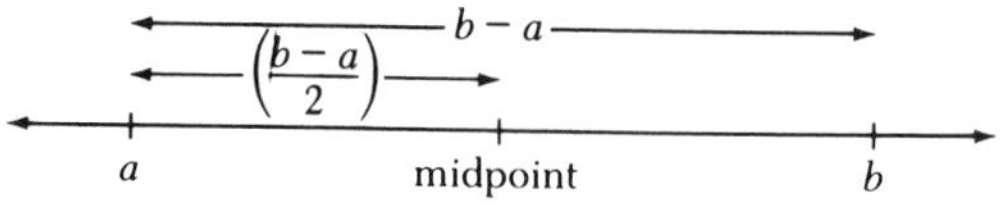

FIG. 20.10

Therefore the midpoint is

$$a + \frac{(b-a)}{2} = \frac{2a}{2} + \frac{b-a}{2} \qquad \text{(adding fractions)}$$

$$= \frac{2a + b - a}{2}$$

$$= \frac{a+b}{2}$$

Midpoints in the Plane

Now we want to find the midpoint of two points lying in a plane. This sounds like it involves using the distance formula, but since that involves tiresome things like square roots, let us do an example and see if we can avoid it.

Suppose we want to find the midpoint between $P = (2, 1)$ and $Q = (8, 3)$. First let us draw a diagram, as in Figure 20.11.

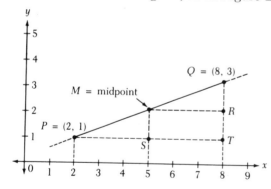

FIG. 20.11

T has the same x coordinate as Q, and the same y coordinate as P, so

$$T = (8, 1)$$

Now look at the triangles PMS and PQT. They are the same shape as one another, so PQT (the large one) is a magnification of PMS (the small one). Since M is the midpoint of PQ, PQ must be twice PM, so the magnification factor is 2. Therefore PT is twice PS, making S the midpoint of PT.

Now P, S, and T lie on a horizontal line, so you can find the x-coordinate of S by averaging the x coordinates of P and T (as on the number line).

But $P = (2, 1)$ and $T = (8, 1)$, so

$$S = \left(\frac{2+8}{2}, 1\right)$$

that is,

$$S = (5, 1)$$

By exactly analogous reasoning with triangles MQR and PQT, we find that R is the midpoint of QT. Since Q, R, and T all lie on a vertical line, you can get R's y coordinate by averaging the y coordinates of Q and T.

Now $Q = (8, 3)$ and $T = (8, 1)$, so

$$R = \left(8, \frac{3+1}{2}\right)$$

that is,

$$R = (8, 2)$$

But M has the same x coordinate as S, and the same y coordinate as R, so

$$M = (5, 2)$$

A general formula for the midpoint, M, between $P = (x_1, y_1)$ and $Q = (x_2, y_2)$ can be derived by a method exactly similar to that used in the example above.

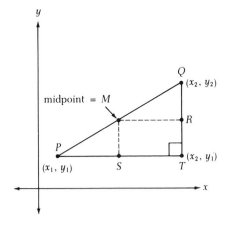

FIG. 20.12

Look at Figure 20.12. As before, S is the midpoint of PT and R is the midpoint of QT. So

$$S = \left(\frac{x_1 + x_2}{2}, y_1\right) \quad \text{and} \quad R = \left(x_2, \frac{y_1 + y_2}{2}\right)$$

M has the same x coordinate as S and the same y coordinate as R, so

$$\text{midpoint, } M = \left(\frac{x_1 + x_2}{2}, \frac{y_1 + y_2}{2} \right)$$

PROBLEM SET 20.2

Find the midpoint of the line segments joining the following pairs of points.

1. $(3, 5)$ and $(1, 1)$
2. $(-2, 3)$ and $(8, -3)$
3. $(6, 4)$ and $(2, 9)$
4. $(1, 1)$ and $(1, 7)$
5. $(1, 1)$ and $(7, 1)$
6. $(\frac{1}{4}, \frac{3}{4})$ and $(\frac{3}{4}, \frac{1}{4})$
7. $(\frac{7}{2}, \frac{8}{5})$ and $(\frac{3}{5}, -\frac{2}{3})$
8. $(-\frac{9}{4}, -\frac{1}{4})$ and $(2.2, -3.7)$
9. $(2.15, -3.98)$ and $(-3.16, 4.02)$
10. $(\frac{2\pi}{3}, \pi)$ and $(\frac{\pi}{6}, \frac{\pi}{4})$

11. Find the midpoint of the hypotenuse of a triangle with vertices $(0, 3)$, $(4, 0)$, and $(0, 0)$.

12. Find the center of the rectangle with vertices $(2, -1)$, $(2, 4)$, $(8, -1)$, and $(8, 4)$ by finding the midpoint of one of the diagonals.

13. Show that for any rectangle, the diagonals bisect one another; that is, show that they have the same midpoint. *Hint:* Align the rectangle on a Cartesian coordinate system parallel to the axes. Label the vertices connected by one diagonal (a, b) and (c, d). What are the coordinates of the other vertices?

CHAPTER 20 REVIEW

1. Show that the points $(-2, 1)$, $(2, 3)$, and $(10, 7)$ lie on a straight line. *Hint:* If they are on a line, then the distance from the first to the second plus the distance from the second to the third should equal the distance from the first to the third.

2. Find the length of the diagonal of a cube of side 3. See Figure 20.13.

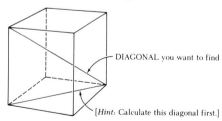

DIAGONAL you want to find

[*Hint:* Calculate this diagonal first.]

FIG. 20.13

3. Will a cardboard square of side 5 fit into a covered cubic box of side 4? Why or why not?

4. A baseball "diamond" is just a 90-foot square, with home plate and the three bases as vertices. The pitcher's rubber is 60 feet 6 inches from home plate, and lies on the line connecting home plate and second base. How far is it from the rubber to second base? From the rubber to first base? *Hint:* It may help to think of the baseball diamond as lying on a Cartesian plane, with home plate at the origin and the first- and third-base lines as the x and y axes, respectively.

5. Using similar triangles, find the two points that divide the line segment between (1, 8) and (10, 14) into three equal parts. These are the trisection points of the segment.

6. Find the trisection points of the segment between (2, 3) and $(-5, -8)$.

7. Find the general formula for the trisection points of a segment between (x_1, y_1) and (x_2, y_2).

For Problems 8–11, consider the triangle in Figure 20.14.

8. Find the midpoint of AB. Call this point C'.
9. Find the midpoint of BC. Call this point A'.
10. Find the midpoint of AC. Call this point B'.

11. The segments AA', BB', and CC' are called the *medians* of the triangle (i.e., a median connects a vertex to the midpoint of the opposite side). Find the trisection points of each of the three medians. What do you notice?

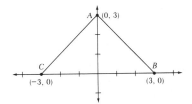

FIG. 20.14

12. Show that the point (4, 4) is a trisection point of each of the medians of the triangle with vertices at (2, 6), (9, 7), and $(1, -1)$. The point (4, 4) is called the *centroid* of the triangle.

13. Find the general formula for the coordinates of the centroid of a triangle with vertices at (x_1, y_1), (x_2, y_2), and (x_3, y_3).

Find the enclosed areas of the following figures.

14.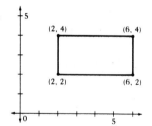

15.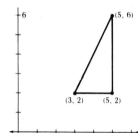

16.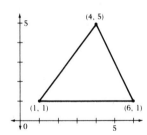

17.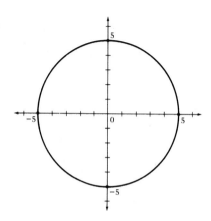

18.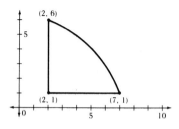

19.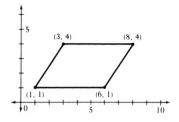

20.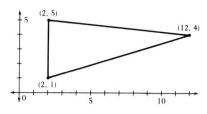

21.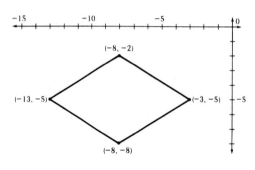

22.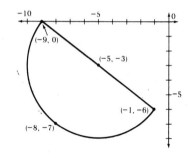

Land is frequently bought and sold on a price-per-square-foot basis. The buyer and seller of the following irregularly shaped tracts of land have agreed on a price of 50¢ per square foot, and surveyors have filed the following maps. What is the total selling price of each tract?

23.

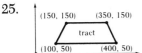

24.

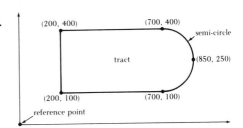

25.

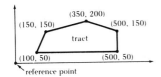

26.

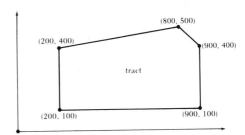

27.

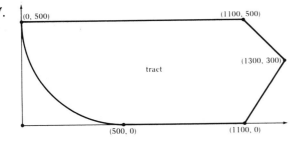

28.

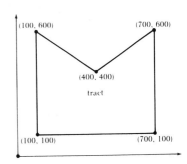

29.

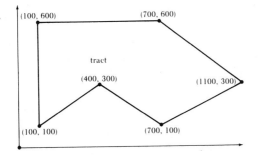

30.

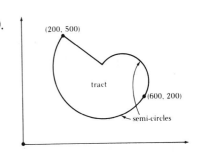

21 LINES

21.1 "STEEPNESS" AND "POSITION" OF A LINE

Imagine that you're trying to tell me—without pointing—how to draw a particular line on the Cartesian plane. What do you need to tell me in order that I may draw exactly the line that you are thinking of? One way is this. First tell me what angle the line is at. Is it a "steep" line, or does it climb slowly? Does it climb from left to right, or from right to left? But knowing what angle the line is at isn't enough, because I can draw lots of different lines at the same angle. For example, see the lines in Figure 21.1. But if you also tell me something about where the line is—a point it goes through or one of the intercepts—then I can draw your line.

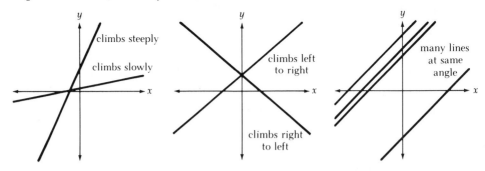

FIG. 21.1

So it seems that a line is characterized by two things: its "steepness" and its "position." Once we know those, we can draw the line, and given a line, we can say how "steep" it is, and what its "position" is.

The point of this section is to look at the equations whose graphs are lines, and to see what in the algebraic equation makes a line "steep" or "not steep," and what determines its "position." Before we can do this, however, we must know what kinds of equations to look at. As it turns out, *any linear equation in x and y has a graph that is a line*—which is why such equations are called linear.

What Determines the "Position" of a Line?

Remember the graphs of $y = x$ and $y = x + 1$ from Chapter 19, which are redrawn in Figure 21.2. Comparing these graphs, you can see that $y = x + 1$ is everywhere one unit higher up than $y = x$. The reason is that $y = x + 1$ has all its y's one greater than the corresponding y's on $y = x$.

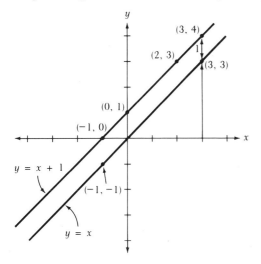

FIG. 21.2

Changing the 1 in $y = x + 1$ will move the graph up or down. For example, look on p. 454 at the graphs of $y = x + 2$, $y = x + 3$, or $y = x - 1$. The first, $y = x + 2$, has all its y's two greater than the corresponding y's for $y = x$, and so the graph of $y = x + 2$ is the graph of $y = x$ shifted up two. (If you don't really believe this, please do the graph of $y = x + 2$ with a table of values.) By similar arguments you can figure out the graphs of all of the lines shown in Figure 21.3. All these lines have the same "steepness," and the only thing that varies is their position. Also, all of these lines have equations of the form $y = x + b$. Looking at the graphs will show you that b tells you where the line cuts the y axis—so b is the y-intercept of $y = x + b$. To check this algebraically, realize that finding the y intercept means finding y when $x = 0$. Substituting $x = 0$ into $y = x + b$ gives $y = 0 + b$ or $y = b$, which is the y intercept.

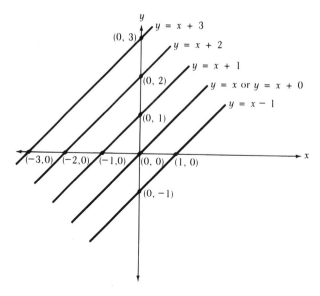

FIG. 21.3

So it is the b in $y = x + b$ that is responsible for the "position" of the line.

What Determines the "Steepness" of a Line?

Now let's look at several equations: $y = 2x + 1$, $y = 3x + 1$, $y = -x + 1$, $y = -2x + 1$, and $y = -\frac{1}{2}x + 1$. Since at the moment we have no idea what changing the coefficient of the x will do to the graph, we will have to use a table of values. For $y = 2x + 1$, the table of values is given in Table 21.1. By doing similar tables of values for the others, we can draw the graphs of the other equations. They are shown in Figure 21.4. Interestingly enough, they all turn out to be lines through $(0, 1)$ but with different "steepness."

Table 21.1

x	$y = 2x + 1$
-2	-3
-1	-1
0	1
1	3
2	5
3	7

The coefficient of x seems to determine the "steepness" of the line. In the equation $y = mx + 1$, if m is positive, the line climbs from left to right; if nega-

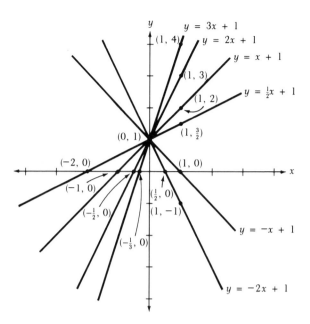

FIG. 21.4

tive, it climbs from right to left. The larger the magnitude of *m*, the "steeper" the line is.

Therefore, *the m in y = mx + 1 is responsible for the "steepness" of the line*.

If you plot $y = mx + 2$ for the same values of *m*, you get exactly the same picture but moved up by one. Each line now crosses the *y* axis at (0, 2) but has the same "steepness" as before. So the *m* in $y = mx + 2$ still represents steepness.

What about changing the coefficient of *y*? What does putting a 2 in front of the *y* do? Luckily, nothing new: The equation $2y = x + 1$ is the same as $y = \frac{1}{2}x + \frac{1}{2}$, whose graph is a line parallel to $y = \frac{1}{2}x + 1$ but moved down $\frac{1}{2}$, as shown in Figure 21.5 on p. 456.

No matter what nonzero coefficient *y* is given, you can always divide through by it and get an equation of the form $y = mx + b$ in which *m* represents steepness.

In general:

> Any equation of the form $y = mx + b$ gives a line where
>
> *b* is the *y* intercept
>
> *m* is the "steepness"

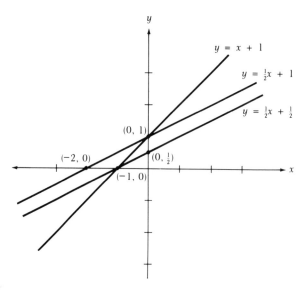

FIG. 21.5

PROBLEM SET 21.1

Graph and label the following lines on the same coordinate system.

   1. $y = x$ 4. $y = -3x$

   2. $y = -x$ 5. $y = \frac{1}{3}x$

   3. $y = 3x$ 6. $y = -\frac{1}{3}x$

Graph and label the following lines on the same coordinate system.

   7. $y = x$ 10. $y = x + 3$

   8. $y = x + 1$ 11. $y = x - 2$

   9. $y = x + \frac{1}{2}$ 12. $y = -x - 2$

Graph the line $y = 2x + b$ when:

  13. $b = 2$ 17. $b = \frac{2}{23}$

  14. $b = -1$ 18. $b = 7$

  15. $b = 0$ 19. $b = \frac{1}{30}$

  16. $b = 1$

Graph the line $y = mx + 3$ when:

  20. $m = 1$ 24. $m = 3$

  21. $m = 0$ 25. $m = 5$

  22. $m = -1$ 26. $m = \frac{1}{100}$

  23. $m = 2$

Graph the line $ay = 3x + 1$ when:

27. $a = 1$ 31. $a = 400$
28. $a = 3$ 32. $a = \frac{1}{10}$
29. $a = 0$ 33. $a = -\frac{1}{10}$
30. $a = -1$

21.2   DEFINITION OF SLOPE

From the previous discussion it must be obvious that the idea of the "steepness" of a line is of great importance. However, it is hard to compare the "steepness" of two lines or see why m should represent the "steepness" without knowing a good deal more exactly what we mean by "steepness." We need a precise definition—although it must certainly be one that fits in with our intuitive ideas.

The "steepness" of a line is meant to measure how fast it is climbing (or descending). If, for example, you wanted to convey to someone the steepness of a road, you might tell them that it rises 1 foot for every 10 feet of horizontal distance. See Figure 21.6. (This is sometimes called a "1 in 10" hill, and is, by the way, an extremely steep one.) The ratio $\frac{1}{10}$, which is called the *slope*, is a measure of the steepness of the hill. A road of slope $\frac{3}{20}$ would therefore be climbing 3 feet for every 20 horizontal feet or 3 meters for every 20 horizontal meters, and so on. See Figure 21.7. As another example, a mountain face of slope $\frac{7}{3}$ would look like Figure 21.8.

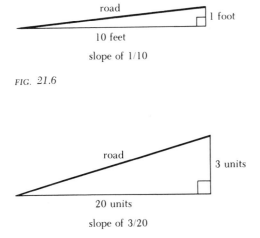

road

1 foot

10 feet

slope of 1/10

FIG. *21.6*

road

3 units

20 units

slope of 3/20

FIG. *21.7*

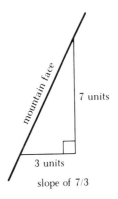

mountain face

7 units

3 units

slope of 7/3

FIG. *21.8*

So we will define the slope of a line as:

$$\text{slope} = \frac{\text{vertical change between two points on the line}}{\text{corresponding horizontal change}}$$

Suppose the points $P = (x_1, y_1)$ and $Q = (x_2, y_2)$ are on the line whose slope we want to find. See Figure 21.9. T is the point (x_2, y_1) because T has the same y coordinate as P and the same x coordinate as Q.

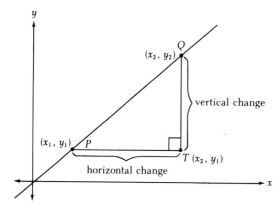

FIG. 21-9

The vertical change between P and $Q = y_2 - y_1$.
The corresponding horizontal change $= x_2 - x_1$.
So:

$$\text{slope} = \frac{y_2 - y_1}{x_2 - x_1}$$

There are three things that you should notice about this formula:

1. We are interested in the *change* in y from P to Q, represented by $y_2 - y_1$, rather than by $|y_2 - y_1|$, which represents a distance. Only by looking at changes rather than distances can we see whether y is increasing or decreasing, and knowing this is essential to seeing whether a line climbs or descends as we move from left to right.

2. *It doesn't matter which point you call (x_1, y_1) and which (x_2, y_2).*

$$\text{slope} = \frac{y_2 - y_1}{x_2 - x_1} = \frac{-(y_1 - y_2)}{-(x_1 - x_2)} = \frac{y_1 - y_2}{x_1 - x_2}$$

Therefore, if you interchange x_1 and x_2, and y_1 and y_2, you still get the same value for the slope. The only thing that is important is to be consistent: x_1 and y_1 must be the coordinates of the same point.

3. *It doesn't matter which two points on the line you pick—the slope will always come out the same.*

Suppose $P = (x_1, y_1)$, $Q = (x_2, y_2)$, $R = (x_3, y_3)$, $S = (x_4, y_4)$ are all points on the same line. We will find the slope by using P and Q and by using R and S. See Figure 21.10

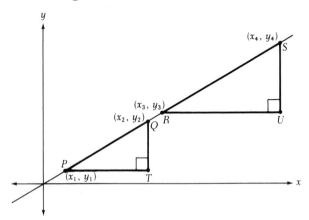

FIG. 21-10

The two triangles PTQ and RUS are similar because both contain a right angle, and angles QPT and SRU are equal—both being the angle the line makes with the x axis. Also, angles PQT and RSU are equal, so the triangles have the same angles and are similar. Therefore,

$$\frac{SU}{QT} = \frac{RU}{PT} = \text{``magnification factor''}$$

Hence,

$$\frac{y_4 - y_3}{y_2 - y_1} = \frac{x_4 - x_3}{x_2 - x_1}$$

Multiplying through by $\dfrac{y_2 - y_1}{x_4 - x_3}$ we get

$$\frac{(y_2 - y_1)}{(x_4 - x_3)} \cdot \frac{(y_4 - y_3)}{(y_2 - y_1)} = \frac{(y_2 - y_1)}{(x_4 - x_3)} \cdot \frac{(x_4 - x_3)}{(x_2 - x_1)}$$

So

$$\frac{y_4 - y_3}{x_4 - x_3} = \frac{y_2 - y_1}{x_2 - x_1}$$

but the slope from P to Q is $\dfrac{y_2 - y_1}{x_2 - x_1}$

and the slope from R to S is $\dfrac{y_4 - y_3}{x_4 - x_3}$

So the slope using R and S = the slope using P and Q.

EXAMPLE   *Find the slope of the line through the points* $(-2, 1)$ *and* $(3, -1)$.

If we let $P = (-2, 1)$ and $Q = (3, -1)$, the formula tells us that

$$\text{slope} = \frac{(-1) - (1)}{(3) - (-2)} = -\frac{2}{5}$$

Looking at Figure 21.11 gives us an idea what such a slope means. You can see that in going from $(-2, 1)$ to $(3, -1)$, y drops by 2 and x increases by 5. The fact that y drops while x increases gives us the minus sign, the fact that y changes only two-fifths as fast as x gives us the $\frac{2}{5}$.

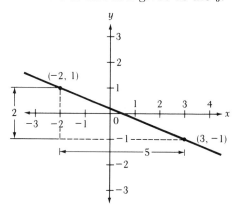

FIG. 21-11

EXAMPLE   *Find the slope of the line* $y = \frac{1}{3}x + 2$.

In order to use the slope formula we must find two points on the line, which can be done in the same way that we create a table of values—by picking x values and plugging them into the formula to get the corresponding y values. If $x = 0$, $y = 2$, and if $x = 1$, $y = \frac{7}{3}$, so $(0, 2)$ and $(1, \frac{7}{3})$ are on the line.

$$\text{slope} = \frac{\frac{7}{3} - 2}{1 - 0} = \frac{\frac{1}{3}}{1} = \frac{1}{3}$$

If we had picked two different points, say $(2, \frac{8}{3})$ and $(3, 3)$, on the line, the slope would have been

$$\frac{3 - \frac{8}{3}}{3 - 2} = \frac{\frac{1}{3}}{1} = \frac{1}{3}$$

which is the same as before.

As you might expect, a graph shows that the line climbs from left to right and gains 1 unit of height for every 3 in a horizontal direction. See Figure 21.12.

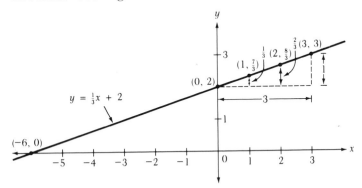

FIG. *21-12*

However, the most noticeable thing about the slope of the line $y = \frac{1}{3}x + 2$ is that it comes out to be the coefficient of x—namely, $\frac{1}{3}$. This is great, because we know that m (here $\frac{1}{3}$) measures the "steepness" of the line and we want slope to measure the same thing. A proof that the slope always turns out to be m is given in the next section.

EXAMPLE   *Find the slope of a horizontal line.*

Along a horizontal line the x coordinate changes, but the y coordinate does not. Suppose the line cuts the y axis at $(0, b)$; then the line must also go through $(1, b)$, as well as $(2, b)$, $(3, b)$, $(-1, b)$, and so on. See Figure 21.13.

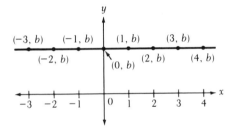

FIG. *21-13*

Using the points $(0, b)$ and $(1, b)$, we get

$$\text{slope} = \frac{b - b}{1 - 0} = \frac{0}{1} = 0$$

So the slope of a horizontal line is 0.

EXAMPLE *Find the slope of a vertical line.*

On a vertical line the y coordinate changes whereas the x coordinate is constant. Suppose the x intercept is k; then the points $(k, 0)$, $(k, 1)$, $(k, 2)$, and so on, are on the line. See Figure 21.14.

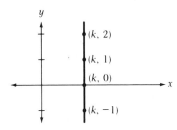

FIG. *21-14*

Using the points $(k, 0)$ and $(k, 1)$, we get

$$\text{slope} = \frac{1 - 0}{k - k} = \frac{1}{0}$$

which is undefined.

So the slope of a vertical line is undefined.

PROBLEM SET 21.2

What is the slope of the line that passes through the points:

1. $(3, 2)$ and $(7, -6)$
2. $(-1, -2)$ and $(6, -2)$
3. $(12, 5)$ and $(-1, -1)$
4. $(4, 2)$ and $(4, 6)$
5. $(-1, 5)$ and $(-3, 0)$
6. $(1, 2)$ and $(19, 2)$
7. $(0, 0)$ and $(4, -5)$
8. $(1, 0)$ and $(\frac{4}{3}, 7\frac{1}{8})$
9. $(-1, 3\frac{1}{2})$ and $(6, -4\frac{3}{5})$
10. $(4, 3\frac{9}{23})$ and $(4, \frac{81}{23})$
11. $(0.03, 7.20)$ and $(0.61, 6.31)$
12. $(1.22, 8.40)$ and $(-2.19, -1.57)$

Graph and label the following lines and find the slope in each case.

13. $y = x + 1$
14. $y = x - 1$
15. $y = -x + 1$
16. $y = -x - 1$

17. $y = 2x - 3$ 19. $x + 2y = 4$
18. $y = -3x + 2$ 20. $3x - 2y = 1$

21. Take the line $y = 2x + 2$ and pick any three pairs of points on the line. For each pair of points compute the slope between them. Show that the slope in each case is the same.

22. Can you draw a straight line through the points:

$$A = (100, \ 106) \qquad B = (107, \ 112) \qquad C = (114, \ 118)$$

21.3   SLOPE-INTERCEPT FORM OF A LINE

An equation of the form

$y = mx + b$ where m and b are constants is the equation of a line in *slope-intercept form*. For any such line, *m is the slope* and *b is the y intercept.*

Since m can be varied to give any slope, and b can be varied to give any y intercept, any line (except a vertical one) has an equation of this form. For completeness, we will now demonstrate officially that m and b really are what we claim they are.

To show the slope of the line $y = mx + b$ is m   Suppose (x_1, y_1) and (x_2, y_2) are two points on the line. This means that the coordinates of these points satisfy the equation; that is, $y_1 = mx_1 + b$ and $y_2 = mx_2 + b$. Now

$$\text{slope} = \frac{y_2 - y_1}{x_2 - x_1} = \frac{(mx_2 + b) - (mx_1 + b)}{x_2 - x_1}$$

$$= \frac{mx_2 + b - mx_1 - b}{x_2 - x_1}$$

$$= \frac{mx_2 - mx_1}{x_2 - x_1}$$

$$= \frac{m(x_2 - x_1)}{x_2 - x_1}$$

$$= m$$

To show the y intercept of the line $y = mx + b$ is b   On the y axis, $x = 0$, so the y intercept is given by

$$y = m \cdot 0 + b = b$$

To show the x intercept of the line $y = mx + b$ is $\left(-\dfrac{b}{m}\right)$ (provided $m \neq 0$)   On the x axis, $y = 0$, so the x intercept is given by

$$0 = mx + b$$

or

$$x = -\frac{b}{m}$$

Note: This assumes $m \neq 0$, that is, that the line is not horizontal.

Examples of the Use of the Slope-Intercept Form of a Line

EXAMPLE   *Find the slope of the line*

$$\frac{x}{2} + \frac{y}{3} = 1$$

and draw its graph.

Rewrite

$$\frac{x}{2} + \frac{y}{3} = 1 \quad \text{as} \quad \frac{y}{3} = -\frac{x}{2} + 1$$

so that

$$y = \left(-\frac{3}{2}\right)x + (3) \quad \overset{b}{\underset{m}{}}$$

Now you can see that the slope is $-\frac{3}{2}$. The y intercept is 3, and the x intercept is

$$\frac{-b}{m} \quad \text{or} \quad \frac{-3}{\left(-\frac{3}{2}\right)} = 2.$$

The intercepts allow you to plot the graph very easily, since you know it goes through the points (0, 3) and (2, 0). The line is graphed in Figure 21.15.

The graph also shows that in moving from P to Q, y decreases by 3, and x increases by 2—as you would expect of a line that has slope $-\frac{3}{2}$.

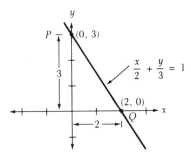

FIG. 21-15

EXAMPLE   *Graph the family of lines* $y = mx$.

Any line of the form $y = mx$ goes through the origin, since $b = 0$. As m varies, the equation $y = mx$ gives a whole collection of lines; this collection is called a family of lines and is shown in Figure 21.16.

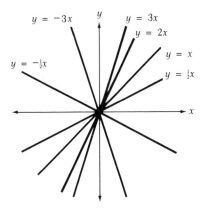

FIG. 21-16

You should notice how m affects the angle of the line $y = mx$. If $|m|$ is large, the line is steep (e.g., $y = 3x$, $y = -3x$); if $|m|$ is small, it is not steep (e.g., $y = \frac{1}{2}x$). If m is positive, the line climbs to the right (e.g., $y = 3x$); if m is negative, it descends to the right (e.g., $y = -3x$).

On any of these lines y and x are connected by an equation of the form:

$$y = (\text{constant})x \qquad \text{where constant} = m$$

and so y is *proportional* to x, and the constant of proportionality is the slope m. Therefore, if you know that one quantity is proportional to another and you draw their graph, it will be a straight line through the origin.

Finding the Equation of a Line Given the Slope and a Point

EXAMPLE   *Find the equation of a line of slope −2 and through the point (−1, 3).*

Since the slope is −2, the equation must be of the form

$$y = -2x + b$$

Now the coordinates of the point (−1, 3) must satisfy the equation since the point lies on the line. So

$$3 = -2(-1) + b$$

This equation can be solved for b:

$$b = 3 - 2 = 1$$

The equation of the line is $y = -2x + 1$. Its graph is shown in Figure 21.17. In this example the slope fixes the direction of the line, and the point (−1, 3) determines its position.

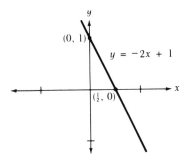

FIG. 21-17

Finding the Equation of a Line Given Two Points

EXAMPLE   *Find the equation of a line through the points (2, 6) and (−2, 8).*

These two points can be used to find the slope of the line between them.

$$\text{slope} = m = \frac{8 - 6}{-2 - 2} = \frac{2}{-4} = -\frac{1}{2}$$

Therefore, the equation is of the form

$$y = -\frac{1}{2}x + b$$

To find b—which determines the position of the line—we use the fact that the point $(2, 6)$ lies on the line. (The point $(-2, 8)$ could equally well be used.) Substituting $x = 2$, $y = 6$ into $y = -\frac{1}{2}x + b$:

$$6 = \left(-\frac{1}{2}\right)(2) + b$$

So

$$b = 7$$

Therefore, the line has the equation

$$y = -\frac{1}{2}x + 7$$

Alternative Method of Finding the Equation of a Line Given Two points

Instead of using the two points to find the slope, we could have used the fact that *both* points must satisfy the equation of the line. This would give us simultaneous equations for m and b.

Specifically, the fact that the points $(2, 6)$ and $(-2, 8)$ lie on the line $y = mx + b$ means that

$$6 = 2m + b$$
and
$$8 = -2m + b$$

We can now treat these as simultaneous equations for m and b: Add the equations to solve for b:

$$14 = 2b$$
or
$$b = 7$$

Subtract the equations to solve for m:

$$-2 = 4m$$
or
$$m = -\frac{1}{2}$$

Therefore, the equation of the line is

$$y = -\frac{1}{2}x + 7$$

as before.

PROBLEM SET 21.3

Find the equation of the lines satisfying the following conditions:
1. Passes through the points (3, 5) and (1, 1)
2. Passes through (−1, 5) and has a slope of 2
3. Has y intercept $= -31$ and slope $= 2$
4. Passes through (−1, −5) and has a slope of $-\frac{1}{2}$
5. Passes through (2, −5) and (−3, 3)
6. Passes through (3, 0) and (0, 3)
7. Has x intercept $= 4$ and slope $= 3$
8. Passes through (3, 1) and (5, 4)
9. Has x intercept $= 5$ and y intercept $= 5$
10. Passes through (0.01, −0.03) and (4.3, 2.02)
11. Passes through (1, −16) and has slope $= m$
12. Passes through (a, 0) and (0, a)
13. Passes through (a, b) and has slope $= 2$
14. Passes through (a, 0) and (b, c)
15. Passes through (a, b) and has slope c

16. Graph and label the lines in Problems 1–10.

Find the slope, x intercept. and y intercept of:

17. $2x + y = 1$
18. $\frac{x}{2} + \frac{y}{2} = 1$
19. $-x - 4 = 5$
20. $0.2y - 0.04x = 1.3$
21. $\frac{2y + 1}{x + 1} = 3$
22. $2x = \frac{3(y - x + 1)}{2}$
23. $3y + x = 2$
24. $4 = y - x$
25. $\frac{7}{x + y} = \frac{2}{x}$
26. $0 = 10y - 7x - 4$
27. $\frac{2}{2x + y} = \frac{3}{x + y}$
28. $\frac{y}{4} - x = \frac{3}{2}$
29. $\sqrt{2}x - y = \sqrt{18}$
30. $ax + by + c = 0$

31. Graph and label the lines in Problems 17–19.

Find the equations of the following lines (32–35).

32.

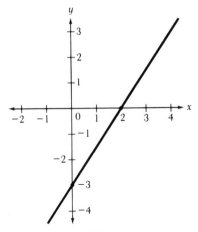

33.

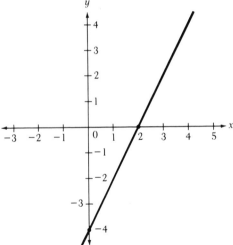

34.

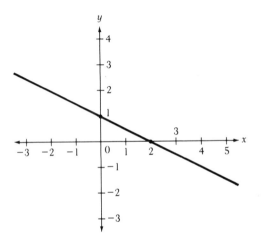

35.

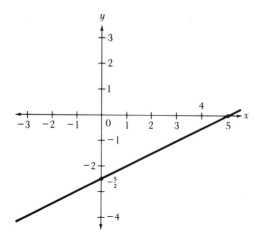

36. Match the following graphs with the equations below.
    (a) $y = x - 4$ (d) $y = -3x - 4$
    (b) $-2x + 3 = y$ (e) $y = x + 5$
    (c) $4 = y$ (f) $y = \frac{1}{2}x$

(A)

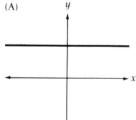

(B)

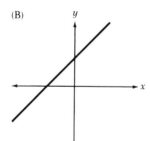

(C)

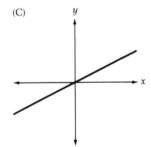

(D)

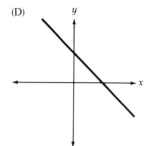

(E)

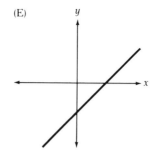

(F)

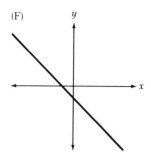

37. At which points do the following lines intersect the x and y axes?
    (a) $x - 2y = 3$

    (b) $1 = \dfrac{x}{4} + \dfrac{y}{5}$

(c) $0 = x + 2y - 7$

38. What is the length of that portion of the line between the x and y axes in each of the lines in Problem 37?

39. In Figure 21.18, if P is the graph of $y = mx + c$, and Q is the graph of $y = nx + d$:
  (a) Which is larger, m or n? Why?
  (b) Which is larger, c or d? Why?

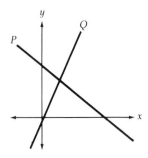

FIG. 21-18

40. Here are three equations:

$$2x + 3y = 12$$

$$-4y = 24x + 3$$

$$x = 2y - 3$$

  (a) Which of the above equations has the graph with the greatest slope?
  (b) Which has the greatest y intercept?

41. Consider a graph of Fahrenheit temperature, y, against Centigrade temperature, x, and assume that the graph is a line. You know that when $y = 212$, $x = 100$ (because 212°F and 100°C both represent the temperature at which water boils). Similarly when $y = 32$, $x = 0$ (water's freezing point). Hence (100, 212) and (0, 32) are two points on the line.
  (a) What is the slope of the graph?
  (b) What is the equation of the line?
  (c) Use the equation to find what temperature in Fahrenheit corresponds to 20°C.
  (d) At what temperature are the Centigrade and Fahrenheit temperatures equal?

42. Mark out on graph paper in addition to the axes the lines $y = 12$ and $x = 6$. This is a pool table. The *pool rule* states that if a ball travels along a line with slope m and strikes the side of the table, it will bounce back along a line with slope $-m$.
  (a) A ball starts at (3, 8) with slope 2 toward the y axis. Where does it

strike the y axis? What slope does it have after bouncing off the y axis? What is the equation describing the new path?

(b) Follow the ball in (a) for two more bounces.

(c) Assume the pockets are located at the corners [e.g., (6, 12)]. Where are the others? If you hit a ball from (3, 8) to $(0, \frac{16}{3})$, will it go into a pocket?

(d) **Start two balls at (3, 8),** one toward the y axis at slope $\frac{3}{2}$ and the other toward $(12, 5\frac{1}{3})$. On the second bounce, which is closer to a pocket?

21.4   HORIZONTAL AND VERTICAL LINES

We have already graphed some horizontal and vertical lines in Section 19.1, as a result of looking for all the points with $x = 3$ or with $y = -5$. Let's look at horizontal and vertical lines again in the light of what we know about slopes and intercepts.

Horizontal Lines

At the end of the section on slope (Section 21.2), you saw that the slope of a horizontal line was 0—as you might expect, since it does not climb at all.

If you want to find the equation of a horizontal line at a height of 4 above the x axis, you know its slope is 0 and its y intercept is 4, so the equation is

$$y = 0 \cdot x + 4$$

or

$$y = 4$$

This equation looks funny because it does not contain an x. However, you are meant to think of it as meaning that x can be anything, but y must be restricted to 4, which gives the points on the horizontal line in Figure 21.19. Therefore the equation $y = 4$ does specify exactly those points that lie on the line.

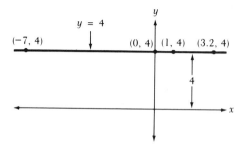

FIG. 21-19

In general, a *horizontal line at a distance b from the x axis has equation* $y = b$. If b is positive, the line is above the x axis, and if b is negative, it is below the x axis. In the special case that $b = 0$, we have the equation $y = 0$, which must be the x axis. If you think about it, this is reasonable because the x axis does consist of all the points whose y coordinates are 0. So $y = 0$ *is the equation of the x axis.*

Vertical Lines

We showed earlier that the slope of a vertical line is not defined. For example, the slope of a line between (1, 2) and (1, 4) involves calculating

$$\frac{4 - 2}{1 - 1} = \frac{2}{0}$$

which is not defined, and so the slope is not defined either. A vertical line is therefore the one and only case that cannot be described by the slope-intercept form of the equation.

Looking at the equation of a horizontal line, however, gives you an idea of what the equation of a vertical line might look like. Since a horizontal line consists of all the points with a certain y coordinate, say 3, and has equation $y = 3$, it would be reasonable for a vertical line that consists of all the points with a certain x coordinate, say 5, to have an equation $x = 5$. And indeed, if you understand the equation $x = 5$ to mean that the y coordinate can be anything, but the x coordinate is restricted to 5, then the equation specifies exactly those points on a vertical line at a distance of 5 to the right of the y axis. See Figure 21.20.

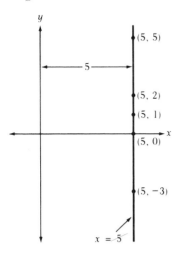

FIG. 21-20

In general, the equation of a *vertical line at a distance k from the y axis is* $x = k$. If k is positive, the line is to the right of the y axis, and if negative, to the left. If $k = 0$ the equation reads $x = 0$ and represents the y axis—which does, after all, consist of all those points whose *x* coordinate is 0. So *x = 0 is the equation of the y axis.*

21.5   PARALLEL AND PERPENDICULAR LINES

Parallel Lines

A set of parallel lines all climb at the same rate and in the same direction (see Figure 21.21), and therefore you will not be surprised to learn that *parallel lines have equal slopes.*

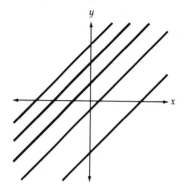

FIG. 21-21

Proof that if $y = m_1 x + b_1$ **and** $y = m_2 x + b_2$ **are parallel, then** $m_1 = m_2$ Suppose the two lines in Figure 21.22 are parallel. In order to go on from here, we have to get some notation straight.

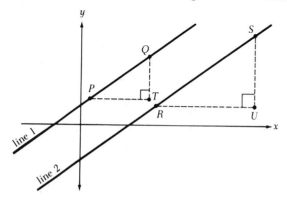

FIG. 21-22

$|TQ|$ will mean the distance between *T* and *Q*. It is always a positive number, and

$$|TQ| = |QT|$$

T and Q are on the same vertical line, and without the absolute value signs TQ will mean the change in y in going from T to Q.

QT therefore means the change in y in going from Q to T, and so

$$TQ = -QT.$$

For two points on the same horizontal line, such as P and T, PT means the change in x in going from P to T.

TP means the change in x in going from T to P and so, again,

$$PT = -TP.$$

Now let's go back to Figure 21.22. Since two parallel lines cut the horizontal at the same angle, the angles QPT and SRU must be equal. In the same way, the angles PQT and RSU must be equal. Therefore the triangles PTQ and RUS are similar and ratios of corresponding sides are equal, giving us:

$$\frac{|TQ|}{|US|} = \frac{|PT|}{|RU|}$$

But, looking at the graph, TQ and US must have the same signs. We have no way of knowing if both are positive or both negative, but we do know that either way the ratio $\dfrac{TQ}{US}$ will come out positive. Therefore,

$$\frac{|TQ|}{|US|} = \frac{TQ}{US}$$

By exactly the same reasoning,

$$\frac{|PT|}{|RU|} = \frac{PT}{RU}$$

Therefore, the equation

$$\frac{|TQ|}{|US|} = \frac{|PT|}{|RU|}$$

can be replaced by:

$$\frac{TQ}{US} = \frac{PT}{RU}$$

Multiplying this through by $\dfrac{US}{PT}$ we get:

$$\frac{TQ}{\cancel{US}} \cdot \frac{\cancel{US}}{PT} = \frac{\cancel{PT}}{RU} \cdot \frac{US}{\cancel{PT}}$$

so

$$\frac{TQ}{PT} = \frac{US}{RU}$$

But

$$\frac{TQ}{PT} = \frac{\text{vertical change along line 1}}{\text{corresponding horizontal change}} = \text{slope of line 1}.$$

Similarly,

$$\frac{US}{RU} = \text{slope of line 2}$$

So

$$\frac{TQ}{PT} = \frac{US}{RU}$$

tells us that

slope of line 1 = slope of line 2

EXAMPLE   *Find the equation of a line parallel to $2y = 3x + 2$ but through the origin.*

Rewriting $2y = 3x + 2$ as $y = \frac{3}{2}x + 1$, you can see that its slope is $\frac{3}{2}$. Therefore the slope of the line that we are looking for is $\frac{3}{2}$, and so its equation must be of the form

$$y = \frac{3}{2}x + b$$

Our line is to go through the origin and so has a y intercept of 0, so $b = 0$.

Therefore the equation of our line is $y = \frac{3}{2}x$.

Perpendicular Lines

It is easy to tell whether or not two lines are parallel to one another—if they are, they have the same slope. Since lines that are perpendicular to one another are important, it would also be useful to know how to tell if two lines were perpendicular, or how to find a line that is perpendicular to some other line. As it turns out, *perpendicular lines have slopes that are negative reciprocals of one another*, or, in other words:

If $y = m_1x + b_1$ and $y = m_2x + b_2$ are perpendicular, then

$$m_1 = -\frac{1}{m_2}$$

(or, equivalently,

$$m_1m_2 = -1)$$

Note: These formulas don't hold for horizontal and vertical lines because, although they are perpendicular, the slope of a vertical line is not defined.

Justification of $m_1 = -\dfrac{1}{m_2}$ If two lines are perpendicular, in general one will have to be climbing as you go from left to right while the other is descending, and therefore they will have to have slopes of opposite signs. The fact that they are perpendicular also means that if one line is climbing (or falling) fast—that is, is nearly vertical—then the other line must not be climbing or falling fast, and so must be nearly horizontal. In other words, if one line has a slope of large magnitude, the other has a slope of small magnitude. These facts certainly fit in with a formula like $m_1 = -\dfrac{1}{m_2}$ Of course, they don't prove that such a formula holds, but they certainly suggest it.

Proof that $m_1 = -\dfrac{1}{m_2}$ for Perpendicular Lines   Suppose that L_1 and L_2 are any two perpendicular lines and that P is their point of intersection. Draw any vertical line cutting them at the points Q and R as shown in Figure 21.23, and draw in the horizontal line PS also.

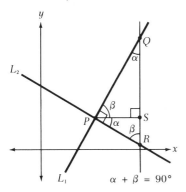

FIG. *21-23*

The angle QPR is a right angle because L_1 and L_2 are perpendicular. This means that angles PQS and RPS are equal because both make 90° when added to angle QPS. Similarly, angles QPS and PRS are equal, since both make 90° when added to angle SPR.

Therefore the triangles PQS and RPS are similar.

This means that the ratios of lengths of corresponding sides are equal:

$$\frac{|SQ|}{|PS|} = \frac{|PS|}{|SR|}$$

or

$$\frac{|SQ|}{|PS|} \cdot \frac{|SR|}{|PS|} = 1 \quad \left(\text{multiplying both sides by } \frac{|SR|}{|PS|}\right)$$

Looking at Figure 21.23, we can see that SQ and SR must have opposite signs. Which is positive and which negative is impossible to say in general, but we can say definitely that

$$SQ \cdot SR = -|SQ| \cdot |SR|$$

since one of SQ and SR must be negative, and $|SQ| \cdot |SR|$ is positive because it is the product of two lengths. Therefore,

$$|SQ| \cdot |SR| = -SQ \cdot SR$$

PS may be positive or negative, but either way it is true to say that:

$$|PS| \cdot |PS| = PS \cdot PS$$

since both sides are squares and therefore positive. Therefore,

$$\frac{|SQ|}{|PS|} \cdot \frac{|SR|}{|PS|} = 1$$

becomes

$$-\frac{SQ}{PS} \cdot \frac{SR}{PS} = 1$$

or

$$\frac{SQ}{PS} \cdot \frac{SR}{PS} = -1$$

But

$$\frac{SQ}{PS} = \frac{\text{vertical change along } L_1}{\text{corresponding horizontal change}} = \text{slope of } L_1 = m_1$$

and, similarly,

$$\frac{SR}{PS} = \text{slope of } L_2 = m_2$$

Therefore, we have

$$m_1 \cdot m_2 = -1 \quad \text{or} \quad m_1 = -\frac{1}{m_2}$$

EXAMPLE   *Find the equation of the line perpendicular to $3y + x = 7$ and through (a, a).*

The equation $3y + x = 7$ can be rewritten as $y = -\frac{1}{3}x + \frac{7}{3}$, and so its slope is $-\frac{1}{3}$. Therefore the slope of a perpendicular line is

$$-\frac{1}{-\frac{1}{3}} = 3$$

and so its equation must be of the form

$$y = 3x + b$$

If the point (a, a) is to lie on the line, its coordinates must satisfy the equation—in other words, b must satisfy

$$a = 3a + b$$

so

$$b = -2a$$

Therefore the equation is

$$y = 3x - 2a$$

PROBLEM SET 21.5

Find the equations of the lines satisfying the following conditions:
  1. A vertical line through $(0, 4)$
  2. A horizontal line 2 units below the x axis
  3. Parallel to the x axis and through $(0, -4)$
  4. Parallel to the y axis and through $(\sqrt{2}, 0)$
  5. Parallel to $3x + 5y = 8$ and through the origin
  6. Perpendicular to the y axis and through $(3, 7)$
  7. Perpendicular to $y + 3x = 7$ and through the origin
  8. Perpendicular to $2y - x - 1 = 0$ and through $(1, -1)$
  9. Parallel to $3x + y = 15$ and through $(1, 4)$
 10. Perpendicular to $x = 5y - 5$ and through $(-2, -3)$
 11. Perpendicular to $y - 1 = -3(x - 4)$ and through $(1, 1)$
 12. Perpendicular to $2x + 3y = 4$ and through $(2, 1)$
 13. Parallel to $2x - 3(y - 2x) - 3 = 0$ and through $(7, 5)$
 14. Perpendicular to $x = zy$ and through $(3, 0)$
 15. Perpendicular to $\dfrac{x + y}{3} - 5 = 0$ and through $(-0.2, 3.3)$
 16. Perpendicular to $y + ax = 0$ and through the origin
 17. A horizontal line through $(4, k)$

18. Parallel to $\dfrac{ay + bx}{c} = n$ and through the origin

19. A vertical line through (a, b)

20. A horizontal line through the point of intersection of $3x + 5y = 8$ and $x = 16$.

21. Sketch the lines in Problems 1–20.

22. Show that the line segments joining the points $(-1, 1)$, $(5, 3)$, and $(1, 5)$ form a right triangle.

23. The equations of the sides of a triangle are $2x - y + 5 = 0$; $y + 5 = 3x$; $x + 2y = 6$. Show that this is a right triangle.

24. Show algebraically that the line through $(2, 3)$ and $(4, 9)$ is perpendicular to the line passing through $(1, 2)$ and $(4, 1)$.

25. Let A be the line through $(1, -2)$ and $(4, 2)$. Let B be the line perpendicular to A and through $(1, -2)$. If a square were made with line A as one side, line B as another side, and the two points $(1, -2)$ and $(4, 2)$ as vertices, what would be the coordinates of the other two vertices?

26. The vertices of a quadrilateral are $(-4, 1)$, $(0, -2)$, $(6, 6)$, and $(2, 9)$.
   (a) Show that this figure is a parallelogram.
   (b) Show that this figure is actually a rectangle.

27. An ant is crawling along the line $y = 2x - 2$ in the direction of increasing y. When he comes to the point $(3, 4)$ he takes a left turn (90°) and keeps going in a straight-line path. What is the equation of his new path?

28. Show that the diagonals of a square are perpendicular. *Hint:* Draw a general square on a coordinate system as in the figure.

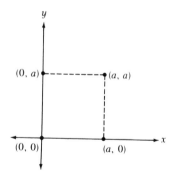

In Problems 29–32, give the equations of line L and line M.

29.

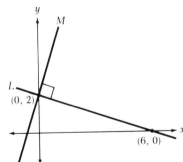

30.

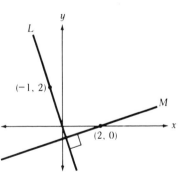

31.

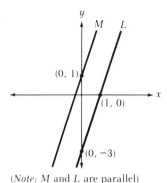

(*Note:* M and L are parallel)

32.

33. Find the equation of the line tangent at the point $\left(\dfrac{\sqrt{2}}{2}, \dfrac{\sqrt{2}}{2}\right)$ to a circle of unit radius centered at the origin. *Hint:* **A tangent and the radius at the same point are perpendicular.**

34. The line segment from $(1, -2)$ to $(-2, 3)$ in the figure is the diameter of a circle.

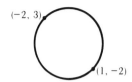

(a) Find the coordinates of the center of the circle.

(b) Find the area of the circle. (You may leave π in your answer.)

(c) Find the equation of the line through the point $(-4, -5)$ and perpendicular to the line connecting the points $(-2, 3)$ and $(1, -2)$.

For Problems 35–43 you will need to find the point of intersection of two lines. If a point lies on two lines, its coordinates satisfy the equations of both lines. Therefore, to find a point of intersection, solve the equations of the lines simultaneously.

35. Write the equation of the line that passes through the origin $(0, 0)$ and through the point of intersection of the following two lines:

$$2x - y + 5 = 0$$

$$y + 5 = -3x$$

36. An electric current is running through a wire along the line $y = 2x + 1$. An electron at $(4, 4)$ is attracted toward the wire and moves toward the closest point on the wire (i.e., it moves along a line perpendicular to the wire).

    (a) What is the equation of the line along which the electron travels?

    (b) Where does it hit the wire?

    (c) How far does it travel before hitting the wire?

37. You know you can find the distance between two points by using the formula

$$D = \sqrt{(x_1 - x_2)^2 + (y_1 - y_2)^2}$$

It is also possible to find the distance between a point and a line, and between two parallel lines. For example, take the line $y = 2x - 1$ and the point $(-2, 5)$. The shortest distance between the point and the line is the length of the line segment through $(-2, 5)$ and perpendicular to $y = 2x - 1$. The procedure to follow is this:

    (a) Find the equation of the line perpendicular to $y = 2x - 1$ and through $(-2, 5)$.

    (b) Find the point at which the two lines intersect, say (a, b).

    (c) Find the distance between $(-2, 5)$ and (a, b). This is the distance we want.

38. What is the distance between $(1, 6)$ and $y = 2x + 3$?

39. What is the distance between $(6, 9)$ and $y = -\frac{3}{4}x + 1$?

40. What is the distance between $(11, 0)$ and $y = 2x + 3$?

41. Given the two parallel lines $y = -3x - 5$ and $y = -3x + 2$, find the distance between them. *Hint:* Pick a point on one line and proceed as in Problem 37.

42. Find the distance between $y = 2x + 5$ and $y = 2x - 5$.

43. Can you find a general formula to express the distance between any point (x_0, y_0) and any line $y = mx + b$?

CHAPTER 21 REVIEW

Find equations for the following lines:
1. The line passing through $(-2, -3)$ and $(7, 9)$
2. The line passing through $(-3, -2)$ with slope $-\frac{1}{3}$
3. The line passing through $(0.02, -0.01)$ and $(-0.02, -0.04)$
4. The line with x intercept $\frac{1}{5}$ and y intercept -3
5. The line through $(4, 5)$ with slope 0
6. The line passing through $(-\frac{2}{5}, 3)$ and $(1, -6)$
7. The line through $(0, \frac{5}{8})$ with slope $\frac{98}{89}$
8. The line with x intercept -4 and y intercept $-\frac{1}{4}$
9. The line with slope 2 passing through (a, b)
10. The line with slope $\dfrac{1}{m}$ passing through $(x_0 + t, x_0 - t)$

Graph the following:
11. $\dfrac{x}{4} + \dfrac{y}{-6} = 1$
12. $2x - \frac{1}{3}y = 4$
13. $0.2(x - 3) = (y + 1)$
14. $y = -\frac{1}{2}x - 1$
15. $\dfrac{x - y + 3}{4} = \dfrac{2x + y + 1}{6}$
16. $\frac{3}{8}x = \frac{2}{3}y$
17. $2x = -42$

18. Find the equation of the line perpendicular to the line in Problem 2 but passing through $(3, 2)$. Graph it.

19. Find the equation of the line perpendicular to the line in Problem 3 but passing through $(0.02, -0.01)$. Graph it.

20. Find the equation of the line parallel to $2y - 7x = y + 4$ but passing through $(-1, 2)$. Graph it.

21. If a line passes above the point $(-2, 3)$ and below the point $(4, -2)$, what is the greatest slope a line perpendicular to it could have?

22. Start at $(-1, 2)$ and head directly for $(2, -2)$.
    (a) Where do you hit the y axis?
    (b) What is the equation of the line travelled on?
    (c) Two lines might be described as skewed if the ratio of the products of their slopes and the sum of these slopes $= -1$. At $(2, -2)$, what is the line skewed to the one that you are on?

23. In an attempt to move into the sports field, the Gallup poll compiled Table 21.2 about the players for a well-known football team. It shows the

number of times each player got to carry the ball (this is the number of attempts), the total distance he carried it (this is the yards), and the average distance he carried it, per attempt.

Table 21.2

Rushing Statistics—Five Game Totals			
Player	*Attempts*	*Yards*	*Average*
Ron	60	334	5.6
Jon	39	201	5.2
Al	62	192	3.1
Tom	36	120	3.3
Burke	18	49	2.7

(a) Putting "attempts" along the x axis and "yards" up the y axis, plot a point for each player, and draw a line from that to the origin.

(b) What is the equation of the line joining Al to the origin?

(c) What is the slope of the line perpendicular to that joining Burke to the origin?

24. I. A. Rich owns an estate on the corner of Ivy Street (which runs east–west) and Vine Avenue (which runs north–south). The driveway on the estate runs in a straight line from a point on Vine Avenue 300 yards north of the corner directly to the house. If this driveway runs northeast–southwest and is 1500 yards long, how far is the house from Ivy Street?

25. A wire is laid on a piece of graph paper in such a way that it forms the graph of $y = 3x + 2$. If this wire is moved vertically by 2 units, where does it then cross the x axis?

26. Able is a crook, fleeing from a crime by motorboat going at a constant speed in a straight line across Axis Lake. He moves from $(-5, 10)$ to $(-3, 10)$ in 10 minutes. Just at the time he reaches $(-3, 10)$ Baker, a policeman, leaves the Axis Lake Police Station on an island at $(3, 2)$ to catch him. Baker wants to catch Able as soon as possible, but the maximum speed of a police boat is 10 units per hour. Give the equation of the straight-line path Baker should follow in order to catch Able.

27. Suppose that

(i) Two lines are "reciprocal" to each other if their slopes are reciprocals (for example, $y = 3x + 2$, $y = \frac{1}{3}x - 1$)

and

(ii) Two lines are "negative" to each other if their slopes are of the same magnitude but opposite in sign (for example, $y = 2x + 1$, $y = -2x + 3$)

then:

(a) Can a line ever have the same slope as a line "reciprocal" to it? Give an example, or explain why not.

(b) Is it possible that a line could be both perpendicular *and* "negative" to another line? Give an example, or explain why not.

(c) Is it possible that a line could be both "negative" and "reciprocal" to another line? Give an example, or explain why not.

28. Two lines, A and B, intersect at the point $(2, 3)$. Both lines have positive slopes, and the slope of line A is greater than the slope of line B.

(a) Which line will have the greater y intercept?

(b) Which line will have the greater x intercept?

(c) Which line will have its x intercept closer to the origin $(0, 0)$?

(d) If the problem had stated that both lines had *negative* slopes, would your answers have been different? (Be careful!)

Problems 29–30 require you to find the point of intersection of two lines by solving the equations simultaneously. See note before Problem 35, in Problem Set 21.5.

29. Given a triangle with vertices $(0, 0)$, $(0, 4)$, and $(4, 0)$, show that the three lines connecting each vertex to the midpoint of the opposite side meet in a single point.

30. A radio beacon is located at $(1, -1)$ and another at $(-\frac{2}{3}, \frac{3}{4})$. Your navigation equipment tells you that the line joining you to the first beacon has slope $-\frac{1}{2}$ and the line joining you to the second has slope $\frac{1}{2}$. Where are you?

Appendix: Geometry

Two-Dimensional Shapes

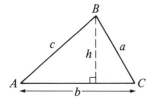

TRIANGLE

Perimeter $= a + b + c$

Area $= \dfrac{b \cdot h}{2}$

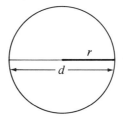

CIRCLE

$d = 2r$

Circumference $= \pi d = 2\pi r$

Area $= \pi r^2$

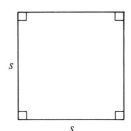

SQUARE

Perimeter $= 4s$

Area $= s^2$

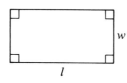

RECTANGLE

Perimeter $= 2\,(l + w)$

Area $= lw$

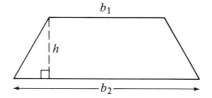

TRAPEZOID

Area $= \dfrac{(b_1 + b_2)\,h}{2}$

Three-Dimensional Shapes

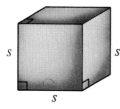

CUBE

Surface Area $= 6s^2$

Volume $= s^3$

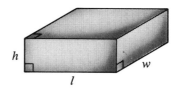

RECTANGULAR BOX

Surface Area $= 2(lw + lh + wh)$

Volume $= lwh$

SPHERE

Surface Area $= 4\pi r^2$

Volume $= \dfrac{4}{3}\pi r^3$

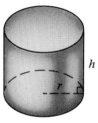

CYLINDER

Surface Area $= 2\pi r^2 + 2\pi rh$

Volume $= \pi r^2 h$

CONE

Volume $= \frac{1}{3}\pi r^2 h$

Triangles: Types and Properties

RIGHT

Pythagorean
Theorem:
$a^2 + b^2 = c^2$

ANY

$\alpha + \beta + \gamma = 180°$
(The sum of the
angles of any
triangle is 180°.)

EQUILATERAL

All sides equal
All angles equal

ISOSCELES

Two sides equal
Two angles equal

 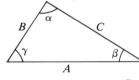

SIMILAR TRIANGLES

Similar triangles have the
same shape and the same
angles but are not neces-
sarily the same size.

Ratios of corresponding sides are equal:

$$\frac{A}{a} = \frac{B}{b} = \frac{C}{c} \quad \text{"Magnification factor"}$$

Changing Units

To change (convert) units, multiply by a suitably chosen ratio (which must be
equal to 1) so that all units cancel except the ones you want. For example:

$$1 \text{ ft} = 12 \text{ in, so } \frac{12 \text{ in}}{1 \text{ ft}} = 1 \quad \text{Therefore } 5 \text{ ft} = 5 \text{ ft} \cdot \frac{12 \text{ in}}{1 \text{ ft}} = 60 \text{ in}$$

Note that:

$1 \text{ ft} = 12 \text{ in} \quad \text{but} \quad 1 \textit{ square} \text{ ft (ft}^2) = 144 \text{ in}^2 \quad \text{and}$
$1 \textit{ cubic} \text{ ft (ft}^3)_| = 1728 \text{ in}^3$

$1 \text{ cm} = 10 \text{ mm} \quad \text{but} \quad 1 \text{ cm}^2 = 100 \text{ mm}^2 \quad \text{and} \quad 1 \text{ cm}^3 = 1000 \text{ mm}^3$

ANSWERS TO ODD-NUMBERED PROBLEMS

Chapter 1 Exercises

	Real	Complex	Whole	Natural	Rational	Integral	Irrational
1.	°	°	°	°	°	°	
3.	°	°	°		°	°	
5.	°	°			°	°	
7.	°	°			°		
9.	°	°			°	°	
11.	°	°			°		
13.	°	°			°		
15.	°	°	°	°	°	°	
17.	°	°			°		
19.	°	°			°		

21. 1, 2, 3, 4, 5, 6, 7

23. 6, 12, 18, 24, 30

25. None

27. Finite

29. Infinite

31. (i) $2n$ (ii) $2n + 1$

35. Odd

37. Impossible—the sum of even integers is even.

39. Either none, two, or four

41. All

43. No

45. Some

47. Some

49. Some

Chapter 2 Exercises

1.

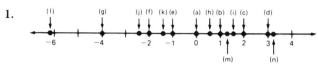

3. $H, \dfrac{-27}{8}$; $I, -\sqrt{8}$; $J, -\sqrt{2}$; $K, \dfrac{9}{10}$; $L, 2.8$; M, π

5. 5	**17.** 8	**29.** -5	**41.** 9	**53.** 4
7. -3	**19.** 0	**31.** -1	**43.** 4	**55.** 0
9. 3	**21.** -1	**33.** -7	**45.** 10	**57.** 4
11. -5	**23.** 1	**35.** 1	**47.** 4	**59.** -6
13. 7	**25.** -12	**37.** -6	**49.** 9	**61.** -12
15. -14	**27.** -1	**39.** -1	**51.** 5	**63.** 0

65. −5	79. −18	93. 27	107. 64
67. 2	81. −96	95. 64	109. No, $3 million
69. 0	83. 2	97. 16	111. (a) East (b) 3 miles
71. −3	85. 2	99. 0	113. (a) 22°C (b) 347°C
73. 11	87. −24	101. −32	115. 7 pounds loss
75. −6	89. 1	103. 64	117. 42
77. 6	91. 1	105. 64	

Chapter 3 Exercises

1. 20	11. 3	21. $70
3. −3	13. 4	23. $1540
5. 12	15. 3	25. 660 miles
7. −18	17. −2	
9. 9	19. 0	

Chapter 4 Exercises

1. $2 \cdot 3^2 \cdot 5 \cdot 11$	13. 432	25. $\frac{3}{7}$	37. $\frac{23}{24}$
3. $2^2 \cdot 3^2 \cdot 5^2$	15. 252	27. $-\frac{1}{14}$	39. (a) $\frac{1}{40}$ (b) $\frac{39}{40}$
5. $3^3 \cdot 5$	17. 1848	29. $\frac{9}{10}$	41. (a) 0 (b) $\frac{1}{4}$
7. $2^3 \cdot 19$	19. 1008	31. $\frac{3}{20}$	43. (a) $\frac{1}{3}$ (b) No
9. $2^2 \cdot 3 \cdot 13 \cdot 17$	21. $\frac{7}{180}$	33. 478	45. (a) $\frac{1}{3}$ (b) $\frac{1}{12}$ (c) $\frac{1}{4}$
11. 24	23. $\frac{1}{24}$	35. (a) $\frac{1}{4}$ (b) 55	

Chapter 5 Exercises

1. 2.3	33. 30	65. 25%	97. 1
3. 0.0025	35. 2.9	67. 16.$\overline{6}$%	99. is less than
5. 0.001	37. −0.0509	69. 12.5%	101. is less than
7. 1.053	39. 0.33$\overline{4}$	71. 10%	103. is more than
9. 0.024	41. 1.610	73. 2%	105. is more than
11. 2.6	43. 0.018	75. 75%	107. is equal to
13. 0.2	45. 24.15	77. 175%	109. is less than
15. 1.$\overline{6}$	47. 0.065%	79. 0.1236	111. is more than
17. 0.8$\overline{3}$	49. 0.706%	81. 57.5	113. 0.25%
19. 0.$\overline{3}$	51. 23.0%	83. 75	115. (a) $108.90 (b) 11.$\overline{1}$%
21. 0.2	53. 1200%	85. 39.2	117. (a) 107.8 (b) 4.2%
23. 0.$\overline{142857}$	55. 0.00003	87. 18.8094	(c) less than 163.9
25. 0.$\overline{1}$	57. 0.0075	89. 0.08236	119. (a) 4.0 (b) one home
27. 0.04	59. 3.21	91. 16.683	run (c) 0.55
29. 0.$\overline{6}$	61. 0.16	93. 62.307	121. (a) 9% (b) $246,600
31. 0.8	63. 50%	95. 70.933	

Chapter 6 Exercises

1. $3\sqrt{5}$
3. $(\frac{1}{4})(\sqrt{\frac{1}{5}})$
5. $20\sqrt{210}$
7. $2\sqrt[3]{3}$
9. 8
11. $(\frac{2}{3})\sqrt[4]{4}$
13. 0.12
15. 2
17. $\dfrac{\sqrt{3}}{6}$
19. $\dfrac{\sqrt[4]{3}}{3}$
21. $\dfrac{\sqrt[3]{4}}{4}$
23. $\dfrac{\sqrt{6}}{24}$
25. $\sqrt{5}$
27. $\sqrt{15} + 2\sqrt{5}$
29. 2
31. $2\sqrt{7}$
33. $15 - 10\sqrt{2}$
35. 0.6
37. No. You can't add unlike square roots.
39. No. $\sqrt[3]{8 + 27} = \sqrt[3]{35} \neq 5$
41. No. You can't combine a cube and a square root.
43. is less than
45. is less than
47. is less than
49. $3\sqrt{5}$
51. $\sqrt{3} + 1$

Chapter 7 Exercises

1. $1.567 \cdot 10^6$
3. $7.0 \cdot 10^{-2}$
5. $2.3 \cdot 10^{-2}$
7. $6.5 \cdot 10^{-4}$
9. $1.1 \cdot 10^{-1}$
11. $3.8 \cdot 10^{-8}$
13. $7.06 \cdot 10^{-3}$; 0.706%
15. 2.23; 223%
17. $5.0 \cdot 10^{-1}$; 50%
19. $2.7 \cdot 10^{-4}$; 0.00027
21. $3.0 \cdot 10^{-6}$; 0.000003
23. 1.013; 1.013
25. $3.\overline{3} \cdot 10^{-1}$; $0.\overline{3}$
27. $2 \cdot 10^4$
29. $9 \cdot 10^{-4}$
31. $5.9 \cdot 10^{199}$
33. $7.8 \cdot 10^{-4}$
35. $3.0 \cdot 10^{-9}$
37. $2 \cdot 10^{-2}$
39. $1.42 \cdot 10^{-12}$
41. 10^{-1}
43. $1 \cdot 10^{-22}$
45. $4.7 \cdot 10^3$ kwh/person
47. (a) $9.46 \cdot 10^{17}$ cm
    (b) $7.2 \cdot 10^{21}$ atoms/hr
49. 1
51. 3
53. 1
55. 4
57. 2

Chapter 8 Exercises

1. two terms; first degree
3. one term; zero degree
5. four terms; not a polynomial
7. 1. 14 3. 41 5. 0
9. 22
11. $\frac{6}{7}$
13. 0.6
15. 50
17. -66
19. two; $\left(\dfrac{p + q}{p - q}\right)t; p$
21. three; $3a(x^2 + 1)$; $\dfrac{b}{x^2 + 1}$; $(a + b + 2)\dfrac{x^2 + 1}{x^2 + 2}$
23. $14; -3; -1; -\dfrac{1}{2}; -5$
25. $a; (a + b); \dfrac{1}{b}$
27. a^2b and $-3ba^2$ and $5aba$; $2ab^2$ and $(2ab)b$
29. $x(x + 2)$ and $\dfrac{(x + 2)}{3}x$

Problem Set 9.1

1. $8t$
3. $-7xy$
5. $-as^2 + 9s^2$
7. $4p^2t^3$
9. $4x^2 - 6xy + y^2$
11. $-x^3 + 5x^2 + 7x - 2$
13. $7t^2 + 4rt - 13r^2$
15. $-11x^2 - 5x + 5$
17. $10m^2 - 6m + 9$
19. $3a^2 - 7ab + b^2 - c$

Problem Set 9.2

1. $2a^5 - 4a^3$
3. $6ab^2c + 2a^2b^2c^2$
5. $2y^2 + 5y - 12$
7. $a^2 + 2as + s^2$
9. $a^2 + b^2 + c^2 + 2ab + 2bc + 2ac$
11. $6x^2 + 19x - 7$
13. $p^2r + p^2st + q^2r + q^2st$
15. $x^3 + 6x^2 + 12x + 8$
17. $x^3 + x^2 - 14x - 24$
19. $6b^3 - 29b^2 - 17b + 60$
21. $2r^2 - 38$
23. $6r^5s^2p^{12} - 18r^2s^2p^3 + 3r^3s^2p^2$
25. $2y^4 + 4y^3 - 11y + 3$

Problem Set 9.3

1. $y + 1$
3. $1 + st$
5. $x - 3$
7. $10a^2 - 12a + 1$
9. $1 - 3x$
11. $x - 2, R = -4$
13. $7y - 19, R = -6$
15. $y^2 - 7y + 28, R = -118$
17. $2x + 3$
19. $3x + 11, R = 7$
21. $x + 6, R = x + 14$
23. $a^2 + ab + b^2$
25. $6x - 5y$
27. $2a^2 - 3ab - 2b^2$
29. $2a - 5b$

Problem Set 9.4

1. $5 - 2r$
3. $5x - 7$
5. $5p + 9$
7. $2d - 7$
9. $0.11p^2$
11. $23c - 10d$
13. $2a^2 - 2r^2z$
15. $a - c$
17. $-5a + 20$
19. $0.3z + 0.52t$
21. $20k - 20$
23. $-10b + 16$
25. $x(y + 1) - x(y - 1) - 2y = 2(x - y)$

Problem Set 9.6

1. xy
3. $4x$
5. $4x^3yz^2$
7. $7xz + y$
9. $x + 1 - z$
11. $x(x + 10)$
13. $3a(b - 2c)$
15. $5x(3b + 4a)$
17. $2a^3(3a^2 + 1)$
19. $3T(4RT - Q)$
21. $\dfrac{LT}{\pi}(L^2 - 9T^2)$
23. $(x + y)(x + y - 3)$
25. $11zy(y + 2t^2 + at)$
27. $3a(ab + 3b^3c + 9ac)$

29. $ad(a^2bc^3d + 4c - 3bd)$
33. $(9x - 2r)\,(a - b)$
35. $(2x - 1)\,(y - 2)$

37. $(3a - 1)\,(x - y)$
39. $(x + 2)\,(2x - 2) = 2(x + 2)\,(x - 1)$

Problem Set 9.7

1. $(2x + 1)\,(3x + 5)$
3. $(z + 7)\,(z + 2)$
5. $(x - 6)\,(x + 2)$
7. $(t - 5)\,(t - 2)$
9. $(r + 1)\,(2r - 3)$

11. Unfactorable
13. $(2b - 5)\,(5b - 9)$
15. $(8 - x)\,(x - 2)$
17. $(3a + 1)\,(5 - a)$
19. $2(3x - 1)(x + 3)$

21. $(3k - 1)\,(k - 1)$
23. $(a - 100)\,(a + 1)$
25. $(2x - 3)\,(x + 5)$

Problem Set 9.8

1. $(b + a)\,(b - a)$
3. $2(x + 2)\,(x - 2)$
5. $(d + 3)^2$
7. $(3p + 2)\,(3p - 2)$
9. $(y + 4)^2$

11. $8(1 + d)\,(1 - d)$
13. $(ab + 4)\,(ab - 4)$
15. $3(2y - 5)^2$
17. $(4t + 7)^2$
19. $(x^2 + 9 - 6x)\,(x^2 + 9 + 6x) = (x - 3)^2\,(x + 3)^2$

Problem Set 9.9

1. $(x + 2y)\,(x + y)$
3. $(x^2 + 2)\,(x^2 - 2)$
5. $p^3(1 + r)\,(1 - r)$
7. $(2x - 1)\,(x + 2)$
9. Unfactorable
11. $(4c - 4d + 1)\,(3c - 3d - 2)$
13. $(a + b + 1)^2$
15. Unfactorable
17. $(r + s - 3x)\,(r - s + 3x)$
19. $x(x + 1)\,(x - 1)$
21. $(7 + 2y)\,(7 - 2y)$
23. $a^2(a + 3)\,(a - 2)$
25. Unfactorable
27. $(3a + 5b)\,(2a - 3b)$
29. $(5c - 2b + 10x)\,(5c - 2b - 10x)$
31. $(t + 7 + s + 4)\,(t + 7 - s - 4) = (t + s + 11)\,(t - s + 3)$
33. $2(h + 1)\,(h - 1)\,(h^2 + 1)$
35. $x(x + 1)\,(x + 2)$
37. Unfactorable

39. $\frac{2}{3}(2x + 1)\,(x + 2)$
41. $(ax - by)^2$
43. $(s - a + 1)\,(2s - 2a - 3)$
45. $(12x - 5y)\,(x + 2y)$
47. $5\pi(t + 2\pi)^2$
49. $(a + b)\,(a - b)\,(a^2 + b^2)$
51. $(p + q + 1)^2$
53. $(\sqrt{x} - 1)\,(\sqrt{x} - 2)$
55. $5\left(2 + \dfrac{p}{q}\right)^2$
57. $\dfrac{u}{2\pi}\,(r + s)^2$

Problem Set 9.10

1. $(pq^2 + 8)(pq^2 - 8)$
3. $xy(x + 3y)(x - 3y)$
5. $(f + g)^2(f - g)^2$
7. $2t(1 - 2t + 4t^2)$
9. $a(a^2 - ab - bc - ac)$
11. $(x - 2)(x + 3)(x + 5)$
13. $(y^2 + 2)(1 + y)(1 - y)$
15. $(2y - x^3)(x + 3)$
17. $\left(\dfrac{4ab}{c^3} + 3d^2\right)\left(\dfrac{4ab}{c^3} - 3d^2\right)$
19. $(x - y)(x^2 + y^2)$
21. $(4x^2 - 3)(2x^2 - 1)$
23. $-2m^2\left(2 + \dfrac{m}{r}\right)\left(2 - \dfrac{m}{r}\right)$

25. $x(x + 5)(x + 7)$
27. $(2y - 3)(y + 2)(y - 2)$
29. $(d + 4)(d - 3)(d - 2)$
31. $(t - 3)(t^2 + 3t + 9)$
33. $2(3b - 2)(9b^2 + 6b + 4)$
35. $(a + b)(a - b)(a^2 + ab + b^2)(a^2 - ab + b^2)$
37. $(x + 5t)^3$
39. $(4x - 3)(4x^2 + 3x + 5)$
41. $(2r - 3s)^3$
43. $5(x + 2)(x - 3)(x - 1)$
45. $a^n(7a + 2b)(2a - 3b)$

Problem Set 9.11

1. $y^2 + 4y + 4 = (y + 2)^2$
3. $x^2 - 3x + \frac{9}{4} = (x - \frac{3}{2})^2$
5. $3r^2 + 6r + 3 = 3(r + 1)^2$
7. $2y^2 - 4y + 2 = 2(y - 1)^2$
9. $b - 2b^2 - \frac{1}{8} = -2(b - \frac{1}{4})^2$
11. $r^2 + 6r + 9 = (r + 3)^2$
13. $3x^2 - x + \frac{1}{12} = 3(x - \frac{1}{6})^2$
15. $x^2 + ax + \dfrac{a^2}{4} = \left(x + \dfrac{a}{2}\right)^2$
17. -7
19. -13
21. $-\frac{1}{2}$

23. 3
25. $c - \dfrac{b^2}{4a}$
27. minimum $= -44$; maximum $= 245$
29. minimum $= -\frac{49}{9}$; maximum $= \frac{325}{3}$
31. minimum $= -1$; maximum $= 362$
33. minimum $= -\frac{25}{48}$; maximum $= \frac{297}{2}$
35. minimum $= -\frac{49}{4}$; maximum $= 120$
37. November 13; D.J. $= 428$
39. Distance 50 meters; 50%
41. 100 feet

Chapter 9 Review

1. $\dfrac{21b}{2}$
3. $-y^2 + 3y - xy^2 + x^2y^2$
5. $BCD^2 + \frac{1}{3}BC^2D + AC^2D$
7. $5ab(a + 5b)$
9. $-7t^2(1 + 7t)$
11. $x^3 + 5x^2 + 9x + 9$
13. $M^2 + 2MN^2 + N^4 + 2Mp + 2pN^2 + p^2$
15. $4x^2 + 4xz - y^2 + z^2$
17. $\frac{9}{16}c^2 - 4d^2$

19. $3x^2 - 2x\sqrt{6} + 2$
21. $625a^8c^4d^{12}$
23. $rs + 1$
25. $7x - 2$
27. $8a + 1$
29. $y^2 + 3$
31. 0
33. t
35. x
37. $19A^2DG$

39. $1 - 2a$

41. $(x - 5)(x + 5)$

43. $(y - 100)(y + 1)$

45. $(3x - 1)(x + 2)$

47. $(3q - 2p)^2$

49. $(7 - 8c)(7 + 8c)$

51. $x^2 - 14x + 49 = (x - 7)^2$

53. $2\left(z^2 + \dfrac{z}{2} + \dfrac{1}{16}\right) = 2\left(z + \dfrac{1}{4}\right)^2$

55. $-2(p^2 + 2p + 1) = -2(p + 1)^2$

57. $x^2 - \tfrac{1}{2}y - 2xy + 2x + 1$

59. $x - xyz + y^2z^2 - xyz^2 - xz + x^2yz^2 - xy^2z^3 - xyz^3 - xy^2z^2 + y^3z^3 + y^2z^3$

61. $4ABC^2(2B + A)(2AB + C)$

63. $L(K^2 + KL + N + KN)$

65. $[2(x - y) - b][2(x - y) + b]$

67. $(C + D + 1)^2$

69. $\dfrac{G}{6}(G^2 + 6GH + 3H^2)$

71. $[(P - 2Q) - (R - 2S)][(P - 2Q) + (R - 2S)]$

73. $5(3q - 20)(q - 30)$

75. $(x - 2)(x - 3)(x + 4)$

77. $(X + V + W)^2$

79. $(x - y + 1)(x + y + 5)$

81. $(4R - 4C - 1)(3R - 3C + 2)$

83. $\dfrac{21R^2S^2}{2} + 7R + 3RS^3 + 2S$

85. $2m - 5$

87. $D^2 - \tfrac{3}{4}D + \tfrac{9}{64} = (D - \tfrac{3}{8})^2$

89. 3

91. -4

93. Unequal

95. Unequal

97. 16

99. $200; $2000

Problem Set 10.2

1. $\dfrac{6}{5}$

3. 0

5. $\dfrac{ab}{b^2 + 3b + 3a + ab}$

7. $\dfrac{64a^{14}b^{15}}{5}$

9. $\dfrac{-2x^2 + 11x - 5}{6x^2}$

11. $\dfrac{-2x + 2y}{x + y}$

13. $\dfrac{6b^7c^3}{d^2}$

15. $\dfrac{6z + 2x}{-z^2 + 2zx - x^2}$

17. $\dfrac{b - a}{a^2 + 2ab + b^2}$

19. $\dfrac{-x^2 + 2xy - y^2}{2p^5r^2}$

21. $\dfrac{12a^8}{p^5 + p^4}$

23. $\dfrac{5a^3 + 15a^2x + 15ax^2 + 5x^3 + 2a + 2x}{c - d}$

25. $\dfrac{-d^2 + 2cd - c^2}{c^2d + d^3}$

Problem Set 10.3

1. $\dfrac{6}{35}$

3. $\dfrac{23}{13}$

5. $b - 2$

7. $\dfrac{a + 5}{a + 4}$

9. $\dfrac{2g + h}{3g}$

11. $\dfrac{k}{n}$

13. x

15. $\dfrac{2ab + 1}{2ab - 1}$

17. $\dfrac{x - 3}{4 - x}$

19. $2sk + 6s - 9k$

21. $-\dfrac{a + c}{d + c}$

23. $\dfrac{1}{bcpqr}$

25. $\dfrac{-3}{2y - 3}$

27. $\dfrac{-2r^4t^2}{s}$

29. $\dfrac{a + c}{ac}$

31. $\dfrac{x}{y}$

33. $\dfrac{3ab + 1}{ab - 3}$

35. $\dfrac{s - t}{a}$

37. $3x + y - 4y^2$

39. m

41. $\dfrac{(p + 2)^2\,(p - 2)}{p - 4}$

43. $\dfrac{3(2x - 1)}{(a + 3)\,(2x + 1)}$

45. $\dfrac{5mp(q + 3)}{4qn^3\,(r + 3)^3}$

Problem Set 10.4

1. $\dfrac{7}{2m}$

3. $\dfrac{p + 2}{p + 1}$

5. $\dfrac{2x - y}{x^2 y^2}$

7. $\dfrac{2}{(p + 1)\,(p - 1)}$

9. $\dfrac{-x - 1}{(x + 2)\,(x + 3)}$

11. $\dfrac{5x - 4y}{120}$

13. 0

15. $\dfrac{st - rt + rs}{rst}$

17. $\dfrac{2l^3 - lc + b}{l^2}$

19. $\dfrac{2c}{x - y}$

21. $\dfrac{y^2 z - x^4}{x^5 y^4 z}$

23. $\dfrac{3x - 4}{(x + 2)\,(x - 2)}$

25. 0

27. $\dfrac{4a^2 + 8ax + x^2}{6(2a + x)\,(2a - x)}$

29. $\dfrac{-4a^3 + a + 2}{(2a + 1)^2\,(1 - 2a)}$

31. $\dfrac{x^2 c^2 + y^2 b^2 - z^2 a^2}{a^2 b^2 c^2}$

33. 0

35. $\dfrac{-2b + 1}{a^2 - b^2}$

37. $\dfrac{2x^2 - 11x + 6}{x^2 - 2x}$

39. $\dfrac{y^2 - 3y - 11}{y + 2}$

41. $\dfrac{-1}{x - 1}$

43. $\dfrac{x^2 z + y^2 x + yz^2}{2}$

45. $\dfrac{-1}{s - r}$

Problem Set 10.5

1. -16

3. $\dfrac{35}{72}$

5. $\dfrac{3b}{7a^2}$

7. $\dfrac{10rs^2}{t^2}$

9. -1

11. $\dfrac{x + 1}{2x}$

13. $\dfrac{6ry}{s^2 x}$

15. $\dfrac{6 - 3r}{4}$

17. $\dfrac{3x + 9}{2}$

19. $\dfrac{4}{x + 4}$

21. $\dfrac{(2\pi m - n)^2}{\pi}$

23. $\dfrac{2(1 - z)}{z + 7}$

25. $\dfrac{1}{5x^2 - 25x}$

27. $\dfrac{a^2 - ab}{b}$

29. 1

31. $\dfrac{2 - x^2}{3xy}$

33. $-(a + b)$

35. $\dfrac{r(b - 1)}{r - a}$

37. $\dfrac{3a^4 - 27b^4}{(2b^2)\,(b + 3)}$

39. $\dfrac{(2 - r)\,(9 + 3r + r^2)}{r^2 - 2r + 4}$

Problem Set 10.6

1. $\dfrac{7}{10}$

3. $\dfrac{p}{3}$

5. $\dfrac{4 - x^2}{3x}$

7. $\dfrac{(1 + x)\,(x)}{x^2 + 1}$

9. $\dfrac{-a}{c}$

11. $\dfrac{rs}{r + s}$

13. 2

15. 1

17. $\dfrac{q}{r}$

19. $\dfrac{x}{y}$

21. $\dfrac{1}{c}$

23. $\dfrac{2(x^2 - 1)}{-7x^2 + 10x + 5}$

25. $\dfrac{3j - 24f}{6j + 4f}$

27. $\dfrac{9}{8}$

29. $\dfrac{4a^3 - 3a}{3ab - 5a^3 + 25b}$

31. $\dfrac{x + y}{x - y}$

33. $\dfrac{1}{x + y}$

35. $\dfrac{4}{(2x + 1)(2x - 1)}$

37. $\dfrac{1}{m - 1}$

39. $\dfrac{-(x - 5)^2}{(x - 1)(x + 1)^2(x - 2)^2}$

41. 1

43. x

45. $\dfrac{(x + 1)(2x^2 + x + 1)}{x^3(2x + 1)}$

Chapter 10 Review

1. $\dfrac{5x}{3y}$

3. $\dfrac{a^2\sqrt{2}}{6\sqrt{3}}$

5. $\dfrac{6(1 - b)}{b - 7}$

7. $\dfrac{7}{2E + 1}$

9. $\dfrac{1}{2x - 1}$

11. $\dfrac{1}{a - 1}$

13. $(a + b)(a - b)$

15. $a + b + c$

17. $\dfrac{-2(x + 5)}{(x + 4)(x + 2)}$

19. $\dfrac{(x + 8)(x - 6)}{(x + 3)(x - 2)}$

21. $\dfrac{-(7a + 1)(a - 3)}{4(a + 1)(a - 1)}$

23. $\dfrac{-2(x - 5)}{(x - 3)(x - 2)}$

25. $\dfrac{y^2(2x + y)}{(x + y)^3}$

27. $\dfrac{x - 2}{x - 3}$

29. $\dfrac{-a}{a + b}$

31. 1

33. $1 + \dfrac{y}{x}$

35. $\dfrac{4b}{(1 - ab)(1 + ab)}$

37. $\dfrac{s^2}{s - 1}$

39. $\dfrac{x + A}{3(2x + A)}$

41. $\dfrac{x}{x^2 - x + 1}$

43. $\dfrac{3(x + 1)}{2(x + 3)}$

45. $\dfrac{-1}{5 + a}$

47. $\dfrac{4}{(1 - 4A)(1 + 4A)}$

49. $\dfrac{R^3 - R^2 + 1}{R^3}$

51. $\dfrac{(A + B)^2}{(X + Z)^2}$

53. Equal

55. 0

57. $2y - 5$

59. $A^2 + B^2 = 1$

61. $\dfrac{n + 4}{n}; \ 1 + \dfrac{1}{n + k - 1}; \ \dfrac{n + k}{n}$

Problem Set 12.1

1. -3
3. $\frac{2}{13}$
5. $\frac{5}{2}$

7. $\frac{6}{5}$
9. $\frac{7}{5}$
11. -5

13. 1
15. 2
17. $-\frac{5}{2}$

19. 13
21. 2000
23. 29

25. $\frac{9}{7}$
27. 5
29. 1.7

31. $\frac{1}{2}$
33. $\frac{71}{112}$

Problem Set 12.2

1. -2
3. 3
5. 1
7. 2

9. -8
11. $-\frac{3}{5}$
13. -1
15. $\frac{11}{2}$

17. $\frac{60}{13}$
19. $\frac{5}{11}$
21. $\frac{17}{39}$
23. $-\frac{4}{3}$

25. $\frac{17}{2}$
27. $-\frac{15}{14}$
29. $-\frac{4}{3}$
31. $\frac{1}{3}$

Problem Set 12.3

1. $M = \dfrac{ax}{r-a}$

3. $x = \dfrac{y-b}{m}$

5. $a = \dfrac{2A-hb}{h}$

7. $p = \dfrac{aq}{1-a}$

9. $p = \dfrac{3m}{2m+3}$

11. $t = \dfrac{A}{p-Apr}$

13. $E = \dfrac{nr\,(b-a)}{a-b-n}$

15. $x = \dfrac{-3a+2b-c}{a+b-c}$

17. $d = \dfrac{a(a-b-2a^2)}{b(b+2a^2)}$

19. $a = \dfrac{r^2+sp-q}{s-t}$

21. $p = \dfrac{a^2-6a^3-3b}{b(2a^3+b)}$

23. $y = \dfrac{sc-ta}{ds+tb}$

25. $x = \dfrac{-9a}{5}$

27. $q = 0$

29. $r = \dfrac{9z+6az+2}{3z+2az-8}$

Chapter 12 Review

1. $Q = \frac{8}{3}$

3. $a = \frac{21}{5}$

5. $B = -3$

7. $R = \dfrac{P-E^2}{s(E+1)}$

9. $m = 58$

11. $Y = \dfrac{-Ax}{B} - \dfrac{C}{Bx}$

13. $Y = -0.6$

15. $Q = \frac{62}{17}$

17. $P = Q - \dfrac{AR}{B}$

19. $P = \dfrac{QM+M}{1+3Q}$

21. $z = \frac{10}{9}$

23. $k = \frac{11}{30}$

25. $t = -\frac{1}{2}$

27. $N = \dfrac{P^2-QS+P^2Q}{PS+PQS^2}$

29. $R = \frac{181}{30}$

31. $x = -2$

33. $P = -\frac{22}{23}$

35. $S = \dfrac{3B^2+3Bax+rR-2Rx}{2x-ax-r-B}$

37. $S = -\frac{5}{2}$

39. $X = 43.9$

41. $M = -\frac{40}{11}$

43. $x = -\frac{5}{7}$

45. $x = 2$

47. $C = \dfrac{K+6J-2B}{23J+2JB-6J^2}$

49. $\dfrac{A}{B} = \dfrac{1+TD}{1+TE}$

51. $X = \frac{77}{116}$

53. $X = \frac{11}{12}$

55. $W = -\frac{67}{24}$

57. $\dfrac{m}{R^3} = \dfrac{1}{GT^2}$

Problem Set 13.3

1. 5

3. 10

5. 5

7. 9, 11, and 13

9. 33 and 38

11. $-5, -3,$ and -1

Problem Set 13.4

1. 20

3. 8

5. 50

7. 9

9. 36

11. 30